# 255
# Topics in Current Chemistry

# Topics in Current Chemistry

Recently Published and Forthcoming Volumes

**Anion Sensing**
Volume Editor: Stibor, I.
Vol. 255, 2005

**Organic Solid State Reactions**
Volume Editor: Toda, F.
Vol. 254, 2005

**DNA Binders and Related Subjects**
Volume Editors: Waring, M.J., Chaires, J.B.
Vol. 253, 2005

**Contrast Agents III**
Volume Editor: Krause, W.
Vol. 252, 2005

**Chalcogenocarboxylic Acid Derivatives**
Volume Editor: Kato, S.
Vol. 251, 2005

**New Aspects in Phosphorus Chemistry V**
Volume Editor: Majoral, J.-P.
Vol. 250, 2005

**Templates in Chemistry II**
Volume Editors: Schalley, C.A., Vögtle, F., Dötz, K.H.
Vol. 249, 2005

**Templates in Chemistry I**
Volume Editors: Schalley, C.A., Vögtle, F., Dötz, K.H.
Vol. 248, 2004

**Collagen**
Volume Editors: Brinckmann, J., Notbohm, H., Müller, P.K.
Vol. 247, 2005

**New Techniques in Solid-State NMR**
Volume Editor: Klinowski, J.
Vol. 246, 2005

**Functional Molecular Nanostructures**
Volume Editor: Schlüter, A.D.
Vol. 245, 2005

**Natural Product Synthesis II**
Volume Editor: Mulzer, J.
Vol. 244, 2005

**Natural Product Synthesis I**
Volume Editor: Mulzer, J.
Vol. 243, 2005

**Immobilized Catalysts**
Volume Editor: Kirschning, A.
Vol. 242, 2004

**Transition Metal and Rare Earth Compounds III**
Volume Editor: Yersin, H.
Vol. 241, 2004

**The Chemistry of Pheromones and Other Semiochemicals II**
Volume Editor: Schulz, S.
Vol. 240, 2005

**The Chemistry of Pheromones and Other Semiochemicals I**
Volume Editor: Schulz, S.
Vol. 239, 2004

**Orotidine Monophosphate Decarboxylase**
Volume Editors: Lee, J.K., Tantillo, D.J.
Vol. 238, 2004

**Long-Range Charge Transfer in DNA II**
Volume Editor: Schuster, G.B.
Vol. 237, 2004

**Long-Range Charge Transfer in DNA I**
Volume Editor: Schuster, G.B.
Vol. 236, 2004

**Spin Crossover in Transition Metal Compounds III**
Volume Editors: Gütlich, P., Goodwin, H.A.
Vol. 235, 2004

**Spin Crossover in Transition Metal Compounds II**
Volume Editors: Gütlich, P., Goodwin, H.A.
Vol. 234, 2004

**Spin Crossover in Transition Metal Compounds I**
Volume Editors: Gütlich, P., Goodwin, H.A.
Vol. 233, 2004

# Anion Sensing

**Volume Editor: Ivan Stibor**

With contributions by
E. V. Anslyn · S. R. Bayly · P. D. Beer · S. D. Collyer · F. Davis
S. P. J. Higson · R. J. T. Houk · P. Lhoták · F. P. Schmidtchen · I. Stibor
C. Suksai · S. L. Tobey · T. Tuntulani · P. Zlatušková

Springer

The series *Topics in Current Chemistry* presents critical reviews of the present and future trends in modern chemical research. The scope of coverage includes all areas of chemical science including the interfaces with related disciplines such as biology, medicine and materials science. The goal of each thematic volume is to give the nonspecialist reader, whether at the university or in industry, a comprehensive overview of an area where new insights are emerging that are of interest to a larger scientific audience.

As a rule, contributions are specially commissioned. The editors and publishers will, however, always be pleased to receive suggestions and supplementary information. Papers are accepted for *Topics in Current Chemistry* in English.

In references *Topics in Current Chemistry* is abbreviated Top Curr Chem and is cited as a journal.

Visit the TCC content at springerlink.com

Library of Congress Control Number: 2004112072

ISSN 0340-1022
ISBN-10 3-540-23247-8 **Springer Berlin Heidelberg New York**
ISBN-13 978-3-540-23247-6
DOI 10.1007/b101055



**Springer is a part of Springer Science+Business Media**
springeronline.com

Printed in Germany



Cover design: KünkelLopka, Heidelberg/design & production GmbH, Heidelberg
Typesetting: Fotosatz-Service Köhler GmbH, Würzburg

Printed on acid-free paper 02/3141 xv – 5 4 3 2 1 0

## Volume Editor


Prof. Ivan Stibor

Praha Institute of Chemical Technology
Department of Organic Technology
Technická 5
166 28 Prague 6, Czech Republic
*stibori@vscht.cz*


## Editorial Board

## Topics in Current Chemistry also Available Electronically

For all customers who have a standing order to Topics in Current Chemistry, we offer the electronic version via SpringerLink free of charge. Please contact your librarian who can receive a password for free access to the full articles by registration at:

springerlink.com

If you do not have a subscription, you can still view the tables of contents of the volumes and the abstract of each article by going to the SpringerLink Homepage, clicking on "Browse by Online Libraries", then "Chemical Sciences", and finally choose Topics in Current Chemistry.

You will find information about the

- Editorial Board
- Aims and Scope
- Instructions for Authors
- Sample Contribution

at springeronline.com using the search function.

# Preface

It has often been deplored that anion complexation is a less developed area of supramolecular chemistry. Many recently published reviews and several books illustrates the current high interest in the field as well as its rapid expansion. In contrast to most cations, anions (and organic anions in particular) are usually structurally very diverse group of guests in Cram's host–guest notation. A relatively complicated structure, strong solvation and sometimes narrow pH window of anion existence – all these features make anion sensing relatively complicated topics with many chalenges to inorganic, organic as well as analytical chemists. Anion sensing is a typical multifaceted subject that can be treated from several directions. This was exactly our intention when structure of the present Volume was being prepared.

Several anion binding proteins with precisely defined crystal structure reported recently. Nevertheless, by far the most attention is still being devoted to synthetic host. This is the reason that the first chapter of the present volume written by *F. P. Schmidtchen* reviews the basic principles imposed on artificial host molecules used for the sensing of anions. This is followed by chiral recognition of anions written by *I. Stibor and P. Zlatušková*. It can be better expressed as enantioselective sensing of chiral anions. This subject to the best of our knowledge has not been treated in literature so far. In fact it is usually hidden in reviews on enantioselective recognition of chiral organic acids. Changing pH one can easily switch from free organic acids to their anions. Most of receptors, however, have been designed for recognition of free chiral acids which means the completely different structure of binding site in comparison with that for anions. Much less is known on the principles of recognition of chiral anions. The following chapter is devoted to one famous molecular scaffold – calixarene. It has been chosen and reviewed as structural basis for design, synthesis and study of abiotic anion receptors by *P. Lhoták*. The fourth review is covering the analytically oriented topic – construction and operation of anion sensors written by *F. Davis, S. D. Collyer and S. P. J. Higson*. Synthetic metal-based receptors have been reviewed by *P. D. Beer and S. R. Bayly*. The bottom-line of all these artificial molecules is the presence of metal ion(s) in the form of organometallic and coordination complexes that is useful as a reporter group. The most recent advances in the field are also detailed including the use of metalloreceptors in dendrimer, functionalised nanoparticle and surface-bound anion sensor. The extremely broad subject of chromogenic anion sensors has been reviewed by *C. Suksai and T. Tuntulani* followed by abiotic guanidinium receptors for anion recognition and sensing written by *R. J. T. Houk, S. L. Tobey, and E. V. Anslyn*. These receptors have found a number of very interesting applications as sensors for anions in practice.

Prague, January 2005 Ivan Stibor

# Contents

# Contents of Volume 218
# Host-Guest Chemistry
# Mimetic Approaches to Study Carbohydrate Recognition

**Volume Editor: Soledad Penades**
ISBN: 3-540-42096-7

Top Curr Chem (2005) 255: 1–29
DOI 10.1007/b101160


# Artificial Host Molecules for the Sensing of Anions

Franz P. Schmidtchen (✉)

Department of Chemistry, Technical University of Munich, Lichtenbergstrasse 4, 85747 Garching, Germany
*Schmidtchen@ch.tum.de*



**Abstract** The foundations of supramolecular complexation of anions by abiotic host compounds are evaluated, with the purpose of determining their qualitative and quantitative preference over competing species. Starting from a critical assessment of binding models, the perspective goal in terms of molecular thermodynamics is defined and discussed. General consequences for the strategic design of artificial anion hosts are delineated and their translation into receptor structures is suggested. The usage of the available fundamental interaction types for anion binding is illustrated with the help of prototypical, yet concrete, receptors of proven utility.

**Keywords** Anion host · Host–guest chemistry · Supramolecular association · Statistical thermodynamics · Binding models

## 1 Introduction

Anion sensing refers to the qualitative detection – or even better – to the quantitative determination of a negatively charged molecular species by a device to translate its mere presence into a physically manageable signal [1, 2]. Common experience tells us that it may be a great feat to conceptually design and engineer devices that generate signals in time-dependent (e.g. in luminescence) or equilibrium modes (e.g. in ion-selective electrodes). The main obstacle to overcome is not so much to come up with a suitable instrumental response; most analytes, and in this vein also anions, would answer an initial stimulus in some processable way. Rather, the problem resides in the discrimination of the principally integral response from the effects turned up by competing analytes. Ensuring selectivity in the readable response mandates a differentiation of species on the molecular level, a process commonly termed molecular recognition. However, two options to cope with the requirement of selectivity can be distinguished that differ from one another with respect to the recovery of the sensing device. The first option uses an irreversible transformation involving the target analyte (the specific anion) to generate an easily quantifiable product. Selectivity in this case can emerge from the transformation itself, but can also be improved later on by subsequent manipulations to remove concomitant contaminants, because the desired product will not be lost. This dosimeter approach is particularly suited to the accumulative determination of analytes, but is less favourable if rapid and/or high-amplitude changes in analyte concentration are to be monitored. The latter situation is better served by the alternative option for implementing selectivity in sensing devices: it calls for the employment of supramolecular interactions to scan and probe the variety and concentration of species present in the system under scrutiny. In essence, associative complexes of the anionic analyte and the device are formed which, irrespective of the actual nature of the binding interaction (i.e. covalent or non-covalent), are weak enough to undergo reversible exchange reactions. In general, it is this binding reversibility established in a certain time domain that ultimately governs the selectivity and the time response of the device. Although some sensor devices relying on the concerted response from an ensemble may require a different treatment, it is appropriate in this context to subdivide the events into two components addressing on the one hand the fundamental binding process, and on the other hand the transformation of this basic binding event into a readable signal (Fig. 1). Of course, this procedure is purely arbitrary and there is no physical foundation for it, as in most concrete cases both aspects are intimately interconnected. In this way, however, we can focus our discussion on that part of a sensing device that is acting as a molecular host without regard to the physics of signal generation.

An additional approximation concerns the conversion of binding information into the elicited signal, which is assumed to be unaltered irrespective of the chemical species bound. This is quite a crude premise that is valid only in the limiting case when the response of the device solely depends on the concentration of the unbound species (as e.g. in certain ion-selective electrodes). In all other scenarios the chemical nature of the specific analyte, in conjunction with and relative to the flock of potential competitors, will influence the response not only by the fractional occupation of the binding site but also due to its particular effect on the physical response. If, for instance, an optical sensor reports the appearance of an absorption band at a certain wavelength owing to the complexation of the target anion, the signal may be easily perturbed by a contaminant present in minute amounts only, but interfering by virtue of a

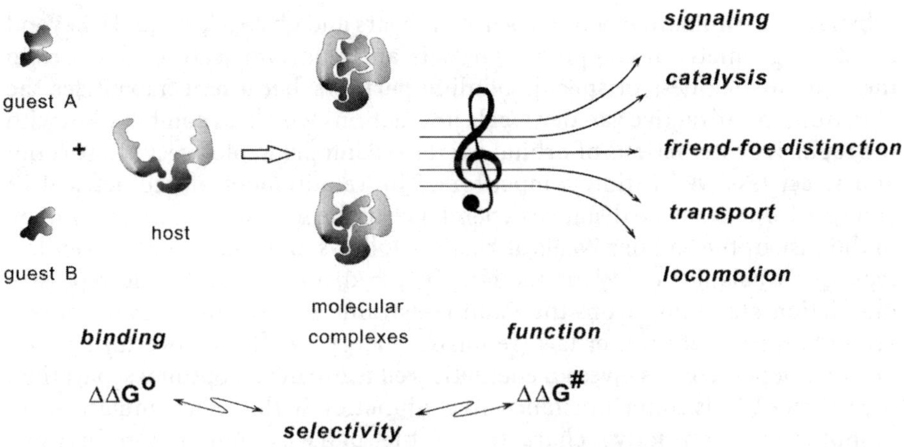


**Fig. 1** Schematic dissection of the sensing process (signalling) into two successive sections: host–guest ground-state complexation and subsequent generation of function.

higher extinction coefficient of its complex at the respective position in the spectrum. Interferences of this type are quite common in cation detection. Since the molecular interactions of anions with their corresponding hosts – as a rule – are weaker than seen in the cation-sensing field, the variability in this influence on signal generation between competing anions is expected to be narrow. For the sake of simplicity in our further discussion, we ignore the relatively small differences that must certainly exist and concentrate our attention on the residual factor that determines sensor selectivity: the host–guest-type reversible association of the anion with a receptive structure and its optimal layout to enforce utmost discrimination of potential binding competitors.

## 2 Conceptual Models for Binding

Molecular binding interactions are fundamental properties of all matter. Though this statement is trivial at first sight and does not deserve special mention, it sets a bold limit to all attempts in the design of specific interactions towards the construction of selective molecular host compounds. The most promising approach towards this goal relies on idealized models that reflect just one prominent aspect of the entire interaction, making it illustrative and as such easily comprehensible. However, owing to the persuasive power of pictorial arguments, the premises of the models may be overlooked and their applicability is then stretched too far. Notorious in this respect is the famous metaphor of the lock-and-key fit in host–guest binding, coined by Emil Fischer more than a century ago to cite geometrical complementarity as the origin for substrate discrimination between various sugars and glycosidases [3,4]. Beyond question, geometrical complementarity is an important feature to ascertain the mutual stickiness of specific binding partners, because it maximizes the help from the attractive van der Waals interactions which, in combination with the repulsive interactions of orbital overlap (Pauli principle), serve to sort out and reject less well-fitting competitors. The requirement of geometrical fit emerges from the steep distance dependence of the interacting surfaces which, in the case of the van der Waals attraction, follows an inverse sixth power law (van der Waals attractive interaction $\sim r^{-6}$; $r$=distance), whilst the repulsive interaction stemming from the Pauli constraint adheres to an even more e xtreme inverse 12th power law (repulsive energy $\sim r^{-12}$). The overlay of both distance dependencies gives an energetic well featuring an optimal separation, the van der Waals contact distance. In conjunction with the accumulative and monotonously attractive character of the pairwise interaction between distinct positions in host and guest, these features constitute the driving urge to form extended interface areas of minimal separation in order to maximize binding.

In quantitative terms the complementary fit model of host and guest binding refers to the energetic difference at constant temperature and pressure $\Delta H_{\text{assoc}}$ (also called the exothermicity) of just two states: one mole each of host and guest molecules totally separated as opposed to the associated complex of the same components in its most favourable configuration. The model has been expanded in various directions to include other types of interaction, notably the participation of functional groups which supplemented the model with Coulomb-, dipolar/multipolar and hydrogen bonding terms, as well as contributions from the intrinsic distortions, bending and deformations affecting the distributions of electron density within the individual host and guest molecules. Though considerable improvements on the original idea were implemented over the years, and even gained in finesse by the inclusion of scaling factors to account for environmental influences (e.g. distance-dependent dielectric permittivity), the fundamental concept of enthalpy-based two-partner two-state binding remained unchanged and still dominates most attempts aimed at understanding molecular host–guest relations today.

In the description and explanation of real experimentally testable situations, the lock-and-key model claims a success story; however, in many cases – and among them are the majority of the more interesting biological examples – it fails to explain host–guest binding affinity and selectivity with reasonably ambitious satisfaction. For instance, anion binding in water very frequently shows endothermic rather than exothermic enthalpies of association, an observation that is incompatible with the naive complementarity model.

The reasons for the mediocre predictive power of simple models are fundamental and obvious: above all they suffer from the blunt comparison of binding enthalpies that emerge from the model treatment with free energies of binding $\Delta G_{\text{assoc}}$ that derive from the readily accessible affinities by straightforward recalculation ($\Delta G_{\text{assoc}}=-RT\ln K_{\text{assoc}}$). Such relations are equivalent to the neglect of the entropic contributions to binding that is justified only in the rare case if the interest in associations is restricted to the temperature regime near zero Kelvin (furnishing $T\Delta S \sim 0$) or the entropy of association (as $T\Delta S_{assoc}$) is very small to render the enthalpy the dominating component. In reality this situation is considerably less frequent than commonly assumed, and in addition also depends dramatically on the polarity of the environment. Entropic factors tend to be of greater importance in intermolecular binding in the protic solvents typically required to generate anionic species as free entities in solution. For the calculation of anion binding in polar solvents around ambient temperature it is mandatory to employ free energies $\Delta G_{assoc}$, although they are less accessible than plain enthalpies and eventually require the more sophisticated and lengthy calculation methods of molecular dynamics.

A second cause for the weakness of the complementarity model arises from the misconception that the binding process of host and guest is essentially the same whether it occurs in vacuum or in a condensed phase. For the latter this idea leads to the frequently voiced opinion that there is no physical contact between the host–guest partners in solution (they stay separated) if the binding free energy is zero. It follows from simple inspection of the relation $\Delta G^0 = -RT \ln K_{assoc}$ that at $\Delta G^0 = 0$ the association constant $K_{assoc}$ equals 1, i.e. for non-zero concentrations of host and guest there is also a finite concentration of the 1:1 associate complex present. Is this binding? The case has been investigated in the context of protein denaturation by low molecular weight additives and has recently been lucidly unfolded by Schellman [5] and Timasheff [6]. The interaction between host and guest molecules, to the extent that the occupation of a binding site on the host matches the concentration of the guest in bulk solution, corresponds to the case of random collision and invokes the replacement of solvent from this site. Unlike the binding scenario in vacuo, the host–guest interaction in liquids is a genuine exchange process with solvent molecules that may favour or disfavour the uptake of the guest. Thus, the measurable free energy of binding $\Delta G^0_{assoc}$ is negative, and only ordinary binding isotherms are found if the interaction of the site with the incoming guest is more exergonic than the interaction of the same substructure with solvent ($\Delta G^0_{solv}$; $\Delta G^0_{assoc} = \Delta G^0_{guest} - \Delta G^0_{solv}$). An exergonic interaction of host and guest in the absolute sense is insufficient to bring about binding.

The very same concept of preferential binding over mere solvation is equally suited to accounting for the effects that rest on preferential exclusion of the guest from the binding site resulting in enhanced solvation. For proteins, osmolytes exerting stabilizing effects on protein structure in water (glycerol, sucrose, trimethylamine oxide etc.) belong to this category of guests. Meticulous thermodynamic analysis reveals that: "When the interaction is strong it is identical to the usual molecular definitions of binding. When the interactions are weak or negative they can differ significantly from the molecular description" [7]. As anion binding in protic solvents frequently comes with small or even positive association enthalpies, indicative of the similar magnitude of the interaction of the host site with guest or solvent, we can expect unorthodox binding phenomena that are not readily understandable in terms of a naive stoichiometric molecular binding model. Instead, the concept of a single binding site and along with it the idea of a unique host–guest complex structure breaks down as the interaction enthalpies with guest and solvent become more and more alike, which gives room for a greater heterogeneity and a statistically more blurred interaction type. This trend is not abrupt and does not show up in the analytical expression of binding. In fact, simulation calculations of a multiple-site protein system demonstrated that the individual binding constants characterizing the affinity at each site could be varied by a power of ten without noticeable impact on the binding isotherm [8]. As no use of the specific nature of the protein was made, the results must apply to all supramolecular host systems interacting weakly enough with their guests. One is urged to conclude that under these energetic conditions even a fine and well-fitting representation of the experimental data by a stoichiometric model may not have a true physical meaning. The binding parameters deduced from the model may well constitute a weighted average of individual binding constants and thus conceal the underlying diversity of the system. As a corollary, it is rather unlikely – though not totally excluded – that host and guest form a complex of well-defined topological correspondence of individual moieties, i.e. with a fixed distance and orientation between anchor groups of both binding partners. Rather, a great structural fluctuation in the complexed state comprising a huge number of modes (in proteins one would call them microstates) at any position within the host reachable by the guest must exist, and many of these substates will be thermally populated. This also applies to the temporary sequence of complex arrangements occurring at one particular position.

The important perception is that this scenario is a direct consequence of the participation of a third player, the solvent molecules, in the host–guest binding process. The role of solvation, not so much as a diffuse influence on the environment than as a molecular factor, went unaccounted for in the bilateral complementarity models. The magnitude of the solvation energies, at least in the more polar solvents, is the same or even greater than that of the direct interaction of the guest anion with a receptor site, enabling a multitude of structural arrangements of similar energy that are all populated at ambient temperature and by virtue of this diversity contribute to the positive entropy of binding. On the other hand, structural uniqueness is lost in this situation owing to the scrambling effect of solvent participation. This weakens the attainable discrimination of guest species and must be regarded when designing abiotic receptors.

Still another aspect with a bearing on the selectivity of anion binding of real host–guest systems, that is commonly not addressed when using complementarity models, stems from the crude assumption that the association of the anion host with its target guest is the only relevant and accountable process in solution. This implies that all other ingredients of the system are inert and stand innocently aside. Since at least the anionic guest cannot be added to the system without an equivalent amount of countercation, one tacitly assumes that neither the negatively charged guest nor the complex with the host will interact with the countercation significantly, rendering all these species strong electrolytes under the solvation conditions applied. The validity of this premise is even more dubious if the receptor itself carries a charge. The risk for breakdown scales inversely with the magnitude of the specific interaction. In favourable cases the interference from simultaneous ion-pairing equilibria can be included in the analytical data treatment and then may furnish a diagnostic tool to assess the complexity of the system, as has been demonstrated by Gibson et al. in the binding of organic cations to crown ethers [9, 10]. In most concrete cases, however, one must accept the curtailment that the experimentally accessible binding constants are entirely apparent and do not represent true thermodynamic entities. Recognizing and appreciating this caveat is not only important to avoid wrong conclusions as it touches on host design, but it may also resolve some discrepancies between theoretical calculations and experimental facts. An informative example was reported by Orozco et al., who studied the binding of anions to polypyrrole receptors by calculation [11] subject to the influence of counterions and various solvents.

The involvement of simultaneous equilibria is not restricted to parallel reactions like ion pairing. Consecutive processes like the stepwise formation of higher order complexes may also add to the complexity. Some spectacular examples in anion binding have been unravelled by Gale et al. [12]. Unfolding the intimacies in the sequence and strength of binding events requires a fortunate disposition of several prerequisites. Above all one needs a traceable probe acting in a suitable concentration domain (which in turn also depends on the method of analysis used) to observe the representative distribution of complex species (the speciation) and an appropriate and unchanged regime for the dynamics. Taken together these restrictions leave but a small window open to look at and investigate the situation when a guest species acts as either a competitor or a promoter of binding. The latter case features positive cooperativity in which the later bound guest molecule(s) associate with stepwise higher affinity than the preceding ones. Depending on the extent of coupling between these steps, i.e. on the affinity enhancement, this upward regulation may result in an on/off switch response signalling the presence or absence of the guest in a small concentration range. Generally, we should expect these sophisticated systems to mix up the effects on signalling due to the change in the concentration of the complex species with their certainly distinguished ability to elicit the response necessitating quite complicated models, which ultimately might not be differentiated because of the experimental limits. At an elevated level of modelling allosterism in abiotic receptors, we may adapt and get inspired by examples developed to explain biological phenomena (e.g. the Wyman–Monod–Changeux (WMC) or Koshland–Nemethy–Filmer (KNF) models). However, even these sophisticated models at best give a xylographic picture of higher order complexation.

## 3 Thermodynamic Selectivity

When we accept the primitive standpoint that anion sensing reflects a host–guest binding phenomenon (see above), we have already laid the foundation of a conceptual model that can correlate macroscopically observable quantities like the equilibrium constant (affinity, $\rightarrow \Delta G$), the heat effect observed when mixing solutions of host and guest compounds ($\Delta H$) or the energy required to raise the temperature of that mixture by a certain span ($\Delta C_p$) with molecular events. The stringency of this link is subject to the subtleties and finesse of the particular model and on its quantitative reliability; however, its mere existence provides us with the possibility of molecular design. The translation of structure (whatever its definition) and interaction between structures into thermodynamic quantities and vice versa, enabling in principle the evolutionary optimization of an artificial receptor, nowadays appears only natural and rational from the viewpoint of a chemist. The fundamental framework of thermodynamics, which still rules many engineering applications, has been developed entirely without reference to the atomistic nature of matter. The breakthrough that opened the field to chemistry derived from the statistical treatment and the summation of elementary events to make up macroscopic observables. A cool and unbiased inspection of the impact made by this merger of ideas on molecular design must acknowledge quite a disparity in anchoring the state functions of enthalpy $H$, free (Gibbs) enthalpy $G$ and entropy $S$ in the conceptual thinking of molecular architects. Whilst the influence of the latter is rarely regarded as a tuning handle in the construction of abiotic host compounds, the former two are in the focus of all design attempts. The goal is to maximize the difference in free enthalpy $\Delta\Delta G$ of binding the target guest (let it be A) versus some interfering competitors ($^1B$, $^2B$,…$^nB$). The energetic situation is sketched in Fig. 2.

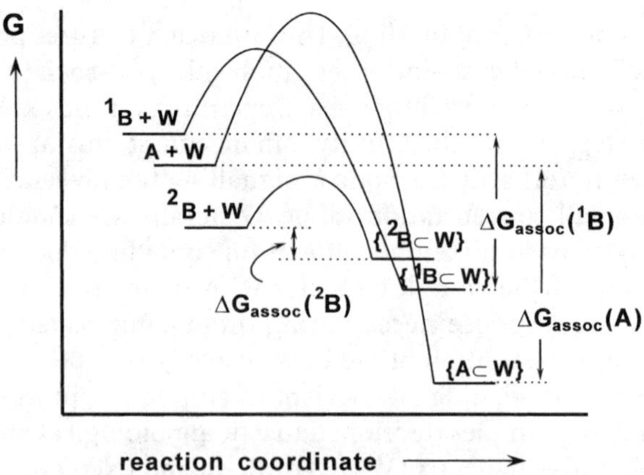


**Fig. 2** Free enthalpy diagram for the complexation of the specific guest A to host W in relation to the competitive substrates $^1B$ and $^2B$.

In the starting state (this is not a state in the thermodynamic sense!) comprising the unbound species the energy levels of the individual host–guest pairs vary as the guest species differ in internal energy and solvation. The complexed state features the bound species encompassing their intrinsic mutual interaction and the solvation of the individual host–guest ensemble. Neither of these "states", which are connected in the diagram by lines of conversion characterizing a reaction profile (in this case the most primitive profile containing just one transition state of arbitrary height was chosen), is experimentally accessible. They are purely fictitious and may only serve to illustrate the derivation of the measurable quantity $\Delta G_{assoc}(A, {}^{1}B, {}^{2}B\ldots)$ emerging as the energetic difference of these states. As a consequence, the selectivity as mentioned above arises as a difference of differences with no special anchoring or reference to one particular state. From this perspective it is senseless or even counterproductive to tailor host–guest interactions solely by the design of the complex structure. The refinement thereby introduced into the host structure may or may not show up as an increased free enthalpy of binding depending on its effect on the starting state, which definitely will not remain untouched. Moreover, even if the refinement was beneficial in binding the target in an absolute sense the structural modification may raise the free enthalpy level, e.g. of the host + competitor $^{1}$B pair (cf. Fig. 2), resulting in augmentation of competitor complexation with no overall improvement in $\Delta\Delta G$. It is this multifaceted interplay of consequences on each of the participating states, along with the comparatively small energetic differences left over from the trade-off of huge amounts of direct interaction/solvation energies, that renders the outcome of molecular design so hard to predict.

The scenario illustrated in Fig. 2 is completely artificial and does not reflect the true conditions where selectivity must be expressed, because the binding events are considered pairwise (host + guest A versus host + guest $^{1}$B etc.) in the *absence* of the competitor and the results are then related to one another. In all real cases thermodynamic binding selectivity is the composite outcome of complexing the target guest in the *presence* of all presumptive competitors, which may vary in a wide range with respect to chemical nature and quantity. In an exact analysis one must quantify the preference of a molecular host W for the desired target species A over its competitors $^{1}B, ^{2}B, \ldots ^{n}B$ by means of constants of mass action (Eq. 1) comprising all stoichiometries possible and populated $^{1}K_A$, $^{2}K_A\ldots$, $^{1}K^{1}_{B}$, $^{2}K^{1}_{B}\ldots{}^{m}K^{n}_{B}$ that in addition must be scaled with respect to the actual concentrations. Generally, there are two ways to assess the thermodynamic selectivity $R_t$. The first (Eq. 5) relates the concentration of the desired complexes $[A_n{\subset}W]$ (Eq. 3) produced from the host and the specific guest A in equilibrium with the free host [W] and guest [A] to the entirety of all host complexes present (Eq. 4).

$$^{1}K_A = \frac{[A \subset W]}{[W] \times [A]}; \qquad ^{n}K_A = \frac{[A_n \subset W]}{[W] \times [A]^n} \tag{1}$$

$$[A \subset W] = {}^{1}K_A [W] \times [A]; \qquad [A_n \subset W] = {}^{n}K_A [W] \times [A]^n; \tag{2}$$

$$[A\subset W] + [A_2\subset W] + \ldots [A_n \subset W] = [A] \times [W] \sum_1^n {}^nK_A \times [A]^{n-1} \quad (3)$$

$$[{}^1B\subset W] + [{}^1B_2\subset W] + \ldots [{}^1B_n \subset W] = [{}^1B] \times [W] \sum_1^n {}^nK_{1_B} \times [{}^1B]^{n-1} \quad (4)$$

$$R_t = \frac{[W] \sum_1^n {}^nK_A [A]^{n-1}}{[A] \sum {}^nK_A [A]^{n-1} \div \sum_1^m \sum_1^n {}^nK_{m_B} [{}^mB]^n} \quad (5)$$

$$R_s = \frac{[A]\,{}^1K_A}{[{}^1B]\,{}^1K_{1_B}} \quad (6)$$

In this instance the selectivity can vary between zero and one and it is immediately apparent that the numerical value of the thermodynamic selectivity is not constant in an ensemble of potential competitors, but depends on its quantitative composition. Selectivity, thus, in any real system is not a molecular property of the host alone, but a function of all components participating. On this basis, however, an evaluation of the merits of design in a series of hosts taking selectivity as a scale for achievement is nearly impossible. Simplifications can be introduced by restricting the number of competitors to just one and further limiting the case to the consideration of 1:1 stoichiometric interactions only. In addition, the concentrations of uncomplexed target analyte and its competitor are set identical. If thermodynamic selectivity is then expressed in a second mode ($R_s$) as a ratio of desired (target) to undesired (competitor) complexes, one arrives at Eq. 6. Under the premises given this expression condenses to the familiar ratio of association constants obtained in the absence of each other.

Clearly, the constraints entered above enable an exact but rather hypothetical comparison of selectivity features, which may fail to live up to the expectations advanced on reasonable grounds in a real application. However, selectivity as defined in the second mode holds the charms of allowing the decoration of the various host designs with high and thus hopefully appealing numbers of merit. Both definitions convey the fundamental point that selectivity in no way is an absolute quantity attributable to a certain host structure like a molecular property. Nevertheless, it is a graded measure characterizing the quality and suitability of a host design with respect to the particular molecular environment (which must be mentioned), and as such can report on the progress in the understanding and realization of molecular recognition.

Thermodynamic selectivity $\Delta\Delta G$ can become large only if the underlying free enthalpies of association $\Delta G$ are even greater. Though maximizing affinity ($\Delta G$) in principle should help thermodynamic selectivity (however, see discussion below), there is an upper limit to this aim due to practical limitations. Taking the popular biotin–streptavidin interaction as a prototypical example of a high-affinity supramolecular complex ($K_{assoc}$~$10^{14}$ $M^{-1}$) and assuming the fastest possible association of host and guest at the diffusion limit (second-order rate constant in water at ambient temperature $\sim 10^9$ s$^{-1}$ M$^{-1}$), one can calculate a monomolecular rate constant for dissociation of the complex of $k_{off}$–$k_{on}$/$K_{assoc}$≅$10^{-5}$ s$^{-1}$ corresponding to a half-life of almost a day, or more than a week for establishing equilibrium. There are not many relevant processes known that depend on thermodynamic selectivity in this time regime.

Biological systems which usually operate on a much shorter time span, but are nevertheless in need of high affinity and selectivity, bypass the problem of slow dissociation by polyvalent binding. This binding mode refers to the simultaneous fixation of several moieties (anchor groups) of one binding partner capable of dedicated interaction with a set of corresponding substructures of the other. The participation of multiple interaction sites results in the accumulation of all free energy contributions compounding the total energy available for differentiating action. Contrary to single-site binding, the adhesion of host and guest can be destroyed in steps by competition with monovalent guest analogues leading to greatly improved dissociation kinetics. In addition, polyvalent binding can be modulated by the mutual influence of the anchor groups on one another. This influence can turn out to be opposing to binding, i.e. the following binding steps addressing the second, third, etc. anchor groups occur with smaller affinities (smaller microscopic $K_{assoc}$ values) than the preceding ones. Though the interaction is progressively weakened in each step, the overall affinity increases with every supplementary site contributing some free energy $\Delta G_i^0$ ($\Delta G_{total}^0 = \Sigma \Delta G_i^0$; $\Delta G_i^0$ is more negative than $\Delta G_{i+1}^0$). Cooperativity between the various anchor groups is negative and in general this situation appears to be more common. However, the alternative scheme in which the affinities in subsequent binding steps are progressively strengthened, giving rise to positive cooperativity ($\Delta G_i^0$ more positive than $\Delta G_{i+1}^0$), holds a peculiar virtue with respect to function: low-affinity binding of the first anchor group to a molecular target can trigger the binding of all subsequent steps, thus providing an overall free energy of interaction that can exceed a certain threshold value defined for initiating some function. Synergistic action here approaches an all-or-nothing switch that is useful on many occasions in supramolecular chemistry, from the clearance of antigens from the blood stream by the immune systemto the formation of kinetically stable molecular capsules from calixarene building blocks.

Another important aspect in the same vein concerns the relationship of selectivity and affinity. Though it seems straightforward to expect that the high affinity of a molecular host for the correct guest will enhance its capacity for differentiation and thus will boost selectivity, there is no strict interconnection between these features. One may well imagine a host compound that rearranges its structure on binding the specific target in an energy-demanding process, rendering just a minute amount of observable free enthalpy. This host displays low affinity, yet is specific because most competitors cannot bring about the conformational change necessary for binding, and consequently in the ideal case will not be complexed at all. This so-called induced fit scenario is most effective in discriminating against non-specific guests when adventitious complexation cannot be avoided due to high concentrations of the competitor or its particular physical properties (charge, hydrophobicity, size etc.). Conversely, if the host was totally rigid and just offered a cavity of well-defined dimensions, only guests smaller than a threshold size could enter and form a complex. For establishing selectivity versus larger guests it is irrelevant whether there is good fit of the small guest capable of invading the cavity, because all the bigger ones cannot penetrate at all. Of course, in most real cases selectivity must also be expressed in a certain concentration regime, which imposes hard restrictions on the affinity level that must be realized, but in principle affinity and selectivity can be manipulated independently from each other.

Tailored shaping of thermodynamic selectivity requires considerable insight into intimate molecular details. In comparison, the blocking of such delicately balanced functions by brute binding force interference should be easier to achieve. This lays the foundations for the best-established application in tailoring thermodynamic selectivity: rational drug design. Guided by complementarity principles, artificial compounds are looked for that would interact best with a target site, for instance, in a protein that is essential for some biological function. Better binding of the artificial guest prevents binding of the natural substrate and interrupts its proper function. The need for selectivity is basically twofold because drug binding has not only to override the natural host–guest process, but also must avoid interfering at any other potential target site (here disregarding the equally important so-called ADME (absorption–distribution–metabolism–excretion) characteristics). The approach chosen most frequently employs the computational maximization of the interaction enthalpy of drug and target, with or without taking solvent participation and the internal flexibility of the binding partners explicitly into account (docking). However, as we have stressed above, the tacit assumption that the enthalpies of interaction $\Delta H$ derived in this way reflect or even parallel binding free energies $\Delta G$ is not founded. Rather it is common experience that, owing to the phenomenon of enthalpy–entropy compensation, dramatic trends in enthalpy variation do not materialize in equilibrium affinities. Nevertheless, in the absence of practicable methods to account for association entropies, and with the proviso of being conscious of their fundamental limitations, docking procedures can provide guidelines in thermodynamic selectivity optimization.

## 4 Host–Guest Binding as Seen from the Viewpoint of Statistical Thermodynamics

To illuminate the binding situation we can retreat to the statistical thermodynamics point of view. Experimentally accessible state functions ($\Delta G$, $\Delta H$, $\Delta S$, $\Delta C_p$ etc.) relate to the differences of *average states* in the ensembles of all constituents of the system including counterions, buffers, solvents etc. before and after the molecular association event. The ensembles possess a number of distinct mutual arrangements of all molecules, each one characterized by an energy value, and for all real cases in solution the number of these arrangements will be very large. This also applies to the differences in the molecular configurations describing the change occurring on host–guest binding, i.e. the difference between the initial and final states of the binding process.

For this discussion it is appropriate to set the free enthalpy $G$ of the uncomplexed ensemble at 0, which makes $\Delta G=G_{complex}-G_{free}$ independent of concentration and corresponds to the choice of a standard state. In any solution phase the energy levels will be very narrowly spaced or even degenerated (i.e. they have exactly the same energy in distinguished arrangements), and one can define numbers $m(E)$ within the number of all physically possible states $M$ having an energy between $E$ and $E+dE$. Statistical mechanics tells us that the entropy can then be calculated from $S(E)=R \ln m(E)$, and thus represents the scaled logarithm of the sheer number of possible binding modes having a particular energy value. The energies of molecular configurations, however, vary from an absolute minimum value, the most stable arrangement of the host–guest complex in its environment, to a maximum value that features the utmost repulsion of all interacting partners within a set of physical constraints. Which in this collection of states is actually populated at thermal equilibrium and to what extent is governed by the Boltzmann distribution. Every value in the energy spectrum $m(E)$ must then be multiplied with the respective Boltzmann factor $\exp -E/RT$ to arrive at the population distribution of the molecular complexes $n(E)=m(E)\times\exp -E/RT$. From this relationship it is immediately clear that the population distribution as a function of the interaction energy must feature a maximum at some higher than minimum energy value, because it is the product of two functions with opposing slopes at moderate energies (in relation to the thermal energies present at experimental temperatures). While the exponential Boltzmann term decreases monotonously with increasing energy, the term representing the number of available energy states $m(E)$ increases steadily at low energies. Conventionally, one can set the most stable molecular configuration (the ground state) at the origin of an energy plot. Since there is only one configuration of minimal energy possible, by definition, the entropy of this state would also be zero ($S(E)=R \ln 1$). At higher energies many more binding modes are possible, corroborating the purely qualitative statement above. At thermal equilibrium all states are populated according to $n(E)$ and their summation yields the partition function $Q$ of the associative process ($Q=\int n(E)dE$). The experimentally accessible free energy is then given by $\Delta F= -RT \ln Q$ (the free enthalpy $\Delta G$ is a little more complicated because the expression contains additional terms to account for the volume work).

For any real binding case in condensed phases the situation is much too complex to allow the synthetic composition of the partition function. Nevertheless, the fundamental equations yield some insight into molecular binding processes that have been ignored or underappreciated in discussions of artificial molecular recognition.

For instance, it is obvious that the free enthalpy of the ground state (which is also the enthalpically most stabilized state because the entropy component is zero) calculated frequently in molecular modelling packages, with or without inclusion of solvent, cannot in any generality be related to the experimentally measured free enthalpy of complexation. The latter refers to an average energy state comprising all populated binding configurations. Unless just one binding mode is enthalpically much more favoured than all others and thus is the only one populated, the tacit assumption of a single interaction pattern of the binding partners has no physical basis (see below).

However, whether or not a molecular interaction rests on one dominant configuration in solution cannot be deduced easily. The problem touches on the fundamental issue of structure–energy correlation, and is heavily loaded with justified and/or completely arbitrary presumptions that in many cases reflect a traditional bias rather than a well-balanced appreciation of scientific arguments.

The design aspect of host–guest chemistry mandates the knowledge-based construction of novel compounds possessing desired properties. The concept of structure seems indispensable in this enterprise, although the term is ill-defined and the International Union of Pure and Applied Chemistry (IUPAC), for instance, consciously avoids the close specification. As a corollary, many fields in science have developed their own interpretation of structure, which sometimes clash with each other and lead to misunderstandings when issues at the common interface of subdisciplines are in need of rationalization.

Structure–energy correlation suffers from such a constellation. While in thermodynamics it needs no special mention that all discussions relate to an initially defined state of the *entire system*, the organic chemistry point of view in general is focused only on the direct mutual interactions of the binding partners. They are looked upon primarily with respect to their molecular skeletons, the proper disposition and orientation of "sticky" functional groups and a purely arbitrary dissection of the overall molecular frameworks into segregated anchor groups. From conformational analysis and in particular from biological host–guest binding studies, the notion transpires that the covalent connectivity is not sufficient to comprehend supramolecular association. Instead a structural hierarchy must be found that is dependent on the molecular environment (solvent, buffers, "inert" salts, phase boundaries etc.) as well as on the time regime taken for study. Among other factors it is the latter restraint that makes the results of the observations dependent on the methods used. For instance, a spectroscopic method probing a special site in the host–guest complex on the microsecond timescale may not report on slower processes happening in the domain of seconds. A method that senses the latter action, too, will necessarily arrive at a different binding constant than the other. If there were agreement of all methods used on the binding constant, however, this serves as a strong argument for true equilibration of all binding modes being established.

The exceedingly high complexity of most molecular associations in solution requires a reductionistic approach that simplifies the real situation to a model.

It is the very nature of a model that only certain aspects of the systems' behaviour are correctly reflected and even this only if provisional constraints are met. Disregard of these restrictions will inevitably lead to erroneous conclusions of host–guest binding intimacies. Model thinking inspired by analogies like the always cited lock-and-key metaphor of molecular recognition, coined by Emil Fischer more than a hundred years ago, emphasize the geometric fit of host and guest but totally ignore the energetic change in the remainder of the system which might even contribute the major fraction to the overall energetics. It is hard to imagine a strong structure–energy correlation, if a respectable if not dominant contribution is constantly omitted.

In a similar way it is widely believed that the mutual interaction of two binding partners in solution involves just one binding mode (the so-called associated–dissociated two-state postulate). Such an assumption appears to be a prerequisite to informed improvement of the molecular design and is based on a plentitude of solid-state complex structures and additionally on the observation of massive exothermicities (negative $\Delta H$) in many, but by no means in all, cases of molecular complex formation.

While X-ray crystal structures of host–guest binding convey a concrete possible binding mode they do not suggest that this is the major, let alone the only one in solution. On crystallization many degrees of motional freedom within the complex are lost and the concomitant reduced entropy is paid for by additional enthalpic interactions which make up the crystal lattice energy. These inter-complex interactions eventually are large enough to squeeze the complex components into positions that seem totally unrealistic when relaxed, for instance, in solution. Of course, the probability of a close match between the crystal and the solution complex structure depends on the rigidity of host and guest, since well-structured binding partners cannot lose much conformational entropy on transfer to the confines of a crystal lattice.

A similar argument holds when the solution structures in a series of solvents of varying polarity are compared to the structure found in a crystal. The less polar solvents are more likely to contain complexes of similar structures to the latter because their smaller solvation energies are less likely to disrupt or rearrange the original. However, there is also no guarantee that a particular host–guest configuration observed in one crystal is actually the most stable one, since it is a common experience that low molecular weight as well as protein–ligand complexes show polymorphism and hold the association partners in different and sometimes rather odd relative configurations in different crystal variants.

A large negative enthalpy of association $\Delta H_{assoc}$, however, is a good sign that there is a prevalent binding mode in solution, because very frequently the increase in exothermicity is counterbalanced, at least in part, due to enthalpy–entropy compensation by a likewise negative entropy component $\Delta S_{assoc}$. This compensatory effect is very general and limits the gain in affinity (free energy) towards a guest species, e.g. when an enthalpically favourable variation is introduced by molecular design into an artificial host compound. Enthalpy–

entropy compensation is the major cause of the difficulty in reading clear trends in the attempt to produce guidelines in host design from deliberate structural modifications introduced in a series of host compounds, if only the association constants of the complexes (i.e. the free enthalpies) are taken as the exclusive observable.

The strongly negative entropy of association $\Delta S_{assoc}$ can be related to an increase in the population of ground states $P_{gs}$ from the dissociated mode to the associated complex by $P_{gs}$(complex)=$P_{gs}$(unbound)×exp $-\Delta S/R$. While zero association entropy does not alter the population distribution, any negative value enhances the fraction of all complex molecules in the ground state. Statistical mechanics does not reveal structural subtleties, but the statement that a unique energy of the ground state relates to a unique structure in this situation seems safe. Of course, the measured entropy change also in this instance refers to the entire system state, but it is useful and customary to divide the total system into two subsystems with necessarily additive energetics. One comprises all the alterations in solvent reorganization that happen on host–guest association, and the other describes the mutual intrinsic energetics of the interacting partners. Traditionally, it is only the latter part that was of interest to the molecular architects and it is a major goal of this article to point out the fallacies in this constraint.

In the concrete case of entropy-opposed association (negative $\Delta S_{assoc}$) the molecular scenario as usual involves the desolvation of the interacting sites. The solvent molecules set free in this process contribute a positive term to the overall entropy, which is accounted for in the solvent reorganization subsystem. In order to make the total entropy change negative as measured, the host–guest intrinsic interaction system must override the desolvation by an even more negative entropy contribution. This is brought about by the association process itself that annihilates six degrees of translational and rotational freedom by making one molecule out of two, and the freezing or constriction of conformational flexibility due to the extended interface of the binding partners. The newly formed associative interaction, however, gives rise to molecular vibrations which in principle add a positive entropy term again. The actual amount depends on the vibrational frequencies. As can be derived in a semiquantitative manner [13], the vibrational entropy around room temperature is appreciable only at frequencies in the vicinity of 400 $cm^{-1}$ and below. Greater exothermicities imply higher force constants and thus higher frequencies, which in turn lead to a smaller contribution in the adverse entropy. More negative enthalpies of association thus entail more negative entropies (another case of enthalpy–entropy compensation), with the result that alterations in the summation state function, the free enthalpy, may be minute. Although there might be very little or no difference in affinity to alternative scenarios, a molecular association characterized by negative enthalpies and entropies features a more structured associated state that is more likely represented by a single host–guest configuration, and thus lends itself to more ready interpretation along the routes outlined by the lock-and-key analogy.

However, the particular thermodynamic signature discussed above is less widespread than commonly assumed. In particular, in strongly solvating and structured solvents (i.e. in all hydrogen-bonding solvents and especially in water) it is quite common to observe positive entropies of association. These come along with negative or even positive (opposing) binding enthalpies, the latter clearly indicating that the direct mutual interaction of the host–guest partners does not dominate the energetics of the global system. In both cases the stability of the associated state as conveyed by the negative free energy is entropy-driven, and as such is substantially governed by the statistical arrangement of the global system. It is not at all reasonable in this situation to attempt the comprehension of the overall energetics by considering the mutual fit of the interacting partners alone.

As an extreme case it is well conceivable that there were no attractive interactions between the associated partners at all, and their association was exclusively due to the endeavour of solvent molecules to minimize the inner surface created in the solvent to accommodate the complex. Since solvation energies are uniformly negative with respect to the gas phase, the deprivation of the contact surfaces from solvent leads to a positive energy that is not counterbalanced by an energetic gain from the direct contact (because this was deliberately set to zero) and must consequently show up as an endothermic result. Nevertheless, a negative free energy translating into a high binding constant $K_{assoc}$ must be possible if the total entropic gain concomitant to the release of solvent molecules, among other factors, overrides the energy loss at the experimental temperature. Again, the gain in solvent entropy can be attributed to the solvent reorganization subsystem and as such is inaccessible to design. Open to design are the molecular frameworks of the interacting partners and though there might be no mutual attraction discernable the structural layout of the contact sites, i.e. the diversity of populated conformational and configurational states, determines the amount of surface hidden from the solvent and, as a corollary since this is the relevant observable, the affinity constant.

According to this analysis, the high diversity of energetic states emerging from the flexibility and high conformational entropy of host and guest in conjunction with the residual solvation results in numerous binding modes, which are all populated because they refer to very similar surface areas being desolvated in the course of association. The binding modes may vary widely in the pairwise correlation of atomic positions in host and guest, but nevertheless they contribute at the same level to the average free energy. Thus, the greater the multitude of binding arrangements, which is an immediate corollary of the structural flexibility, the more positive is the association entropy and consequently the higher is the affinity.

The same reasoning also applies to all molecules possessing completely different covalent skeletons: as far as these species cover a similar fraction of molecular surface buried on association they will compete with the binding of the original guest. Consequently, molecules capable of ready adaptability of their structure to the needs of a binding site cannot be excluded from binding.

Flexibility in general does little harm to the overall binding capacity but is deleterious to selectivity. Conversely, very rigid and geometrically well-structured surfaces may not give a good fit but can define a unique mutual arrangement that is distinctly better (less host–guest interface desolvated) than any other bilateral configuration. In such a case selective binding of this guest will be observed.

From our present discussion, which deliberately avoids any specific interaction of host and guest, the conclusion is inevitable that there is *no* mandatory energetic requirement for a precise fit (e.g. lock-and-key) that stabilizes the host–guest configuration in solution. It does suffice to exclude any potential competitor from binding in the same low-level energy regime as the specific guest. Of course, this goal is more easily met if the interaction energy is strongly negative instead of zero, as in our fictitious scenario. In addition, in many applications of molecular recognition, especially in the living world, selectivity needs to be expressed in a certain concentration range which frequently mandates high affinities to cope with tiny guest concentrations. Selectivity and affinity in these cases are connected via the function to be fulfilled. But in principle the notion transpires that, from the thermodynamic viewpoint, affinity and selectivity can be realized independently from each other and do not require a close relation.

If selectivity or affinity is taken as the target for molecular design, affinity appears to be the goal easier to reach, because fundamental conditions in any environment are known that would lead to enhanced association of the host–guest partners by the sheer accumulation of basic interaction modules. For instance, the introduction of charged groups of opposite polarity in both partners by virtue of Coulomb attraction will almost inevitably lead to stronger binding in low dielectric media. On the contrary, decoration with hydrocarbon or even more, with fluorocarbon groups will augment their stickiness in water to the extent that they eventually form a separate (pseudo)phase coexistent with the aqueous solution.

Selectivity in turn seems to be tied to the singularity of a low-energy binding mode. This must not be taken literally to mean that just one configuration comprising host and guest, and all solvent molecules that are directly attached, is populated at thermal equilibrium. Rather a family of similar configurations having a common motive exists that may differ in molecular detail and thus forms an ensemble of substates. This ensemble needs to be energetically distinct and well separated from any alternative formed in the interaction with a competing guest compound. While the task in molecular design presented in this way seems straightforward and has been tackled by a great number of rational approaches, experience tells us that many reasonably founded expectations for selectivity could not be verified experimentally.

One important reason for the mediocre performance of many artificial hosts may originate from the ambiguity of the binding modes. If we take host–guest binding of anions in water as an example, the blueprint of host design almost always relies on the creation of polar interactions which should attract the guest in a prescribed manner. Frequently hydrogen bonding is engaged based on its strength and known orientational preference, and corresponding functional groups are embedded into covalent frameworks selected according to their capability of setting up the dedicated interactions in the anticipated fashion and to their synthetic accessibility. Rigid skeletons which do retain a defined three-dimensional structure in aqueous solution, either by virtue of their covalent connectivity or by a distinct folding process, and thus could position the hydrogen bond donor groups in the prospected way are quite hard to prepare. As a compromise less rigid structures are chosen, which by informed conformational twisting can bind the guest anion in an enthalpically favourable mode using multiple points of interaction. However, it is well appreciated that hydrogen bonding in water is enthalpically almost silent [14], because it involves just the replacement of hydrogen bonds in the solvation layer of the respective functional group by the host (respective guest) rather than the formation of new ones. At every single site there is little if any enthalpic well, favouring one-point binding of the host–guest structure relative to the hydration of the dissociated substructures. The partly dissociated host–guest configurations, which do not employ the maximal number of attachment sites in binding, are structurally quite diverse and only marginally less stabilized than the minimum energy configuration. As a corollary, they are heavily populated and rather numerous and therefore do contribute the major share to the average free energy that is translated into the association constant. In this rationale it is the fuzziness of structural definition emanating from the flexibility of the hosts as well as from the high competitive power of water molecules that spoils guest anion selectivity.

The strategic remedy may be read from biological receptors. Sulphate or phosphate binding proteins, for example, sequester their specific guests in preformed cavities hidden deeply in the protein interior to eliminate the influence of competing water molecules [15, 16]. The energetically costly dehydration of the guests that is required for transfer to this site is paid for by convergent and joint hydrogen bonding from at least seven backbone amide or side-chain functional groups that even allow the Brønsted acid–base properties of these guests to be addressed. The result is truly remarkable: both proteins bind their respective substrates in water with specificities exceeding $10^5$ over the very similar alternative oxoanion.

## 5 The Construction of Molecular Anion Hosts

Even though the fundamentals of host–guest binding at least in their crude outlines appear well understood, it remains a formidable task to translate these insights into the concrete covalent structure of a compound that in the end displays the anticipated molecular properties. Looking back on a history of 35 years of striving to accomplish selective anion binding by artificial host molecules, it seems fair to state that there is quite a rich abundance of systems for binding target guests with small to moderate affinities concomitant ($10^1$–$10^4$ $M^{-1}$) with mediocre discriminating power, whilst only very few examples stand out and in this respect recommend themselves for technical applications. This difficulty in constructing anion hosts of suitable selectivity and performance, among other factors, triggered the development of an alternative and quite successful approach in anion sensing using pattern recognition in arrays of abiotic anion receptors, which taken as a single entity display insufficient selectivity [17]. Another line of technological development employs combinatorial strategies to arrive at receptors exhibiting the desired sensing features [18]. Although this is a very promising and perfectly legitimate route to solve the stringent technical problem for a particular case, it does not really help in the academic urge to find out about the origin, scope and limitations of the influential factors governing molecular recognition, and thereby does not aid our fundamental understanding.

In order to generate a high-potential anion-sensing host, four issues must be considered and correspondingly answered. The first refers to the *function* that is to be performed by the sensing device. Here the purpose and the physical background of the sensing process need to be defined, because they will set the rigid outer framework for all pertinent solutions. At this stage the decision must be made as to whether the physics of sensing addresses the uncomplexed target species, as e.g. in ion-selective electrodes, or rather directly depends on its physical presence in the sensor itself, for instance, by interfering with quenching pathways of excited states, as in fluorescence detection.

The second issue concerns the *substrate* of interest, since its molecular properties finally must be recognized and distinguished from all other influences. In the case of anion sensing, the first inspection must address the global structure, because in many analytes of interest the anionic moiety constitutes an important, but not necessarily dominant, part of the entire species. Take phosphate as an example. In principle, the basic interaction pattern of the phosphoryl anionic group with a corresponding host might be very similar for inorganic orthophosphate or nucleotides, phosphatidic acids and even phosphorylated proteins. But it is evident that we are unlikely to succeed in creating an abiotic receptor capable of selectively recognizing the phosphoryl anion in all these different species. If the anionic moiety is but a minor appendix of a larger substrate, it may be a matter of definition and thus quite arbitrary to designate the respective receptor as an anion host. In the present discussion we can restrict ourselves to the consideration of small "typical" anions, in which the net negative charge dominates the molecular properties. Here the net charge size, its density and distribution, along with the overall dimensions, topology, polarizability, stereoelectronics, intrinsic binding preferences etc., must be analysed and exploited in host design. Also in this instance, naive thinking in models might introduce an unintentional bias concerning multiply charged anions: even the mere existence of familiar anions like $SO_4^{2-}$ is connected to its molecular environment. The sulphate dianion can readily be observed in the crystalline state where it is stabilized by the lattice of countercations, or in aqueous solution undergoing extensive hydrogen bonding. However, unsolvated sulphate has never been observed in the gas phase and is most probably thermochemically unstable, since stripping the hydrated sulphate cluster successively from its water molecules results in transprotonation ($SO_4^{2-} \times nH_2O \rightarrow HSO_4^{-} \times (n-1)H_2O + OH^-$) or Coulomb explosion [19].

In the next issue this result directs our focus to the *molecular environment*. Because anions as analytes, for electroneutrality reasons, bring in a stoichiometric number of cations, these are natural competitors in any attempt aimed at supramolecular binding by anion hosts. Unspecific ion pairing to form 1:1 complexes or higher aggregates ($(C_n^+A_n^-)_n$) is widespread even in water, especially with species of elevated charge. As a corollary, it is seldom justified to assume complete dissociation of salts in organic solvents of low dielectric permittivity like chloroform, dichloromethane, pyridine or the like to produce free anionic species, because fundamental Coulomb interactions will keep the concentration of free anions at minute levels. The solvent itself finally is of utmost importance as a competitor on the molecular level, as it defines the threshold screen against which preferential binding must stand up. Solvent molecular size was elegantly shown to be a decisive determinant of affinity in host–guest complexation [20]. The overwhelming influence of solvation is in many cases the ultimate remedy when a less than optimal design renders an artificial host insufficiently suited to perform under the originally prospected polar solvation conditions. Switching to a solvent of minor solvating power may then serve as a final retreat and allow molecular association.

After all, the conceptually best host design would just be an intellectual exercise if the experimental realization fails. Thus, the fourth point must address the preparative *feasibility*. Generally, the synthetic expenditure is kept at minimum although a priori one might suspect a strong correlation between preparative input and the output in terms of guest selectivity. Apparently, the guidelines for design at present do not justify strong confidence in a predicted outcome, making it prudent to limit the primary preparative investment. In addition to ready accessibility and modifiability, which in most cases also reduce the time and cost requirement, major points of concern also refer to the chemical stability and inertness under the sensing conditions. For instance, in spite of optimal synthesizability, preparative variability and sensing characteristics, a potential fluorescent sensor would be soon dismissed if its chromophore is bleached or chemically harmed during measurement.

The issues relevant in the construction of anion hosts cannot be grouped in a hierarchical order as there is an intimate interplay between all of them. Successful design mandates a balanced compromise in weighting the individual influences. Advice can and should be sought from pertinent calculations (molecular modelling/dynamics), from the successes and the very few examples of failure accumulated in the literature and collected in recent reviews [21–24], from inspiration provided by abundant biological examples and crystal structures, and not least from personal experience in the handling of host compounds and their decoration with appropriate groups to ease purification, adjustment of solubility, immobilization and conjugation with additional moieties.

As a rule a suitable compromise of answers raised by these issues will feature a host design as an intermediate between two extremes. On the one hand is the encapsulation host that aims at utmost complementarity of size and functional groups and their fixed preorganization, resembling closely the lock-and-key picture to achieve the desired selectivity. This concept requires complete desolvation of the guest on invasion into the molecular cage provided by the host. In combination with the rigid fixation of the host structure this imposes a substantial barrier which surfaces in slow kinetics of guest equilibration and exchange. This conceptual problem may limit the utility of this strategy for sensing purposes. Furthermore, encapsulation and a rigid preorganization of sticky binding functions require a high connectivity of the molecular framework that hampers easy access and modification of such compounds.

At the other extreme a string of anchor groups is connected covalently in branched or unbranched fashion and it is left to a folding process, either prior to guest complexation or even assisted by it, to arrive at a distinct three-dimensional complex structure. Many biological examples, from proteins to nucleic acids to certain carbohydrates, follow this strategy to bind to specific guests. Of course, lack of preorganization leads to inferior binding given the same type and number of sticky groups in an optimally preorganized host. This price is to be paid for the ordering entropy of folding opposing binding. The enhanced flexibility and adaptability in accommodating guest structures differing from the target one also diminishes selectivity; however, the great benefits of this approach in host design in terms of ease of preparation, modification and reuse after partial dismantlement apparently outmatches the disadvantages in selectivity generation, at least for the biological examples.

For artificial host–guest sensors most concepts rely on hybrids between these extremes. Most profitable in this respect is a building block approach, in which some readily available scaffold is chosen to hold a number of anchor groups capable by themselves of contributing only a fraction of the interaction free energy necessary for discrimination against competitors. It is the combined and concerted action of the entire set of anchor groups held in place subject to the direction of the scaffold that results in binding and differentiation of the specific guest from the competitors. This concept does not mandate simultaneous physical interaction of all anchor groups with their corresponding substructures in the guest, as is seductively illustrated in many pictorial representations. Rather it suffices if – on an average – the interaction with the specific guest furnishes a gain of substantially more negative free enthalpy than with any other competing species. Whether or not a single complex structure occurs that dominates the ensemble of associated binding partners and in addition resembles the usually depicted drawing is thermodynamically irrelevant, and should be viewed with suspicion. Singular binding data in general do not report on structures of associative complexes nor on their options of structural interconversion and exchange. Viewing the diversity of possibilities there is a high risk of misinterpretation and we possess a rather dull blade to distinguish between equally plausible alternatives.

The building block approach to host design outlined above relies conceptually on an incremental additivity principle. The accumulation of anchoring functions in a favourable layout to binding (as learned from 1:1 stoichiometric interactions) is expected to boost the binding and differentiation of the target guest from any analogue. Such additivity principles proved invaluable for the progress and development of chemistry as a science (cf. the concept of functional groups); however, the limitations of this approach are obvious. If two groups each holding a molecular property are squeezed into vicinity, the original character of the individual groups is distorted and eventually lost depending on the extent of mutual perturbation (e.g. the neighbouring group effect), which in turn is a function of the electronic or geometrical distance. In molecular hosts the criterion of convergence of anchor groups towards a binding site implies their mutual contact and impact on one another, resulting in a more or less severe deviation of their regular supramolecular behaviour. Since they may be quite remote in their covalent connection and quite flexible in their relative geometrical orientation, the theoretical prediction of guest binding as a synergetic or interfering influence eludes present-day calculation capabilities. In the experimental field there are but a few series of structurally related artificial receptors giving access to the change of a property experienced on inserting an anchor group into the lattice of neighbouring binding functions. Most revealing in this respect are studies in site-directed mutagenesis of proteins which, however, are confined to the aqueous environment. The transfer and validity of these results to host–guest binding in organic solvents is difficult and ambiguous, respectively, and presently does not alleviate the deduction of guidelines for abiotic host design in these environments.

## 6 Realization of the Basic Binding Concepts in Anion Complexation

The feature that renders a molecular species an anion is the negative charge which can only be addressed by electrostatic interactions. The strongest and farthest reaching among them employ Coulomb forces, i.e. the interaction between integer charges. An anion host can only be designed on the exclusive use of Coulomb attractions if the stochastic fluctuation of positive charges around the anionic target in solution is removed and replaced by a permanent field. The basic idea of replacement of the fluctuating ion cloud by covalently linked cationic charges was used early on in the construction of quaternary ammonium cavity hosts **1** and **2**, which were shown to bind the halides and a number of oxoanions in aqueous solution [25, 26].

The anticipated binding mode in which the guest resides in the molecular cavity was confirmed by a crystal structure of the [iodide$\subset$**1**] complex and by

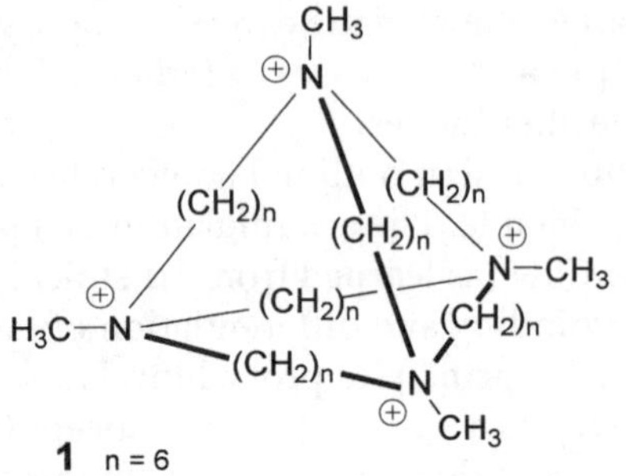

**1** n = 6

**2** n = 8

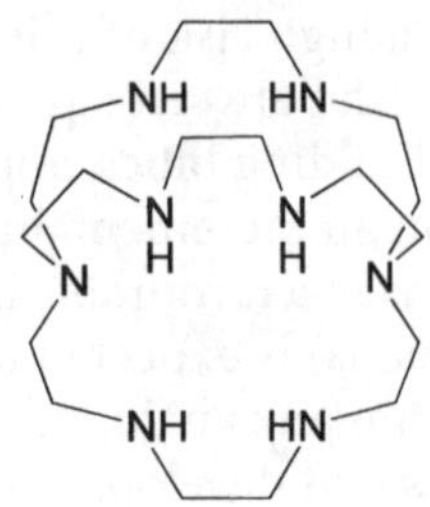

**3**

the non-monotonous sequence of affinities in the halide series, favouring bromide as the best-fitting guest with respect to size. Although devoid of any catalytically active functionality, the larger host **2** proved functional as a molecular catalyst by virtue of its capacity to stabilize highly delocalized anionic transition states as they occur, for instance in nucleophilic aromatic substitutions. The decrease in transition state free enthalpy was shown to derive both from the entropic component of collecting the anionic reaction partners in the confines of the molecular cavity of the host [27,28], as well as from the enthalpic stabilization brought about by the polarizable lining of the cavities' interior surface (i.e. a micro-solvent effect) [29]. The efficiency of ammonium hosts can be much enhanced if, supplementary to the charge, hydrogen bonding can be exploited. Thus the ammonium cryptate **3** was shown to complex fluoride at pH 5.9 in water with log $K_{assoc}$=8.8, displaying discrimination over chloride by a factor of ~$6\times10^7$ [30–32]!

In order to avoid competition by countercations, electroneutral hosts can be designed that may rely on internal charge compensation, as the number of fully charged cationic sites is exactly balanced by covalently attached anionic moieties furnishing zwitterionic structures. Alternatively, oriented electric dipoles can be incorporated into the host structures, which point with their positive ends towards the site where the anion is to be bound. In both cases it is essential that a permanent positive potential is established somewhere in the host structure that is not annihilated by thermal fluctuations. Again the tetrahedral parent amine served as the basic scaffold which was alkylated by α-bromotoluic acid to yield the tetrabetainic compound **4** [33]. Though now this host was electroneutral, it showed little difference in its binding capacity for halide anions compared to the tetracationic counterpart **1**. A van't Hoff analysis allowed determination of the enthalpy and entropy components in the associative process, revealing the surprising result that encapsulation binding to **4** in water is entropy opposed. This is in contrast to ordinary ion pairing in water, typically featuring entropy-driven complexation, owing to the release of firmly bound water molecules from the solvation spheres of the ions. In this case the solvent shell of the guest anion is totally stripped off on penetration into the

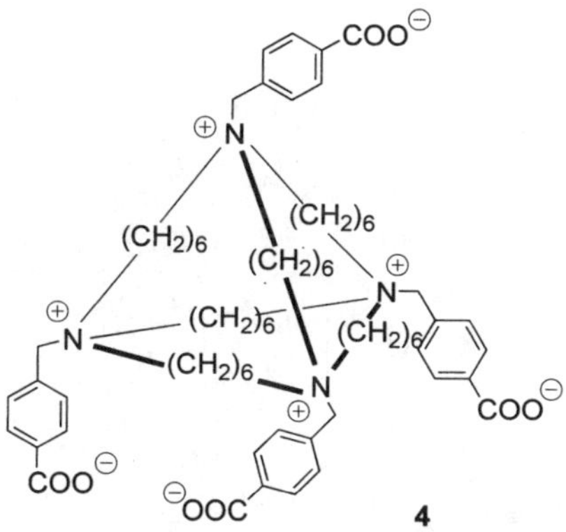
4

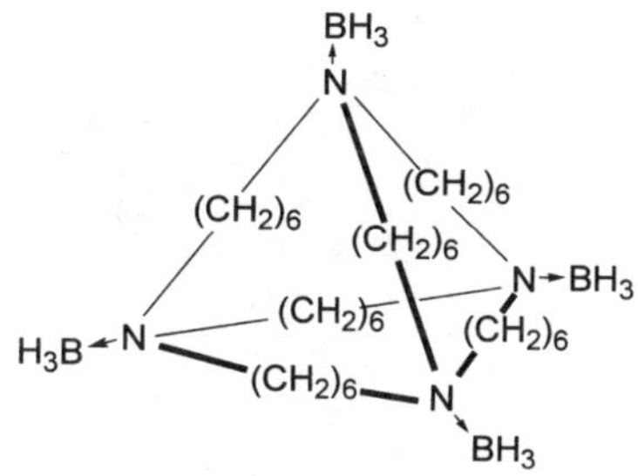
5

cavity, yet the solvent molecules set free apparently cannot compensate for the entropy loss arising from the snug fit and the concomitant rigidification in the host–guest complex.

The amine–borane adduct **5**, formed instantaneously on mixing amine and borane solutions in THF, represents the first example of an anion host that owes its stickiness for anions exclusively to dipolar but non-hydrogen bonding interactions [34]. By virtue of the high connectivity of the molecular skeleton, the dipoles of the dative borane–amine bonds converge with their positive ends head-on to the centre of the molecular cage. This configurational setup is sufficiently strong to bind bromide introduced as the $Ph_4P^+$ salt in $CD_2Cl_2/CDCl_3$ solution, and thereby testifies to the competitiveness of an array of multiple dipoles with a full Coulombic charge in anion binding.

Much more efficient and in general also more readily accessible are electroneutral anion hosts that exploit hydrogen bonding as the major binding tool. A rich variety of anion hosts relying on this principle have been reported over the years comprising in many cases amide [35], urea [36] or thiourea [37] functions, but also simple hydroxyl groups or the more exotic pyrrole [38] or imidazolium heterocycles [39] have been utilized. Spectacular success has recently been achieved in sulphate binding in protic solution (water/methanol mixtures) [40]. The artifical ditopic receptor **6**, composed of two identical cyclopeptide hemispheres interconnected by an adipic acid spacer, organizes these substructures in binding the sulphate guest with positive cooperativity with respect to the monotopic analogue to reach an affinity of log $K_{assoc}$≈5 in aqueous methanol. The free enthalpy of association was found by calorimetry to be more negative than that of iodide or bromide by more than 5 kJ mole$^{-1}$, which is truly remarkable in view of the great difference in the hydration free enthalpies ($\Delta G_{hyd}(SO_4^{2-})$=–1,090 kJ mole$^{-1}$; $\Delta G_{hyd}(I^-)$=–283 kJ mole$^{-1}$), pointing to a peculiar mechanism of stabilization of this guest within the pseudo-cavity of host **6**. One can speculate that it is the set of inwards-directed hydrogen-bond donor sites that allows the guest to retain great motional flexibility while being separated from bulk solvent. As a result the exothermicity experienced from

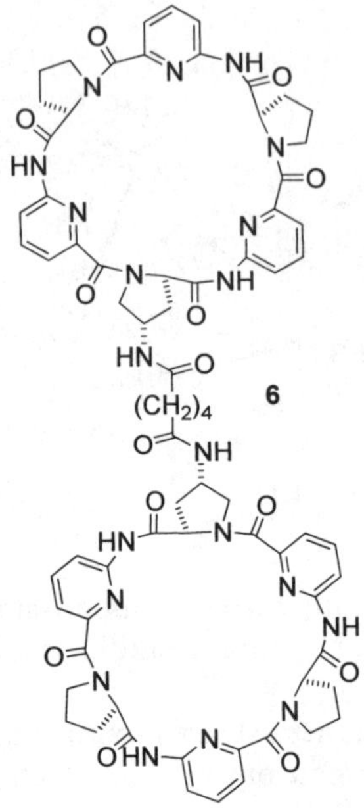

**6**

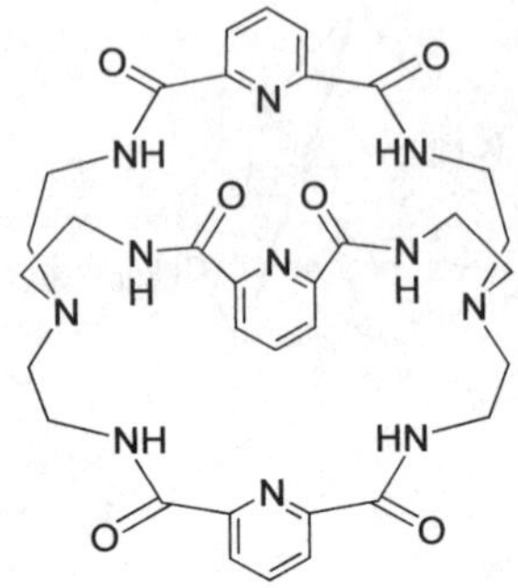

**7**

binding to a spatially restricted and preorganized array of six to eight H-bond donor sites is supplemented by the variety of close contact arrangements (binding modes within the cavity) surfacing as a positive entropy contribution. The combination of both contributions makes up for the high affinity inclusion complex that, on the basis of the thermochemical result, apparently features a collection of bound states rather than the uniquely dedicated interaction between distinct positions in the binding partners.

A contrasting scenario in another neutral hydrogen-bonding receptor is provided by the dipicolinic amide-tris(aminoethyl)amine (tren) cryptand **7** [41]. This host displays a dramatic preference for fluoride over chloride, bromide and a number of oxoanions in DMSO solution (log $K_{assoc}$>5), undoubtedly due to the intense hydrogen bonding of this guest with the pattern of host donor sites. In the $^{19}$F NMR spectrum the fluoride resonance shows as a septet, indicating simultaneous communication of the guest nuclear spin with the six N-H protons of the amide donors. Spin coupling also attests to the partial covalent character of this interaction motif, which also shows in the enhanced lifetime of the particular complex structure visible through slow exchange kinetics.

Based on the fundamental charge–dipole interaction, hydrogen bonding need not necessarily mandate the presence of lone electron pairs in the guest anion. If present, this electronic feature dominates the supramolecular behaviour of anions. As a corollary, it is the most prominent aspect to be removed in the construction of so-called non-coordinative anions ($BPh_4^-$, $BH_4^-$, *closo*-$CB_{11}H_{12}^-$ etc.) that are required in the isolation and study of reactive cationic species. Because lone electron pairs are a constitutive structural feature of most ordinary anions, they might serve as Lewis base anchor points in the interaction with suitably composed poly-Lewis acid compounds. The corresponding electron pair accepting groups may be highly electron-deficient organic moieties like tetrazine heterocycles [42] or oriented aromatic metal π-complexes [43], but broader in scope and potential applicability are compounds incorporating metal centres as direct interaction points. They may be distinguished according to their mode of anchoring the metal in the organic framework. Some metals like tin, boron, mercury etc. form strong and directed covalent bonds that make them useful and kinetically inert tectons in the construction of respective host skeletons, allowing Lewis acidic functionality to be positioned at predetermined locations and intentional directions within the structure.

Alternatively, a more or less sophisticated organic framework can be constructed containing a number of ordinary heteroatomic ligation sites, which can then hold the metal in a Werner-type but coordinatively unsaturated complex via chelating action. Both approaches use the overlap of atomic orbitals in host and guest for mutual adhesion of the binding partners and thus in principle do not conform to the definitions of supramolecular interactions. However, in many cases this distinction is not nearly as clear-cut as it seems. The bonds formed are frequently of low enough energy to allow ready reversible exchange of binding partners, justifying the integration of these examples with regular non-covalent host–guest systems.

A prototypical example to illustrate the covalent incorporation of metal centres is represented by the mercuracarborands, which have been studied extensively and in quite a variety in the solid state and also in solution. The binding of iodide to the [12]mercuracarborand-4 **8** is easily followed in acetone by $^{199}$Hg NMR spectroscopy [44, 45]. Titrating in increasing amounts of iodide diminishes the original broad $^{199}$Hg signal that arises from the perturbation of the solvated nucleus by the butterfly flip of the entire molecule happening in

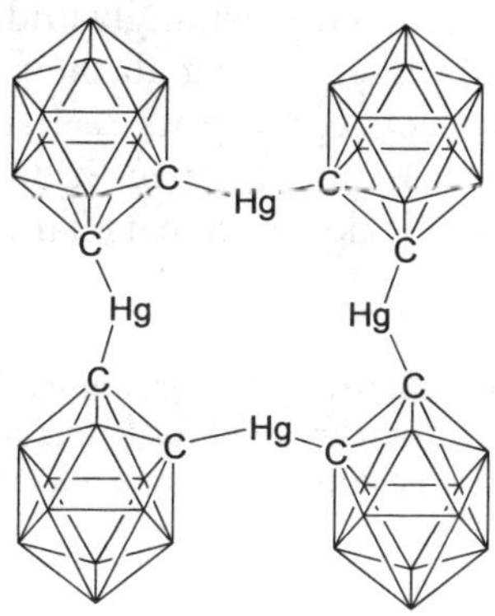


**8**

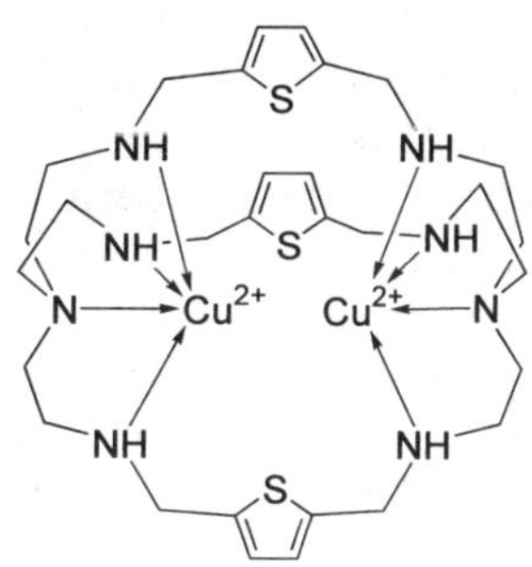


**9**

the same time regime as the timescale sensed by the NMR measurement. A new sharp signal at lower field is observed, representing one unique molecular environment for all Hg nuclei at 1:1 stoichiometry, which is in slow exchange with the unbound host. Further titration brings up another sharp signal indicative of the formation of a 2:1 iodide:host complex, however, with unaltered symmetry with respect to the Hg centres again showing slow exchange with all the other molecular species present.

All these metal organic hosts bear the latent risk of irreversible destruction when put into contact with protic solvents. On the contrary, anion hosts that follow the reversible chelation strategy have been deliberately designed for action in the aqueous environment and thus are much closer to concrete applications. Promising candidates in this respect are again tris(aminoethyl)amine (tren) derivatives, which were also utilized in the amide/ammonium cryptand routes towards anion receptors. Here the spanning of two tren units by 2,5-dimethylenethiophene units leads to a ligand that complexes up to two $Cu^{2+}$ cations in water in an pH-dependent fashion [46]. The resulting complex cation **9** is able to bind azide $N_3^-$ (log $K_{assoc}$=6.75) with 100-fold preference over the structural congener isocyanate $NCO^-$ and $10^4$-fold over isothiocyanate $NCS^-$, whilst competition from the halides or oxoanions like nitrate, acetate, bicarbonate or sulphate is beyond detection. The results from the selectivity studies, along with the UV–Vis absorption characteristics and evidence from crystal structures, all point towards a bidentate bridging complexation of the anion between both metal centres as the cause for the rather high azide selectivity observed.

Apart from one amazing report that documents anion complexation ($Cl^-$, $Br^-$, $PF_6^-$, $Ts^-$) into self-assembled resorcarene capsules in chloroform solution, where the driving force is basically unknown but solvophobic interactions are suspected [47], the examples described represent manifestations of the single-cause use of the variety of binding tools that are at the disposition of the molecular architect for constructing artificial anion hosts. In most real cases he/she will select several of them for implementation into a molecular framework in a proper fashion for their cooperative action. Though we seemingly understand the fundamentals of supramolecular anion binding, putting all the pieces of knowledge together to come up with a novel concept and its realization in anion host–guest binding is an act of creativity requiring in addition a great deal of intuition, some experience and not least a lucky hand that characterizes any piece of art beyond rational arguments.

**Acknowledgements** FPS wants to express his appreciation for the continuous financial support of his research by Hans-Fischer-Gesellschaft, München, and for constructive discussions and advice from many friends and colleagues including E. Anslyn, V. Král and H.J. Neusser in particular.

## References

1. Martinez-Manez R, Sancenon F (2003) Chem Rev 103:4419
2. Suksai C, Tuntulani T (2003) Chem Soc Rev 32:192
3. Fischer E (1894) Chem Ber 27:2985
4. Lichtenthaler F (1994) Angew Chem Int Ed 33:2363
5. Schellman JA (2002) Biophys Chem 96:91
6. Timasheff SN (2002) Biophys Chem 101/102:99
7. Schellman JA (1990) Biophys Chem 37:121
8. Schellman JA (1994) Biopolymers 34:1015
9. Jones JW, Gibson HW (2003) J Am Chem Soc 125:7001
10. Huang F, Jones JW, Slebodnik C, Gibson HW (2003) J Am Chem Soc 125:14458
11. Blas JR, Marquez M, Sessler JL, Luque FJ, Orozco M (2002) J Am Chem Soc 124:12796
12. Camiolo S, Gale PA, Hursthouse MB, Light ME (2003) Org Biomol Chem 1:741
13. Dunitz JD (1995) Chem Biol 2:709
14. Fersht AR (1987) Trends Biochem Sci 12:301
15. Shrestha S, Salins LLE, Ensor CM, Daunert S (2002) Biotech Bioeng 78:517
16. Quiocho F (1996) Kidney Int 49:943
17. Wiskur SL, Floriano PN, Anslyn EV, McDevitt JT (2003) Angew Chem Int Ed 42:2070
18. Linton B, Hamilton AD (1999) Curr Opin Chem Biol 3:307
19. Blades AT, Klassen JS, Kebarle P (1995) J Am Chem Soc 117:10563
20. Chapman KT, Still WC (1989) J Am Chem Soc 111:3075
21. Schmidtchen FP, Berger M (1997) Chem Rev 97:1609
22. Beer PD, Gale PA (2001) Angew Chem Int Ed 40:486
23. Snowden TS, Anslyn EV (1999) Curr Opin Chem Biol 3:740
24. Sessler JL, Seidel D (2003) Angew Chem Int Ed 42:5134
25. Schmidtchen FP (1977) Angew Chem 89:751
26. Schmidtchen FP (1980) Chem Ber 113:864
27. Schmidtchen FP (1981) Angew Chem Int Ed Engl 20:466
28. Schmidtchen FP (1986) Top Curr Chem 132:101
29. Schmidtchen FP (1986) J Chem Soc Perkin Trans II 135
30. Dietrich B, Lehn JM, Guilhem J, Pascard C (1998) Tetrahedron Lett 30:4125
31. Dietrich B, Dilworth B, Lehn JM, Souchez JP (1996) Helv Chim Acta 79:569
32. Reilly SD, Khalsa GRK, Ford DK, Brainard JR, Hay BP, Smith PH (1995) Inorg Chem 34:569
33. Schmidtchen FP, Worm K (1995) Angew Chem 107:71
34. Worm K, Schier A, Schäfer A, Hesse M, Schmidtchen FP (1994) Angew Chem Int Ed Engl 33:327
35. Bondy CH, Loeb SJ (2003) Coord Chem Rev 240:77
36. Bondy CR, Gale PA, Loeb SJ (2004) J Am Chem Soc 126:5030
37. Wu JI, He YB, Zeng ZY, Wei LH, Meng LZ, Yang TX (2004) Tetrahedron 60:4309
38. Sessler JL, Camiolo S, Gale PA (2003) Coord Chem Rev 240:17
39. Ihm H, Yun S, Kim HG, Kim JK, Kim KS (2002) Org Lett 4:2897
40. Kubik S, Kirchner R, Nolting D, Seidel J (2002) J Am Chem Soc 124:12752
41. Kang SO, Llinares JM, Powell D, VanderVelde D, Bowman-James K (2003) J Am Chem Soc 125:10152
42. Garau C, Quintero D, Frontera A, Costa A, Ballester P, Deya PM (2003) Chem Phys Lett 370:7
43. Gawenis JA, Holman KT, Atwood JL, Juriion SS (2002) Inorg Chem 41:6028
44. Wedge TJ, Hawthorne MF (2003) Coord Chem Rev 240:111
45. Yang X, Johnson SE, Khan SI, Hawthorne MF (1992) Angew Chem Int Ed Engl 31:893
46. Amendola V, Fabbrizzi L, Mangano C, Pallavicini P, Zema M (2002) Inorg Chim Acta 337:70
47. Hayashida O, Shivanyuk A, Rebek J Jr (2002) Angew Chem Int Ed Engl 41:3423

Top Curr Chem (2005) 255: 31–63
DOI 10.1007/b101161


# Chiral Recognition of Anions

Ivan Stibor (✉) · Petra Zlatušková

Institute of Chemical Technology, Technická 5, 166 88 Prague 6, Czech Republic
*stibori@vscht.cz*



**Abstract** Chiral recognition of anions is still in its infancy. Despite the fact that chiral analytical separation of different species including anions is a well-developed area of research with ample practical applications, the molecular recognition basis of this technique is still underdeveloped. Both natural and synthetic receptors capable of enantioselective recognition of anions are reviewed.

**Keywords** Anion recognition · Chiral recognition · Synthetic receptors

**Abbreviations**

| | |
|---|---|
| Ac | Acetyl |
| Boc | *tert*-Butoxycarbonyl |
| Cbz | Benzyloxycarbonyl |
| DMSO | Dimethyl sulfoxide |
| ee | Enantiomeric excess |
| h | Hour |
| Me | Methyl |
| mol | Mole |
| NOE | Nuclear Overhauser effect |
| Ph | Phenyl |
| Pr | Propyl |
| s | Second |
| TBDPS | *tert*-Butyldiphenylsilyl |
| TIPS | Triisopropylsilyl |
| Ts | 4-Toluenesulfonyl |

# 1 Introduction

The chemical and more importantly biological activity of any chiral substance depends on its stereochemistry. That is why the design, synthesis, and structure–activity relationships of enantioselective receptors are still very vital areas of research. Chiral synthetic ligands are supposed to open new possibilities in enantioselective catalysis and enantioseparations of racemic chiral compounds, they can be active in different parts of membrane transport and, finally, they can help us in understanding many vital processes in the biological world.

The basis of any chiral recognition event is the formation of diastereomeric complexes composed of chiral receptor and chiral substrate possessing different stabilities. There is a rule of thumb – based on both experience and highly sophisticated calculations valid for the design of new enantioselective receptors: "Chiral recognition requires a minimum of three simultaneous interactions between receptor and at least one of the enantiomers, with at least one of these interactions being stereochemically dependent" [1]. It should be stressed that this three-point rule does not require all three interactions to be attractive (i.e., bonding). In many cases, repulsive steric interactions are invoked, usually in combination with one or more bonding interactions to explain chiral recognition.

In practice it means that the exchange of one enantiomer for another is accompanied by the loss or profound change of at least one of these interactions. The study of enantioselective receptors consists in the measurement of the difference in Gibb's free energy caused by these two processes. The magnitude of this difference is directly proportional to the efficiency of chiral recognition.

Many tools are available for studying the structure and dynamics of multimolecular complexes, with NMR being perhaps the most suited to detailed examination of such complexes in solution. However, it is not yet widely appreciated that liquid chromatography can be extremely useful for the study of solution complexes even though, rigorously, it sometimes takes place not exclusively in solution but rather at an interface. Next to these techniques, UV spectroscopy, fluorescence spectroscopy, and pH titrations are also widely used for this purpose.

Despite the relatively large number of enantioselective receptors of cations and neutral species, reports on the effective chiral recognition of anions are still rare. However, there are a number of reports in the literature describing the recognition of carboxylic acids, and amino acids in particular, performed in aqueous solution at a pH where the carboxylic group has to be at least partly ionized. These reports are also included in this review.

Enantioselective anion sensing is included in a number of reviews covering much broader subjects, namely receptors of anions in general [2–8] and chromophoric sensors of anions [9, 10]. There are also more specialized reviews dealing with receptors for anions based on macrocyclic polyamines [11], amides [12], guanidiniums [13], crown ethers [14], or lanthanoids [15]. This review is also subdivided according to the structure of the receptor host studied.

## 2
## Natural and Modified Natural Receptors

Cyclodextrins, being macrocycles composed of chiral glucopyranose units, are able to bind small molecules in their inner cavity. The addition of cyclodextrins to racemic mixtures of chiral substrates results in the formation of diastereomeric complexes, with different stability for both enantiomers. Moreover, cyclodextrins can be derivatized using a number of well-known methods, including those using chiral structures appended to an already chiral macrocycle. Cyclodextrins have been used either native (uncharged) or charged by the attachment of ionized functional groups, namely amino and carboxy groups. Functional groups with a complexed metal ion have also been described. All these methods were used when looking for chiral recognition of hosts bound inside the cyclodextrin cavity. Interpretation of data thus obtained is not easy as sometimes many chiral elements are operating in conjunction. This type of chiral recognition, caused mainly by steric effects of the substrate bound in the cavity of cyclodextrin, was also observed when anion species were complexed.

Kano has studied extensively the recognition of anions and zwitterions by both native and derivatized cyclodextrins [16]. The native β-cyclodextrin complexation properties were examined with respect to binaphthyl derivatives. For both 1,1′-binaphthalene-2,2′-dicarboxylic acid and 1,1′-binaphthalene-2,2′-diol phosphate almost no chiral recognition was observed. The enantiomers of binaphthyl-2,2′-dicarboxylic acid were better distinguished using a per-*O*-methylated β-cyclodextrin derivative in place of the native β-cyclodextrin.

The same author has reported chiral recognition of α-amino acids by native, anionic, and cationic α- and β-cyclodextrins [17]. Both carboxylates and amines (monosubstituted as well as hexa- and heptasubstituted) were included in this study. The best results obtained were those from a combination of (*S*)- and (*R*)-AcTrp complexed by per-$NH_3^+$-β-cyclodextrin with $K$=2,310 and 1,420 (l/mol). In the detailed study of chiral recognition of substituted phenylacetic acid derivatives by aminated cyclodextrins, these were found to be again only modest with respect to the enantioselection attained [18].

Finally, the recognition of axially chiral 1,12-dimethylbenzo[*c*]phenanthrene-5,8-dicarboxylate (**1**) enantiomers by cyclodextrins was found to be much more efficient [19]. The preliminary $^1$H NMR binding studies in $D_2O$ showed no changes in spectra of the substrate after addition of α-cyclodextrin. However, in the spectra of the (*P*) and (*M*) enantiomers of the tetrahelicene **1**, significant chemical shift changes occurred after the addition of β-cyclodextrin and γ-cyclodextrin. Detailed measurements were performed to determine complexation constants and using the ROESY 2D NMR spectra the structures

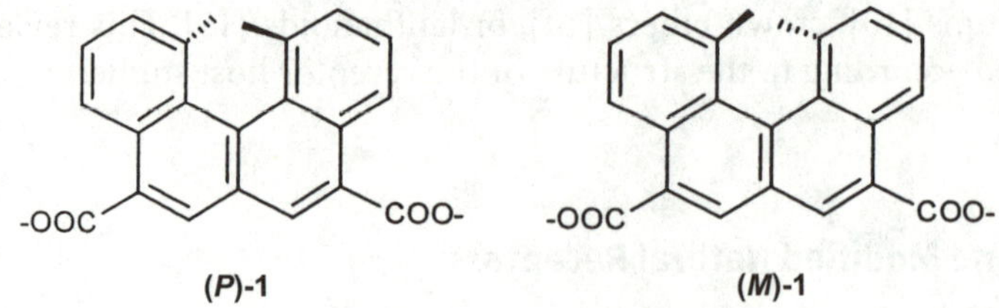

**Table 1** Stability constants of complexes of β-cyclodextrin and tetrahelicenes

| Receptor | Substrate | $K$ ($M^{-1}$) |
|---|---|---|
| β-CD | (*M*)-1 | 18,700±1,700 |
| β-CD | (*P*)-1 | 2,200±100 |
| γ-CD | (*M*)-1 | 3,100±100 |
| γ-CD | (*P*)-1 | 690±20 |

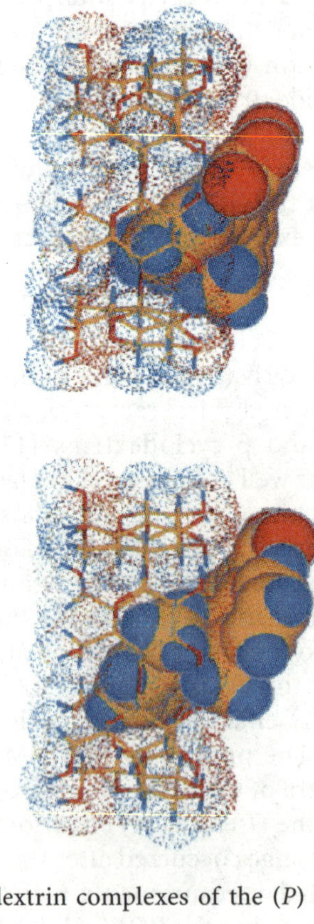

**Fig. 1** CPK models of cyclodextrin complexes of the (*P*) and (*M*) enantiomers of tetrahelicene 1

of the complexes were suggested. The data obtained are summarized in Table 1. The fact that both the receptors prefer the (*M*) enantiomer of **1** can be explained by these models. The (*P*)-**1** enantiomer penetrates deeper into the cavity of the host than its (*M*)-**1** analog, and one of its carboxylate functions gets into the inner part of the cyclodextrin. That causes the enthalpically unfavorable desolvation of this carboxylate group. When the (*M*)-**1** enantiomer is complexed, both carboxylate functions stick out of the cyclodextrin cavity and no desolvation is necessary. This is schematically shown in the CPK models (Fig. 1).

The stability constants and thermodynamic parameters of complexation of 6-amino-6-deoxy-β-cyclodextrin with more than 50 charged and neutral guests have been determined in aqueous phosphate buffer (pH 6.9) by titration microcalorimetry [20]. These data, in comparison with the data of native cyclodextrin, indicated that the chiral discrimination of cationic hosts with anionic guests is governed by a critical counterbalance between the electrostatic interaction of the charged groups in the host and guest and the conventional intracavity interactions of the hydrophobic moiety of the guest, such as hydrophobic, van der Waals, solvation/desolvation, and hydrogen bonding interactions.

Using the native cyclodextrin, the enantiomers of amino acid derivatives were enantioselectively complexed [21]. Further, for a more detailed analysis, zwitterionic tryptophan was employed [22]. For the complexation studies performed on this molecule the α-cyclodextrin was used, as its inner cavity is the smallest. The $^{1}$H NMR measurements showed that (*R*)-tryptophan formed a stronger complex with α-cyclodextrin compared with the (*S*) enantiomer. This is due to the number of hydrogen bonds which can be formed between each enantiomer and the host molecule. The NMR studies showed another very interesting fact: the amino acid is very likely forming no intracavity complex. It has been suggested that it is coordinated near the rim of the cyclodextrin.

Cyclodextrin 6-*O*-monophosphates in both charged and uncharged forms have also been tested for enantioselective recognition of amino acids [23]. The best results were obtained for mono-(6-*O*-diphenoxyphosphoryl)-β-cyclodextrin, giving fairly good enantioselectivity of up to 3.6 for D/L-serine in buffered (pH=7.2) aqueous solution.

A similar complexation study supported by thermodynamic parameters was performed by the same author for mono-(6-anilino-6-deoxy)-β-cyclodextrin and mono-[6-(1-pyridinio)-6-deoxy]-β-cyclodextrin complexation with several amino acids in zwitterionic form [24]. The inclusion complexation was enthalpy driven for the former and entropy driven for the latter host.

Zwitterionic cyclodextrins were designed and synthesized by Tabushi a long time ago as artificial receptors for amino acids in water [25]. Only a very low enantioselectivity was detected for Trp. Inoue also studied the complexation of two new β-cyclodextrin derivatives bearing *m*-toluidinyl and [(9-fluorenyl)-amino]alkylamino groups with various D/L-amino acids by fluorescence spectroscopy in buffered (pH=7.2) aqueous solution. An enantioselectivity as high as 33 was found for D/L-leucine and the former host [26].

Marchelli used the copper(II) complex of histamine-functionalized β-cyclodextrin for chiral recognition and separation of amino acids [27]. The best results were obtained for aromatic amino acids (Trp). Enantioselective sensing of amino acids by copper(II) complexes of phenylalanine-based fluorescent β-cyclodextrin has been recently published by the same author [28, 29]. The host containing a metal-binding site and a dansyl fluorophore was shown to form copper(II) complexes with fluorescence quenching. Addition of D- or L-amino acids induced a "switch on" of the fluorescence, which was enantioselective for Pro, Phe, and Trp. This effect was used for the determination of the optical purity of proline.

Another natural macrocycle used for the chiral recognition of carboxylates is (+)-tubocurarine (**2**) [30]. For complexation studies, the zwitterionic form of this alkaloid was employed. Using $^1$H NMR, fluorescence spectroscopy, and UV spectroscopy in water or $D_2O$, the stability constants of the complexes between tubocurarine and amino acids were determined. The maximum chiral recognition was observed upon the complexation of *N*-acetylphenylalanine and phenylalanine (determined at pH=9). For *N*-acetylphenylalanine, the L-enantiomer was better but weakly bound. Phenylalanine formed a much more stable complex (by approx. one order of magnitude), with that of the D-enantiomer being even stronger. While the enantiospecificity depends primarily on electrostatic interactions, the overall stability is determined by guest hydrophobicity. These conclusions were confirmed by docking calculations for enantiomers of phenylalanine.

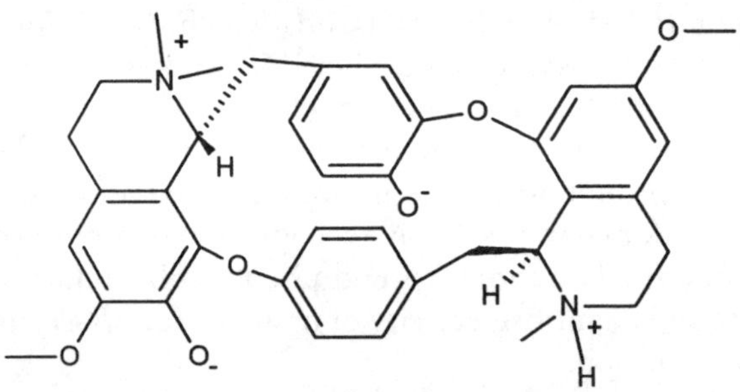

**2**

Very recently, nice recognition of free and *N*-acetylated amino acids (Gly, Ala, Phe) and some structurally related guests by a dicationic cyclophane-type *N,N'*-dibenzylated chiral derivative (**4**) of a bisisoquinoline alkaloid *S,S*-(+)-tetrandine (**3**; DBT) has been studied by NMR titration in water [31]. In contrast to other macrocyclic hosts, DBT shows high affinity and large enantioselectivity ($K(S)/K(R)\geq 10$) toward the smaller *N*-acetylalanine and binds the larger phenylalanine more weakly and nonselectively. The binding specificity of DBT was rationalized on the basis of molecular mechanics calculations.

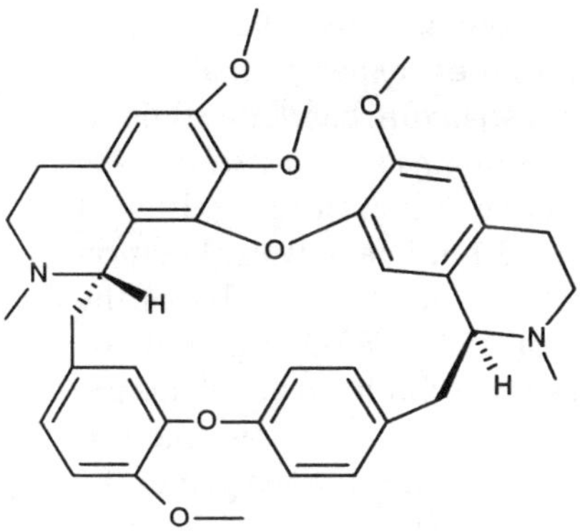

3

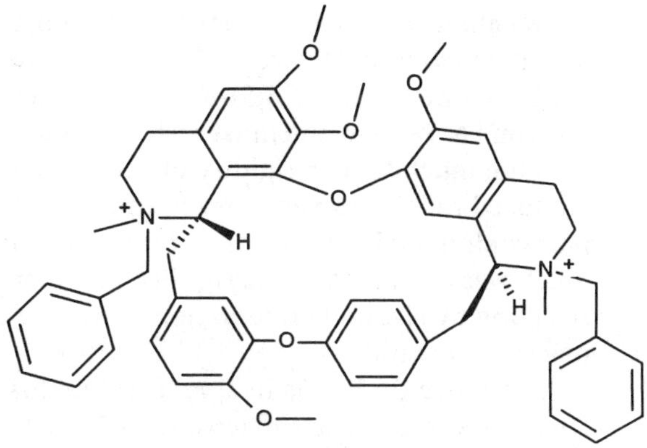

4

## 3 Ammonium-Based Receptors

Oxaazamacrocycles are molecules with combined crown ether and amino group properties; for example, in dependence on the size of macrocyclic ring and the position of heteroatoms stable complexes with various metals can be formed [32]. Furthermore, when protonated, these compounds are excellent anion binders. In fully protonated polyammonium macrocycles, the coulombic attractions and hydrogen bond formation play a dominant role in the anion complexation. The receptors (*R*,*R*)-**5** and (*S*,*S*,*S*,*S*)-**6** were designed to bind dicarboxylates [33].

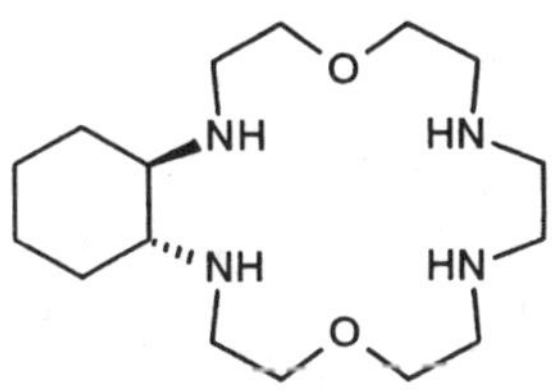

(*R*,*R*)-**5**

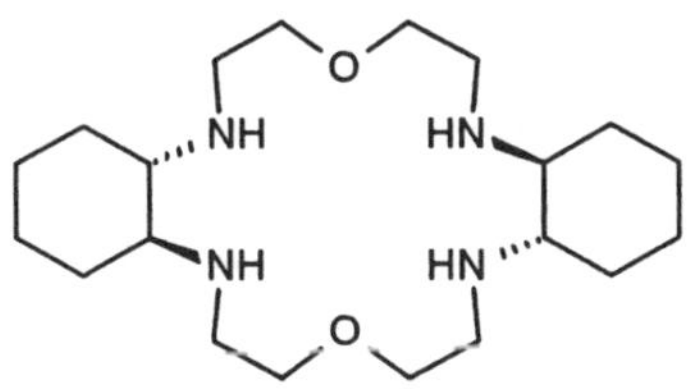

(*S*,*S*,*S*,*S*)-**6**

The protonation constants and the binding constants with different chiral carboxylates have been determined by means of pH-metric titrations. The triprotonated form of (*S*,*S*,*S*,*S*)-**6** showed moderate enantioselectivity with malate and tartrate anions ($\Delta\Delta G$=0.62 and 0.66 kcal/mol, respectively), the strongest binding being observed in both cases with the L-enantiomer. Good enantiomeric discrimination was obtained with tetraprotonated (*R*,*R*)-**5** and

*N*-acetyl aspartate, the complex with the D-enantiomer being 0.92 kcal/mol more stable than its diastereomeric counterpart. All these experimental results allow authors to propose a model for the host–guest structure based on coulombic interactions and hydrogen bonds. The question remains to be answered why the $D_2$ symmetrical (*S*,*S*,*S*,*S*)-**6** was completely inactive in enantiorecognition of the same *N*-Ac-Asp, the most selective anion for (*R*,*R*)-**5**. The only explanation is in the much higher rigidity of the former ligand, which is incapable of the "induced fit" that is very easy to imagine in the case of (*R*,*R*)-**5**. Probably the pair coulombic interactions of protonated amines with both carboxylates are much weaker due to steric rigidity, and that is the reason for the low enantioselection by N–H...O interaction between Ac–NH–Asp and the oxygen of the macrocyclic unit.

The above finding is in agreement with a recent study by Gotor, Lehn, and Dietrich where per-aza macrocycles **7** and **8** were examined [34]. These receptors have a much more flexible backbone and, in addition, three amino groups on "each side" of the macrocycle. Again the protonation features of both ligands have been determined using potentiometry as well as $^1$H and $^{13}$C NMR techniques. This study allows the determination of the basicity constants and the stepwise protonation sites. The presence of cyclohexane decreases the protonation ability, and this effect can be explained in terms of conformational and electrostatic factors. Again, binding of different chiral dicarboxylates has been studied by potentiometry. Here the macrocycle with two chiral cyclohexanes presents higher anion complexation constants. It also exhibits good preference for *N*-Ac-D-Asp. Both receptors have shown an interesting behavior in the complexation of protected *N*-Ac-glutamates. Tetraprotonized forms of the receptors are very selective for the D form of this substrate. Further protonization causes the formation of 2:1 stoichiometry complexes, and the hexaprotonized receptors bind the D-enantiomer with 2:1 stoichiometry while the L-enantiomer is bound with 1:1 stoichiometry.

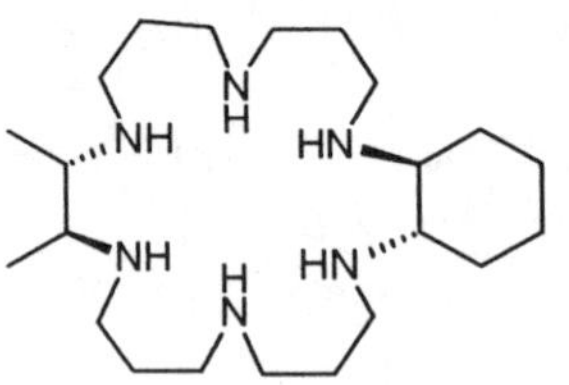


**7**

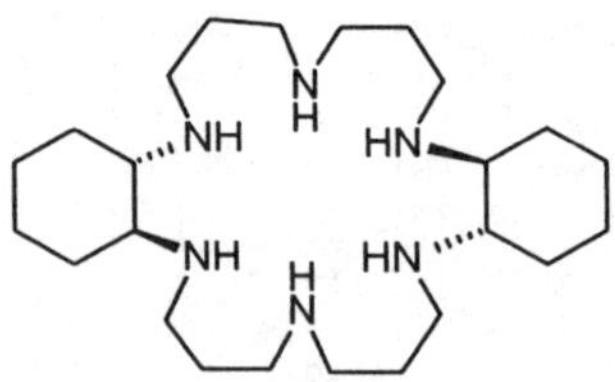


**8**

For the chiral recognition of sodium salts of 2-aminopropionate the imidazolium salt **9** was synthesized [35]. There were four chiral receptors synthesized but only **9** based on (–)-*cis*-myrtanyl imidazole exhibited the yes–no preliminary complexation of both enantiomers of sodium 2-aminopropionates, being able to complex only the (*R*)-enantiomer.

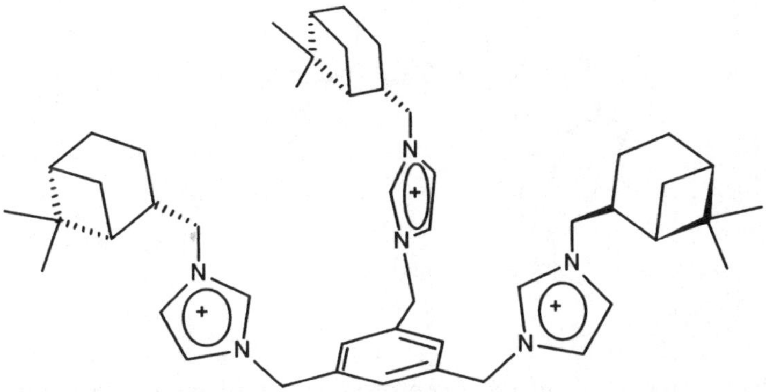

9

## 4
## Guanidinium-Based Receptors

The guanidinium group is known to remain protonated over a wide range of pH and to form characteristic pairs of well-organized, strong hydrogen bonds to carboxylates and phosphates. The receptor **10** was synthesized for the complexation of zwitterionic amino acids. Its chiral structure was based on a bicyclic guanidinium unit, synthesized in optically pure form from L-asparagine. The affinity to aromatic substrates was supported by the attachment of naphthoyl groups to the guanidinium unit [36]. The complexation properties were monitored using single extraction experiments. For the zwitterionic amino acids Phe, Trp, and Val, the enantioselectivity was negligible. Using NMR titration in $CDCl_3$, better chiral recognition was observed upon the complexation of *N*-Ac-try ptophan ($K$=1,051±20% $M^{-1}$ for the L-enantiomer and $K$=534±15% $M^{-1}$ for the D-enantiomer). It represented the first example of enantioselective recognition of anionic species by an abiotic receptor.

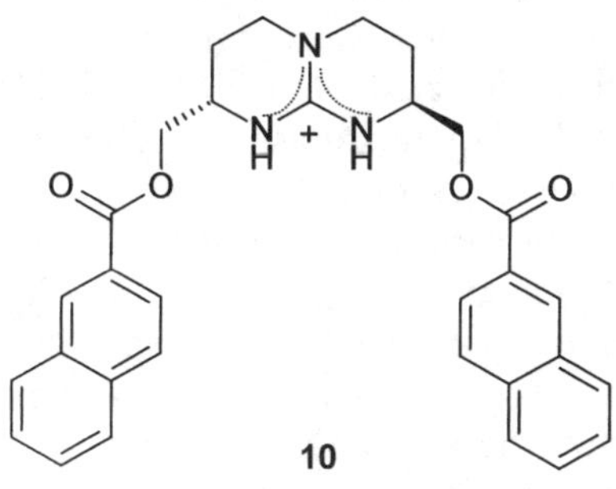

10

Different enantioselectivity has been observed for adenosine monophosphate nucleotides complexed with **11**, where one naphthoyl unit has been sub-

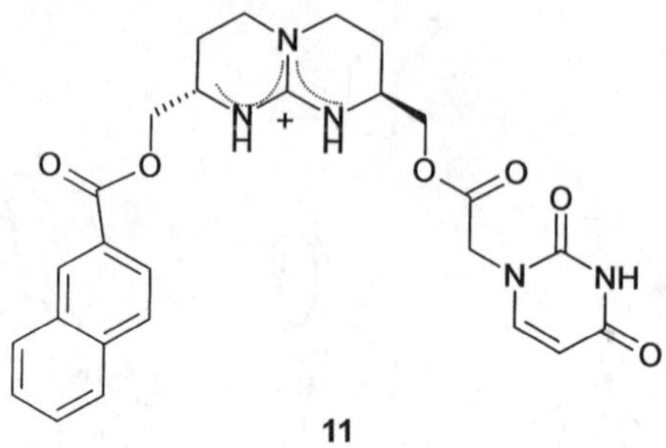

**11**

stituted by uracil [37]. Then, the receptor **12** was synthesized [38]. In the design of this molecule, three structural motifs were used: nonself-complementary binding sites for carboxylate and ammonium, preventing the receptor from internal collapse; an aromatic planar surface for additional selective stacking interaction with the side chain of aromatic amino acids; and a chiral structure for chiral recognition.

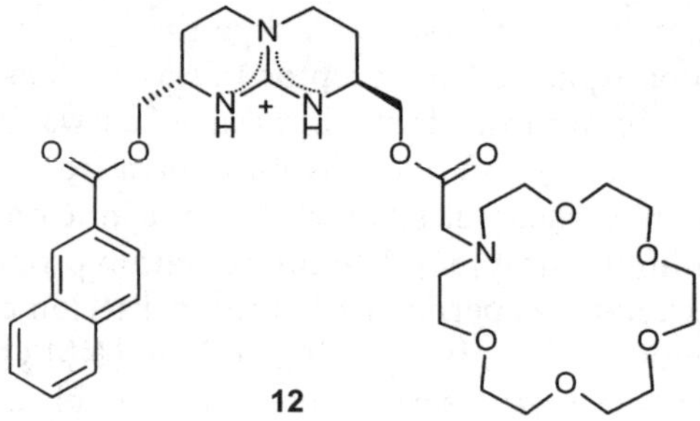

**12**

Despite its ionic structure **12** is almost insoluble in water. The affinity toward amino acids was studied by extraction experiments, where the enantiomers of Phe, Val, and Trp were extracted into $CD_2Cl_2$ from aqueous solutions. The amounts of amino acids extracted were determined by "NMR integration". As expected, the aromatic amino acids were preferred (40% extraction efficiency for L-Phe and L-Trp). In the organic phase, no valine was detected. Moreover, almost no D-enantiomers were extracted by (*S*,*S*)-**12** (D-Trp<0.5%, D-Phe<2%).

The same principle follows for the receptor **13**, designed and synthesized by Schmidtchen [39]. Both recognition structural units have been changed. The aromatic naphthoyl group has been substituted with the bulkier diphenyl-*tert*-butylsilyl group and the aza-18-crown-6 unit by a symmetrical triaza analog that is known to recognize the ammonium cation with high selectivity against alkali metal cations. This compound was capable of discrimination between enantiomers of phenylalanine. The L-enantiomer was extracted into chloroform with 40% ee in a wide pH range. When the analog of **13** without the azamacrocycle in its structure was used, no chiral recognition was observed.

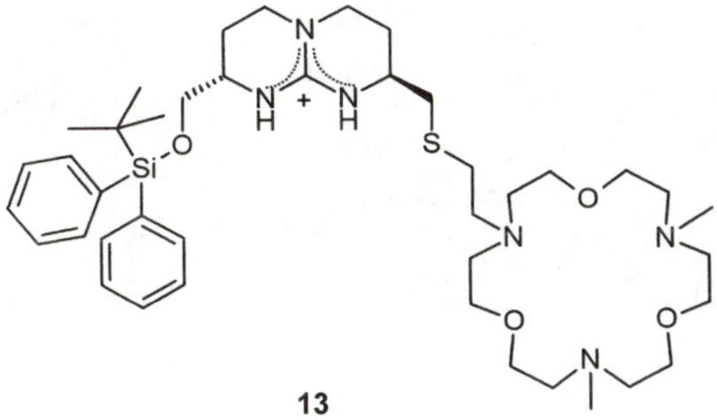

**13**

These results were extended by de Mendoza [40]. A variety of chiral receptors were synthesized and the chiral recognition of zwitterionic amino acid enantiomers upon their facilitated transport across a model membrane was examined. Transport studies were performed in a U-tube cell, consisting of a dichloromethane phase separating two neutral aqueous (source and receiving) phases. Samples of the receiving phase were directly injected into the chiral HPLC column. Ester **14a**, despite being a slower carrier than amide **14b**, showed a higher enantioselectivity. A similar effect was observed with the receptors **15a** and **15b**. Ester **15a**, containing a TBDPS group, surprisingly showed enantioselectivity similar to that of naphthoyl-containing **14b**. The control measurements were performed with receptor **15d**, which contains the nonaromatic triisopropylsilyl protecting group. Since a similar selectivity was found, it has been concluded that the chiral recognition of tryptophan enantiomers was not caused by aromatic stacking interactions. The more probable sources of this discrimination are the hydrophobic and steric effects, which contradicts the primary idea shown in Fig. 2.

When trying to elucidate the structure of complexes **14a** with L-Trp and D-Trp, only ill-defined spectra have been available. No NOE signals could be detected between naphthalene and the indole side chain, although for the complex with L-Trp a few contacts could be attributed to the proximity between the

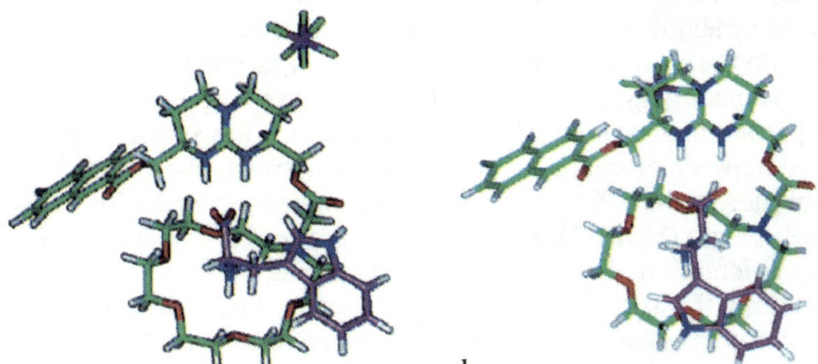

**Fig. 2** Optimized structures of complexes formed by **14** with **a** L-Trp and **b** D-Trp

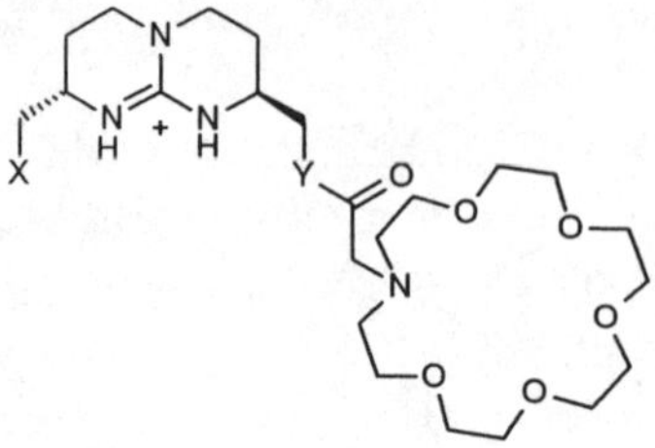

**14a** X=O-2-naphtoyl Y=O
**14b** X=NH-2-naphtoyl Y=NH
**15a** X=OTBDPS Y=O
**15b** X=OTBDPS Y=NH
**15c** X=OH Y=O
**15d** X=OTIPS Y=O

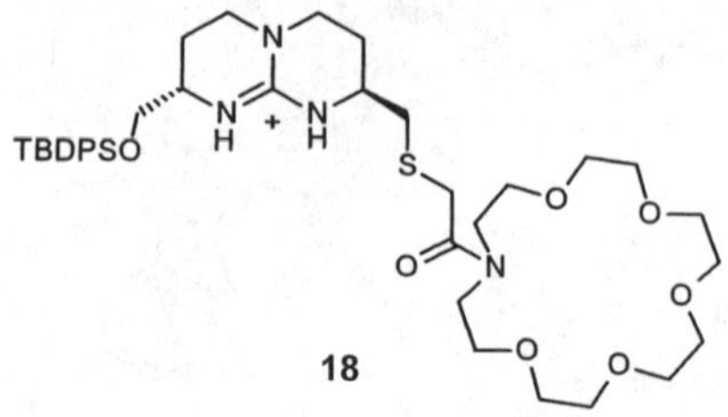

**18**

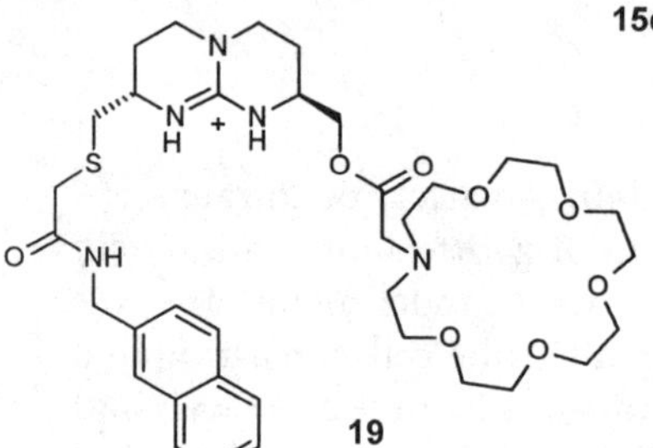

**19**

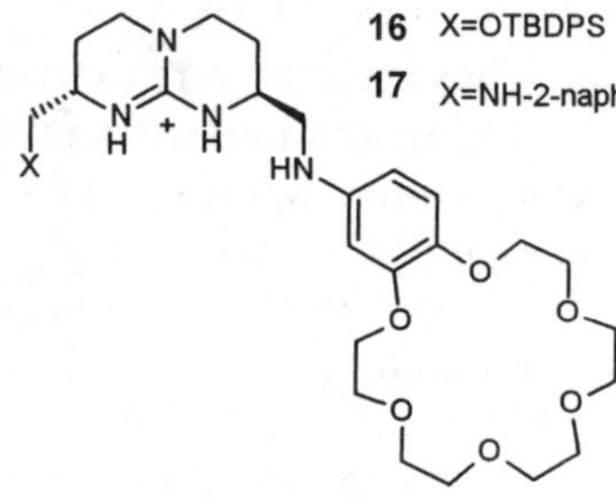

**16** X=OTBDPS
**17** X=NH-2-naphtoyl

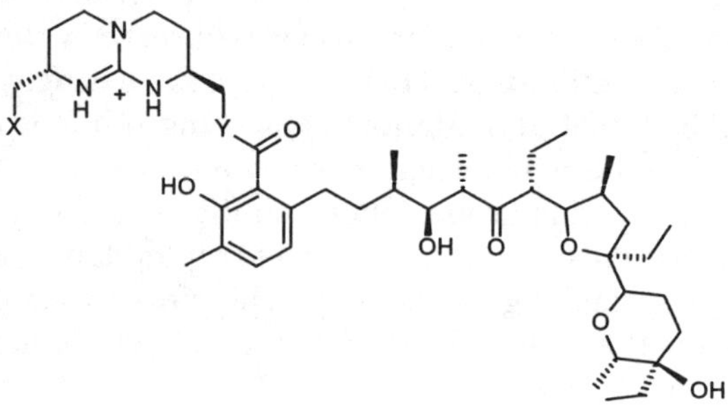

**20a** X=O-2-naphtoyl Y=O
**20b** X=NH-2-naphtoyl Y=NH
**21a** X=OTBDPS Y=O
**21b** X=OTBDPS Y=NH

side chain and the guanidium crown ether. In view of the scarce data obtained from NMR spectra, extensive molecular mechanics and dynamics calculations (1 ns) were performed in a box of chloroform. Suitable fragments were selected from related X-ray structures to build up the initial geometries that were fully optimized by ab initio quantum mechanics, and the amino acid was manually docked into the binding pocket by facing each of its zwitterion functionalities to the two anchoring groups provided by ligand **14b**. The naphthoyl moiety was placed perpendicular to the plane delimited by the aza-crown as no NOE was observed between aromatic hydrogens and aza-crown hydrogens. The overall conformation of the L-Trp was determined from the information provided by the NMR spectra (NOE). This conformation reduced the number of alternative docking orientations and placed the indole ring oriented toward the ester group linking the aza-crown subunit to the guanidium. The optimized structures of complexes between **14b** and both the enantiomers of tryptophan were obtained (see Fig. 2). Surprisingly, the relative distances and disposition of the two anchored pairs are quite similar for both complexes. When representative interaction energies were calculated for each complex during the last 300 ps of the simulation, both the van der Waals and electrostatic contributions to the binding energy were systematically more favorable and less fluctuating for the L-Trp complex. It provided a rationale for the pronounced enantioselectivity of the ligand in favor of L-amino acids, and also strengthened the view that hydrophobic interactions are comparatively of much less importance than electrostatic interactions. This principle must be borne in mind when designing supramolecular systems for specific recognition of amphipathic guests. Balancing both types of interactions represents a difficult compromise and a real challenge. Neither the benzo-crown derivatives **16, 17** nor the thioethers **18, 19** were enantioselective for tryptophan. However, the efficiency of these receptors in amino acid transport was excellent.

The lasalocid derivatives **20a,b** and **21a,b** were less effective enantioselective carriers than the crown-ether-based compounds. Interesting behavior was observed on studying the complexation properties of **21b**. Although this receptor showed a remarkable selectivity for L-Trp, negligible enantioselectivity for Phe was observed.

Chiral rigid guanidinium group(s) have also been attached to the calix[4]-arene scaffold resulting in ligands **22** and **23** [41]. A traditional extraction protocol was used to evaluate the efficiency of these ligands in enantiorecognition of Phe, Trp, and Leu. Both ligands showed preference for L-amino acids while ligand **23** was found to be more effective (Table 2).

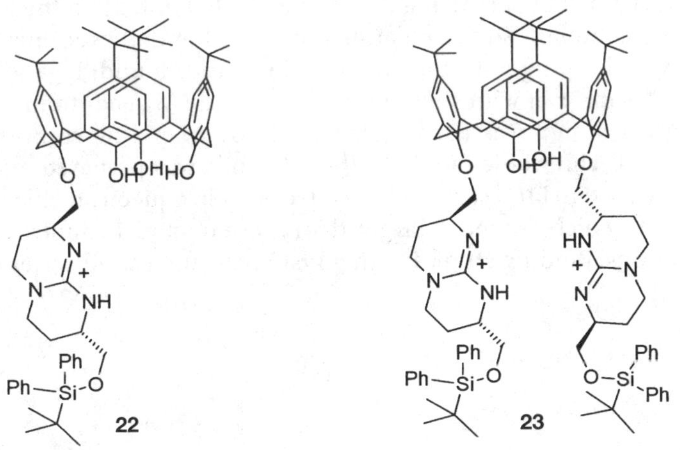


**Table 2** Competeitive extraction – amount of substrate extracted [in %]

| Ligand | L-Phe | D-Phe | L-Trp | D-Trp | L-Leu | D-Leu |
|---|---|---|---|---|---|---|
| 22 | 12.4 | 3.9 | 8.6 | 3.1 | 7.0 | 3.4 |
| 23 | 47.2 | 5.3 | 15.9 | 4.8 | 9.8 | 4.1 |

Davis has shown that the steroid skeleton can be used as a very efficient chiral scaffold [42, 43]. A guanidinium group was attached to a lipophilic steroid skeleton, usually in spatial proximity with other anion-binding active groups such as amides, sulfonamide, or urethane. Very efficient chiral binding sites have been constructed using this approach, resulting in ligands **24–30**.

These ligands were shown to extract many *N*-acyl α-amino acids from aqueous phosphate buffer (pH=7.4) into chloroform and also their preformed tetrabutylammonium salts. L-Enantiomers of the amino acids *N*-Ac-Ala, *N*-Ac-Phe, *N*-Ac-Val, and *N*-Ac-Trp were extracted with sevenfold preference by ligand **24**. Analogously, ligand **25** was found to be even more efficient, as the L-enantiomeric preference was tenfold. Ligands **26** and **27**, however, were much less efficient.

From the so-called unsymmetrical ligands the best one was **30**, recognizing *N*-Ac-Met, *N*-Ac-Phe, and *N*-Ac-Val with an enantioselectivity of 9:1 (L:D). Thorough NMR studies as well as molecular modeling experiments have been applied to explain this efficiency. The structure of the complex of **30** with both enantiomers of *N*-Ac-Val has been calculated using the Monte Carlo method. Whereas the L-enantiomer is bound to carbamoyl NH in position 12 of the steroid skeleton with its acetyl oxygen, the D-enantiomer is bound to the same place by the oxygen of the carboxylate moiety (see Fig. 3). Although the protons linked by the intermolecular NOE are ≈3.8 Å from each other, they can be brought to within 2.2 Å by rotation about the N–Ph bond. The valyl methyls are close to the aromatic rings of the receptor, and are in quite different environments; hence it is not surprising that one is shifted upfield in the NMR spectrum of the complex. The calculations predict slightly lower energy for the L-complex, in keeping with the enantioselectivities found. Ligand **30** is also able to bind naproxen with slight preference for the (*S*)-enantiomer.

A different design was used by Schmuck for guanidiniocarbonyl pyrrole ligand **31** [44]. The whole family of these ligands was prepared and studied. They can be used to bind carboxylates effectively in aqueous media with binding constants which are much larger than those obtained simply with guanidinium cations. Binding studies with a systematically varied series of ligands

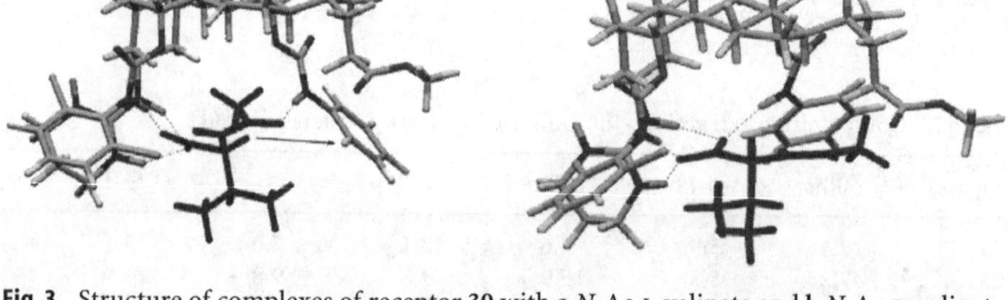


**Fig. 3** Structure of complexes of receptor **30** with **a** *N*-Ac-L-valinate and **b** *N*-Ac-D-valinate

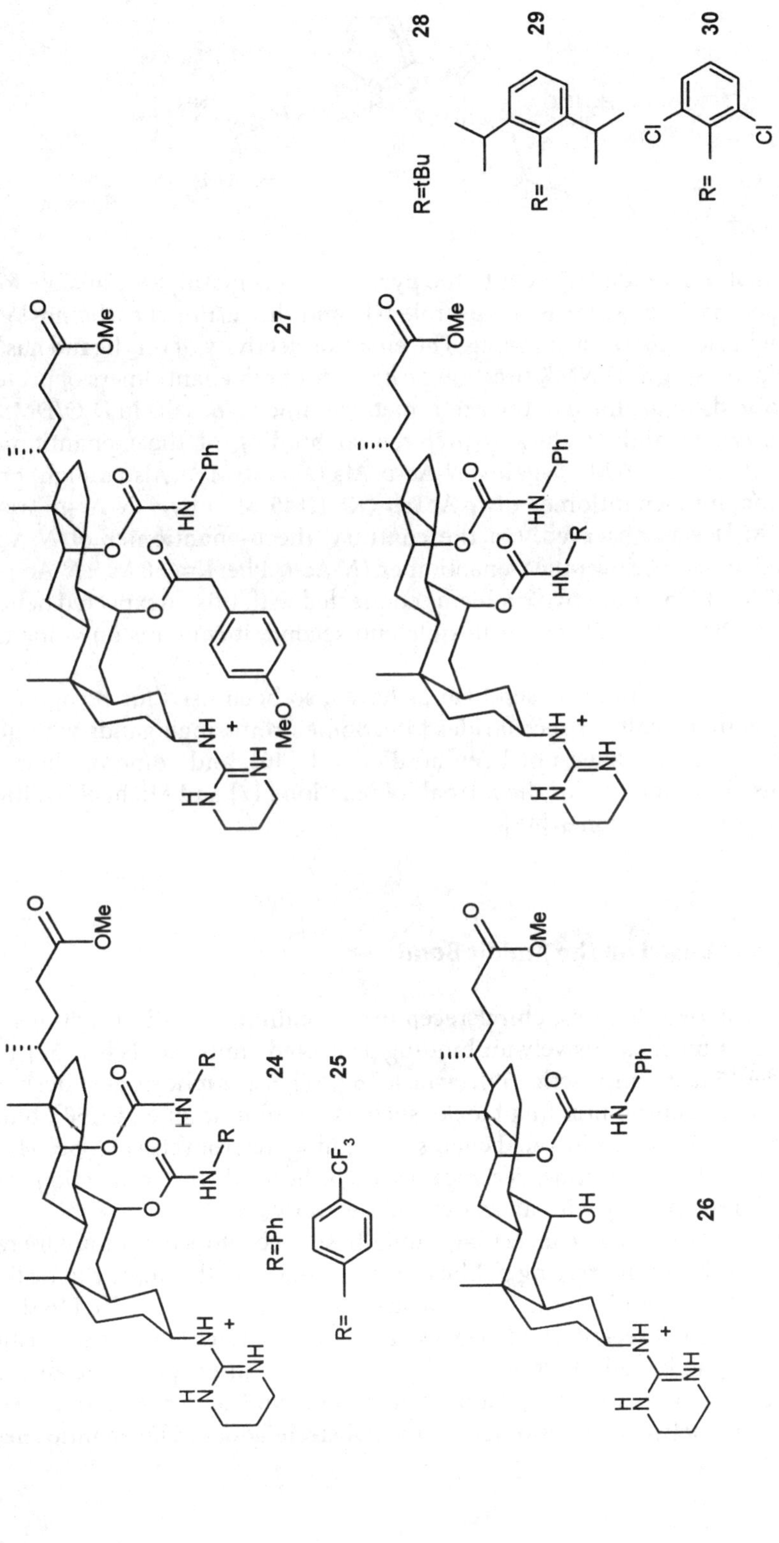
24
25
26
27
28
29
30
R=Ph
R=tBu

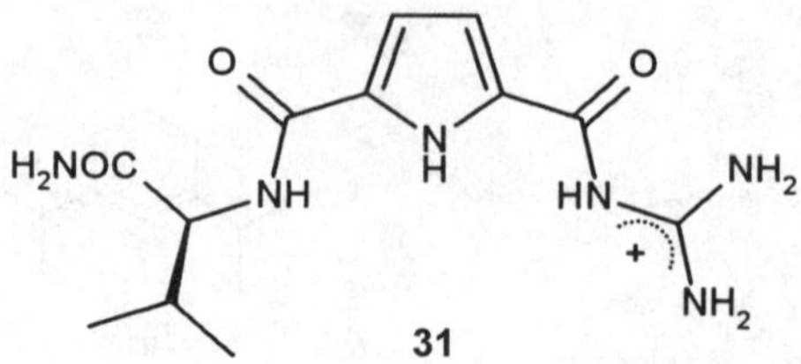

**31**

show that the amide NH next to the pyrrole ring is mainly responsible for the stronger binding, whereas the pyrrole NH and the terminal carbamoyl group contribute less to overall binding. The enantioselectivity of this ligand has been examined using a $^{1}$H NMR titration protocol for both enantiomers of protected amino acids in the form of their tetramethylammonium salts in $D_2O$/DMSO-$d_6$ 40% (v/v). Ligand **31** showed preferential binding of the L-enantiomer of *N*-Ac-Ala ($K$=1,610 $M^{-1}$) against *N*-Ac-D-Ala ($K$=930 $M^{-1}$). Also a slight preference for the L-enantiomer of *N*-AcTrp ($K$=1,145 $M^{-1}$) over *N*-Ac-D-Trp ($K$= 1,005 $M^{-1}$) was observed. On the contrary, the D-enantiomer of *N*-Ac-Phe showed preference over the L-enantiomer, (*N*-Ac-D-Phe: $K$=680 $M^{-1}$; *N*-Ac-L-Phe: $K$=585 $M^{-1}$). There are two problems connected with this unexpected behavior. First, the ligand is still rather flexible and second, it contains only one chiral center.

Chiral guanidinium-based ligands have also been used for recognition of diastereomeric salts of saccharides [45]. Some promising ligands with guanidinium structure have not been studied yet [46], and some of them have been used as catalysts for the nitroaldol reaction [47] and Michael addition to α,β-unsaturated ketones [48].

## 5 Receptors Based on the Amidic Bond

In the last two decades, chiral receptors containing amidic functions were designed almost exclusively for binding protected amino acids [49–57], oligopeptides [54, 58], and lactic [59], tartaric [60, 61] or camphoric acid derivatives [62]. Usually, chiral building blocks such as spirobifluorene [49, 60], binaphthalene [51, 57], or amino acid chains containing macrocycles [52–56, 58] were employed. An interesting receptor was synthesized via connection of the calix[4]arene moiety with an aza-crown derivative [61].

Although the works concerning amide-based receptors for anions are rather rare, several very interesting publications dealing with this topic can be found. One review on peptide and glycocalixarenes has recently appeared [63].

The design of receptors **32–34** was based on the attachment of chiral amino acid chains to benzo-18-crown-6 ether [64]. These molecules were able, upon the addition of cesium salts of amino acids, to form "sandwich" complexes and distinguish between enantiomers of the substrate added. The enantioselective

CE = 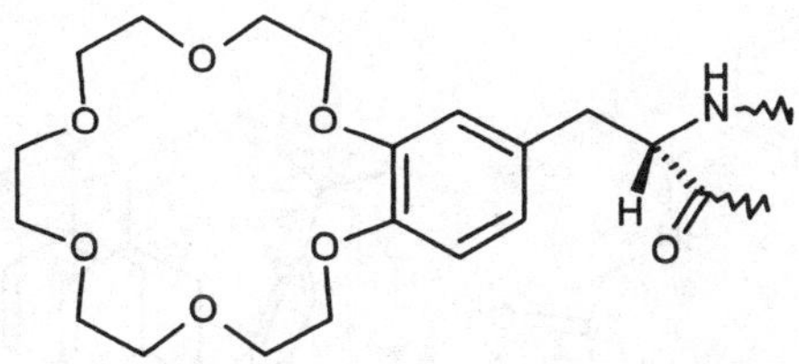

**32** NBoc-Ala-Ala-Ala-CE-Ala-CE-Ala-NH- $CH_2CH_2CH_3$

**33** NBoc-Ala-Ala-CE-Ala-Ala-CE-Ala-NH- $CH_2CH_2CH_3$

**34** NBoc-Ala-CE-Ala-Ala-Ala-CE-Ala-NH- $CH_2CH_2CH_3$

transport of amino acids, facilitated by the receptors **32–34**, was studied in a U-shaped tube using chloroform as a liquid membrane. The amino acid guests (*Z*-Ala, *Z*-Phe, Bz-Ala) were dissolved in cesium hydroxide. The constants of chiral recognition determined from transport ratios of the enantiomers as $k_L/k_D$ are summarized in Table 3.

The discrimination between enantiomers of *Z*-phenylalaninate was observed only using receptors **32** and **33** for the complexation. For **34**, no chiral recognition of enantiomers was detected. The same transport ratios and slow transport of *Z*-phenylalaninate by this receptor can be explained by formation of an "external" complex (*Z*-phenylalaninate is not bound between the amino acid chains of the receptor **34** and no sandwich complex is formed).

Calix[4]arene is a very popular building block for the design of molecules with well-defined mutual positions of substituents. The attachment of amidic residues to this molecule resulted in receptors **35** [65] and **36** [66]. Upon the $^1$H NMR titration of **35** by tetrabutylammonium salts of amino acids in acetone-$d_6$, moderate chiral recognition was observed: *N*-Ac-L-Ala $K$=4,900 $M^{-1}$, *N*-Ac-D-Ala $K$=5,700 $M^{-1}$, *N*-Ac-L-Phe $K$=7,900 $M^{-1}$, *N*-Ac-D-Phe $K$=10,500 $M^{-1}$. The receptors **36a–c** formed stable complexes with *N*-tosyl-L-phenylalaninate in deuteriochloroform (**36a** $K$=1,626 $M^{-1}$, **36b** $K$=4,836 $M^{-1}$, **36c** $K$=6,924 $M^{-1}$). Unfortunately, the measurement using *N*-tosyl-D-phenylalaninate has not been reported.

**Table 3** First- order rates observed for the transport of amino acids fascilitated by **32–34** over a liquid membrane

| Anion | $k$ ($10^4$ mol $l^{-1}h^{-1}$) | | |
|---|---|---|---|
| | **32** | **33** | **34** |
| *Z*-L-Phe | 5.6 | 5.0 | 1.4 |
| *Z*-D-Phe | 4.4 | 6.3 | 1.4 |

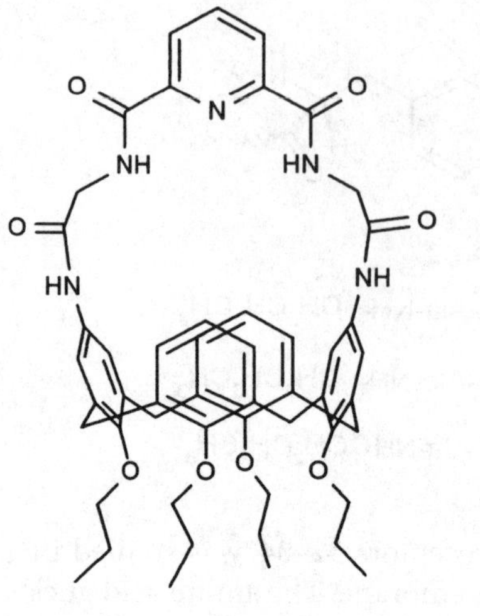

35

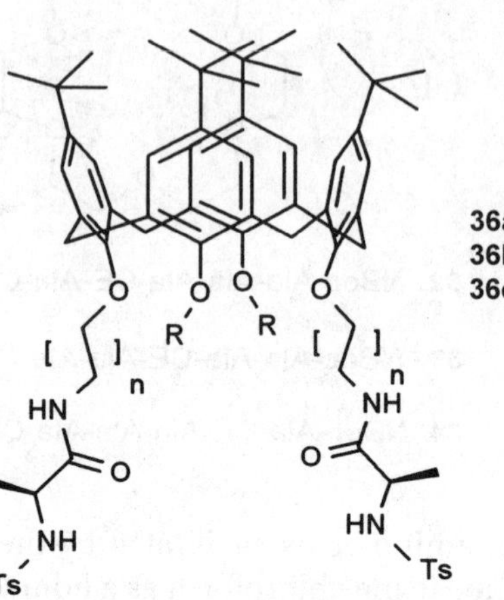

**36a** n = 1, R = H
**36b** n = 2, R = H
**36c** n = 2, R = Pr

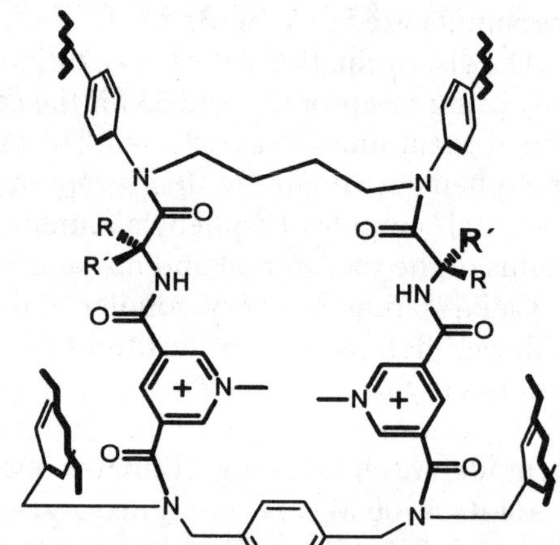

**(+)37** R = iPr, R′= H
**(–)37** R = H, R′= iPr
(only front half of molecule is shown)

Amidic groups were also incorporated into the structure of receptor **37** [67]. This receptor was used for the complexation of substrates capable of quick interchange between their conformational enantiomers in solution. These guests can be fixed in the cavity of **37** predominantly in one of the possible enantiomeric conformations. Using CD spectroscopy, an increase in the amount of (*S*)-helical bilirubin (BR) conformer upon addition of (+)-**37** was observed. The (*R*)-helical conformer was preferred on adding (–)-**37**. The complexation constants were determined from CD spectroscopy data: $6.4\times10^5$ and $6.9\times10^5$ $M^{-1}$

for the complexes of BR with (+)-**37** and (–)-**37**, respectively. A similar behavior of the receptor was observed using pamoic acid (PA) as a guest molecule. Significant changes caused by complexation can be observed in $^1$H NMR spectra too. They were interpreted in terms of probable inclusion of the guest molecule into the cavity of the host.

## 6 Urea- and Thiourea-Based Receptors

In the design of binding sites in receptors for anions, urea and thiourea functions are often employed, due to their ability to form strong hydrogen bonds to organic carboxylates. Chiral thiourea receptors **38** and **39**, based on 6-aminomethylpyridine-2-carboxylic acid, were synthesized by Kilburn [68]. The $^1$H NMR titrations of **38** by tetrabutylammonium salts of various amino acids (*N*-Ac-Ala, *N*-Ac-Phe, *N*-Ac-Asn, *N*-Ac-Gln, *N*-Ac-Trp, *N*-Ac-Ser) were performed in $CDCl_3$. These studies showed the high selectivity of this receptor in recognition of different amino acids – with *N*-Ac-Trp, a complex 30 times more stable than that with *N*-Ac-Ser was formed. However, the chiral recognition of amino acid enantiomers was not so significant. The highest selectivity was observed upon the complexation of *N*-Ac-Gln; the stability constant of the complex of **38** with *N*-Ac-L-Gln was two times higher than that for the D-enantiomer. The complexation constants for *N*-Boc amino acid derivatives were lower than those for *N*-Ac amino acids, but on the complexation of *N*-Boc-Trp an interesting behavior was observed. An improved and reversed selectivity compared with that for *N*-Ac-Trp was found (*N*-Boc-Trp complexation constant ratio was L:D=3:2, but for *N*-Ac-Trp L:D=1:1.5). The receptor **39** was able to discriminate between enantiomers of tryptophan (*N*-Boc-L-Trp $K$=1,925 $M^{-1}$; *N*-Boc-D-Trp $K$=3,785 $M^{-1}$). Upon the complexation of naproxen, a moderate enantioselectivity was observed.

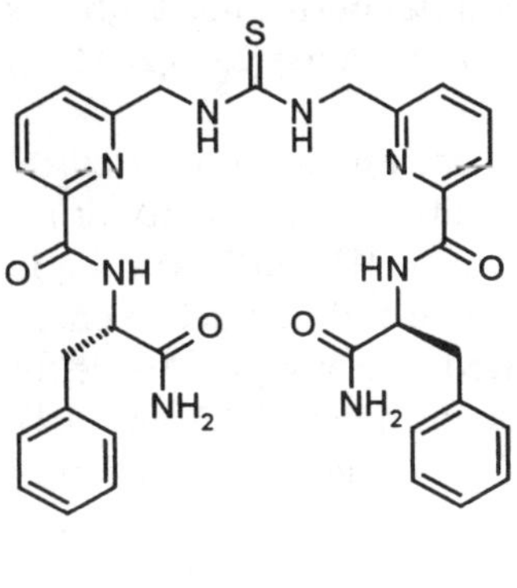

**38**

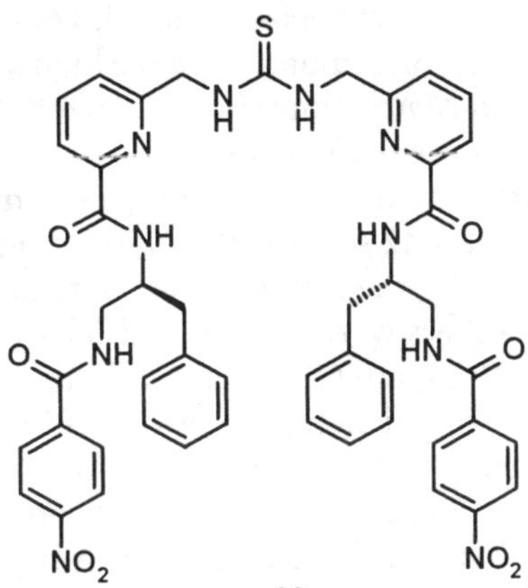

**39**

The bicyclic receptors **40** and **41** were also prepared for enantioselective recognition of chiral carboxylates [69, 70]. The substitution of two benzene rings by pyridines in the receptor **41** enriched this molecule with two additional sites capable of hydrogen bond formation.

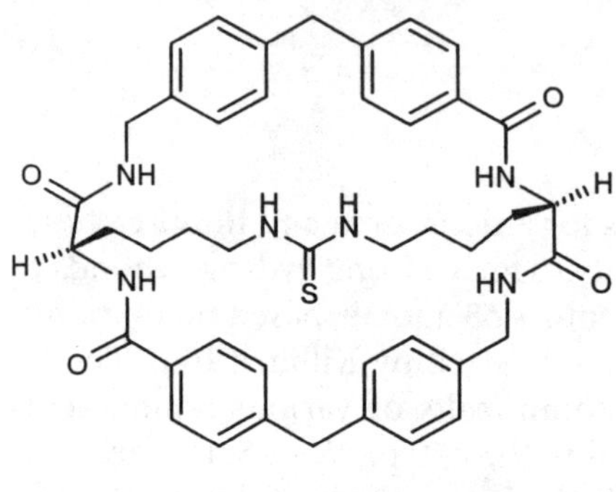

**40**

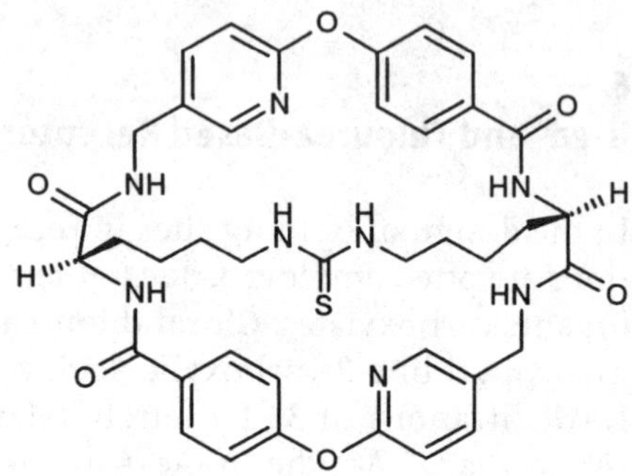

**41**

The stability constants of the carboxylate complexes with the receptor **40** were determined using NMR titrations in $CDCl_3$. The most stable complex observed was formed upon the complexation of the tetrabutylammonium salt of *N*-Ac-lysine ($K$=130,000 $M^{-1}$, that is, ca. ten times higher stability constant compared with the other amino acids). The best chiral recognition was detected during the complexation of enantiomers of *N*-Ac-Phe. Detailed 2D NMR experiments showed the probable reason for this discrimination. While the L-enantiomer of protected phenylalanine is complexed inside the cavity of the host molecule, the *N*-Ac-D-phenylalanine is probably bound outside the cavity via hydrogen bonding to the thiourea group and one of the amidic NH functions.

The ability of receptor **41** to bind enantioselectively to the chiral substrates has not yet been examined. The preliminary NMR studies performed with this receptor demonstrated the high affinity of **41** to *N*-Ac-asparagine ($K$=8,700 $M^{-1}$) and *N*-Ac-glutamine ($K$=55,000 $M^{-1}$).

A very interesting receptor was synthesized by derivatization of a chiral spirobifluorene unit by a urea moiety [71]. The resulting chiral cavity of receptor **42** was capable of binding carboxylates. Using the $^1$H NMR titration in DMSO, the stability constants of complexes with lactate and mandelate enantiomers were determined. These experiments showed a strong preference for (*R*)-lactate ($K$=35,000 $M^{-1}$) compared to the (*S*)-enantiomer ($K$=3,500 $M^{-1}$). A similar behavior of this receptor was observed upon the complexation of mandelates – the more stable complex was formed with the (*R*)-enantiomer ($K$=28,000 $M^{-1}$); (*S*)-mandelate was bound with the constant $K$=1,700 $M^{-1}$. This chiral recognition can be explained by the steric demands of the hydroxyl group of the substrate. This host was also employed to discriminate

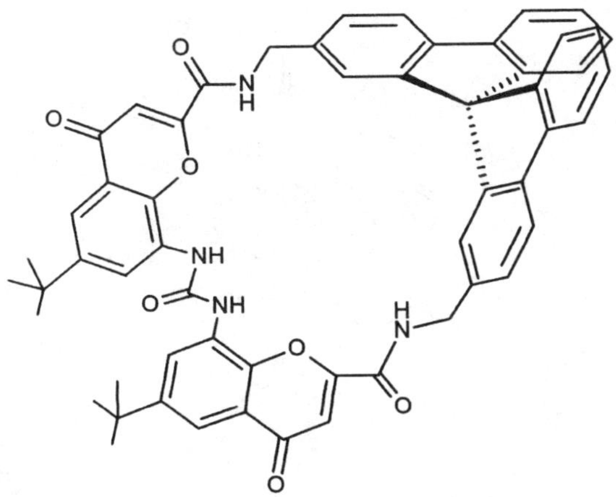


**42**

between the enantiomers of naproxen and lactate, but no chiral recognition was observed.

Thiourea functions were used to attach chiral saccharide units to the molecule of calix[4]arene [72]. The complexation properties of these molecules toward chiral anions have not yet been examined. However, in the preliminary complexation studies ($^1$H NMR titrations in DMSO-$d_6$) the affinity toward acetate and *N*-Ac-L-alaninate was observed.

The synthesis of receptor **43** was based on calix[5]arene [73]. The NMR studies in tetrachloroethane/methanol 2:1 showed the ability of this receptor to bind achiral zwitterionic ε-aminohexanoic acid.

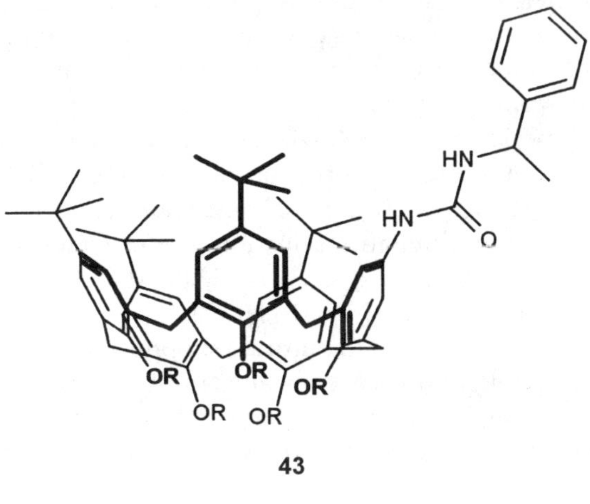


**43**

We have prepared a very simple binaphthalene-based receptor **44** capable of enantioselective recognition of lactate and mandelate [74]. This extremely

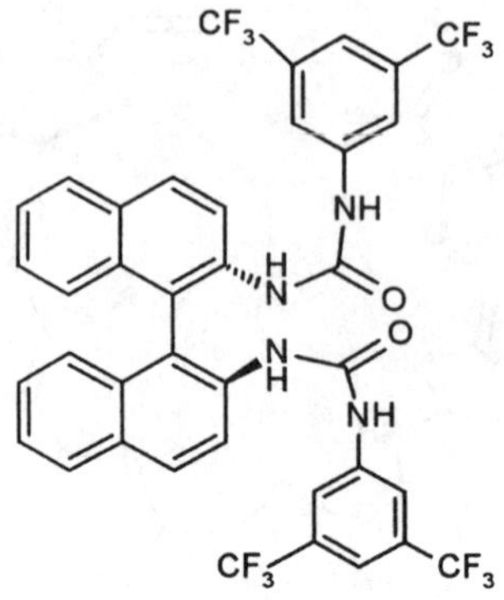

**44**

simple receptor is able to form 1:1 complexes with benzoate and bromide. In addition, 1:2 complexes of **44**:anion were formed with acetate, lactate, and mandelate. Optically pure (*R*)-**44** formed complexes with (*R*)-lactate and (*S*)-lactate in chloroform with different stability (log $K_R$ was 4.60±0.07 and $K_S$ was 3.96±0.06). The enantioselectivity for mandelate is reversed and much weaker, but the complexes formed are very stable with log $K_{assoc}$ over 6.4 in chloroform.

## 7
## Porphyrins and Sapphyrins as Anion Sensing Units

Porphyrins and their derivatives are frequently used in the design of many supramolecular receptors. Sessler has extensively studied many expanded porphyrins including sapphyrins, which are known for their ability to bind anions [75], e.g., fluorides and phosphates [76]. For application in enantioselective recognition the dimeric structures **45–47** have been designed, synthesized, and studied [77]. The complexation profile has been studied using UV–Vis spectroscopy in 5% MeOH–$CH_2Cl_2$. Ligands **45** and **46** preferred *N*-Cbz-Glu against *N*-Cbz-Asp, with very weak enantioselection. Better enantiorecognition was found with the more rigid **47** ($K_{N\text{-Cbz-L-Asp}}$=16,700 $M^{-1}$, $K_{N\text{-Cbz-D-Asp}}$=9,700 $M^{-1}$; and $K_{N\text{-Cbz-L-Glu}}$=3,800 $M^{-1}$, $K_{N\text{-Cbz-D-Glu}}$=16,200 $M^{-1}$).

The sapphyrin unit connected to natural ionophore lasalocid [78] has been used for the design of ligand **48**. Membrane transport was measured for this ligand using a traditional U-tube arrangement. It was found that the ligand is able to transport amino acids in their zwitterionic form with a preference for the L-enantiomer of Phe.

New ligands **49–52** were prepared based on these encouraging results. Unfortunately the enantioselectivity was not improved. For example, compound **52** was found to be unsuitable due to "self-assembled" structures formed in

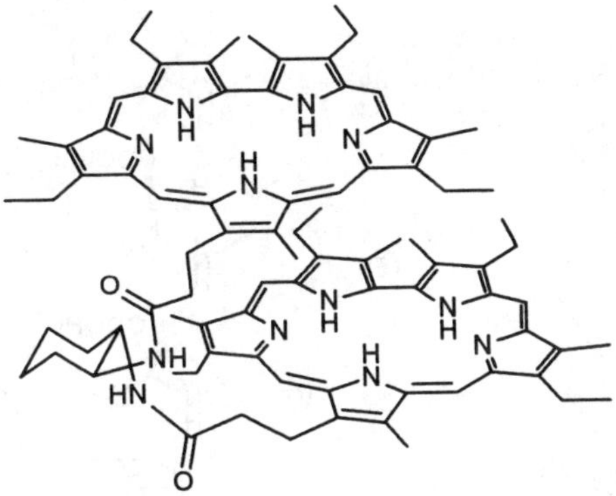

**45**

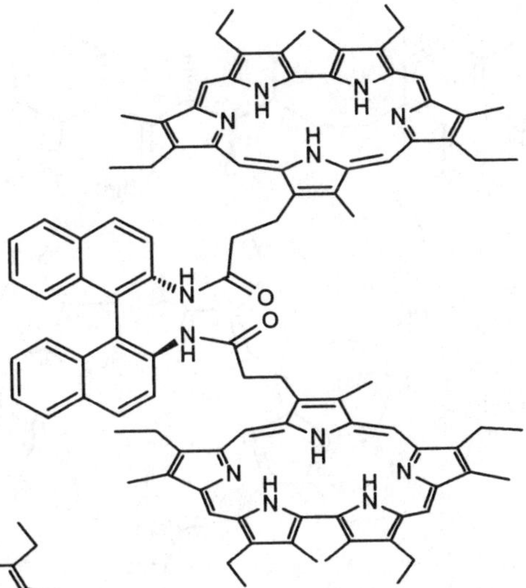

**46**

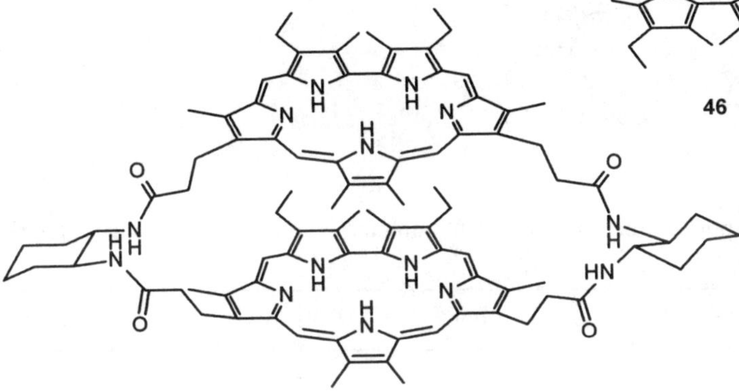

**47**

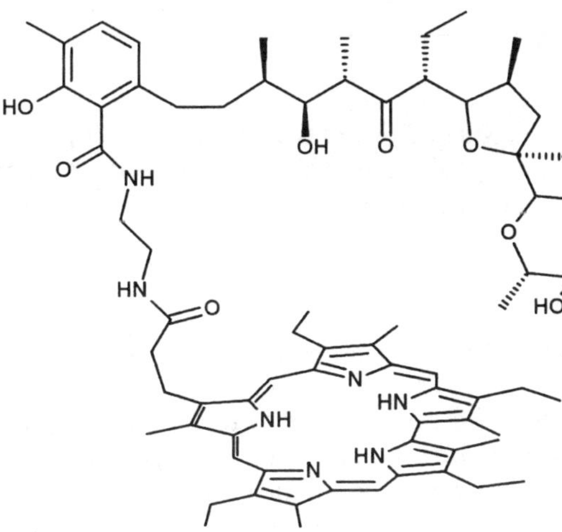

**48**

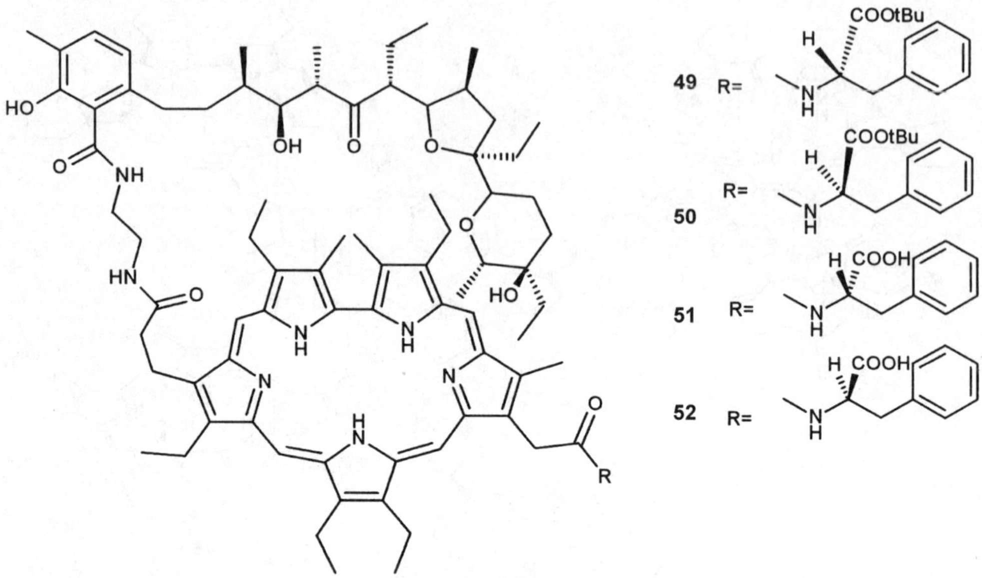


**Table 4** Transport characteristics [$k$ (*$10^{-5}$ mol cm$^{-2}$ h$^{-1}$)] meassured using sapphyrin–lasalocid conjugates

| Receptor | L-Phe | D-Phe | L-Trp | D-Trp |
|---|---|---|---|---|
| **48** | 20 | 12.7 | 5 | 4.2 |
| **49** | 6.4 | 8.2 | 1 | 2.3 |
| **50** | 10.5 | 6.7 | 2.5 | 1.1 |
| **51** | 1.9 | 1.4 | 0.8 | 0.7 |
| **52** | 0.9 | 0.7 | 0.7 | 0.7 |

solution. Better results were achieved with compound **50**, which was capable of transferring Phe and Trp with some preference for the D-enantiomers. These results are summarized in Table 4.

## 8 Metal-Containing Ligands

In principle, there are two possibilities of how to use metal ion(s) for anion sensing. First, the ion can be used for better preorganization of the ligand with respect to anion binding, and second, the cation can be included in the binding site of an anion-sensing ligand in order to explore electrostatic attraction for the binding event.

Ligands **53–57** are representatives of the first approach. Both amides on a bipyridine moiety are brought to spatial proximity by complexation of metal ion between both bipyridine nitrogens [79]. The enantioselective complexation is facilitated by two chiral substituents on amidic nitrogens. Unfortunately, all ligands are relatively flexible and this is probably the reason for their low enantioselectivity for anions. Complexation has been examined using the NMR titration protocol in DMSO-$d_6$, with the tetramethylammonium salt of *N*-Cbz-glutamate. Ligands **53** and **54** showed only negligible enantioseparation of *N*-Cbz-glutamate, while ligands **55** and **56** were much better ($K_{\mathbf{46}\text{-Cbz-L-Glu}}$=160 M$^{-1}$, $K_{\mathbf{46}\text{-Cbz-D-Glu}}$=128 M$^{-1}$). Ligand **57** showed only very weak binding. Coordination on the metal center influenced enantioselection only to a negligible extent.

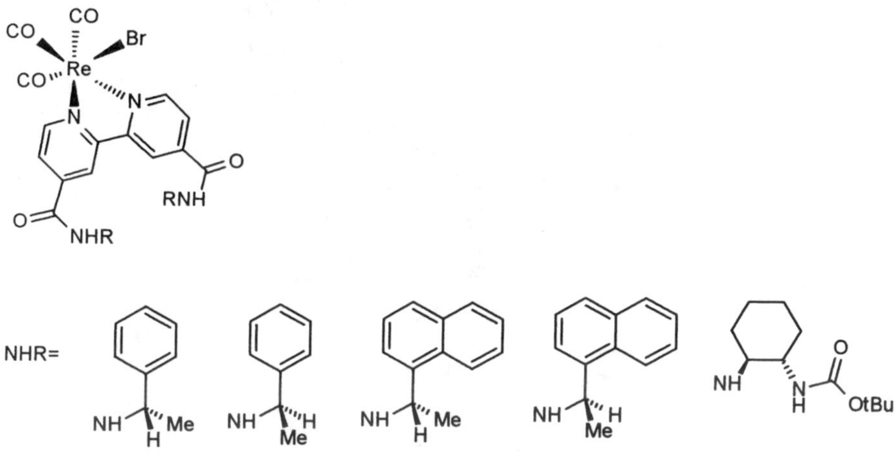


**53** **54** **55** **56** **57**

Ligands **58** and **59** were capable of only very slight enantioselection despite the higher rigidity. Their complexation behavior was followed by NMR titration of sodium salts of L- and D-lactate (ligand **58** only) and tetramethylammonium salts of *N*-Cbz-L- and D-glutamate (both ligands). Moreover, helicity around the metal ion had no influence as both Λ and Δ **58** formed complexes with L-lactate of the same stability within experimental error. Ligand **59** slightly prefered D-glutamate ($K_{\mathbf{50}\text{-Cbz-L-Glu}}$=200 M$^{-1}$, $K_{\mathbf{50}\text{-Cbz-D-Glu}}$=250 M$^{-1}$).

A suitably substituted cyclopentadiene moiety has been used for the construction of ligand **60** using the concept of planar chirality [80, 81]. Only the (+)-(*R*)-enantiomer of **60** was prepared and its structure was proved by X-ray crystallography. Both enantiomers of tetrabutylammonium camphor-10-sulfonate were used for chiral recognition. Only a slight difference was found using NMR titration in $CDCl_3/CD_3CN$ 1:1. (*K*=1,900 M$^{-1}$ for complex with (+)-enantiomer and *K*=1,700 M$^{-1}$ for complex with (–)-enantiomer).

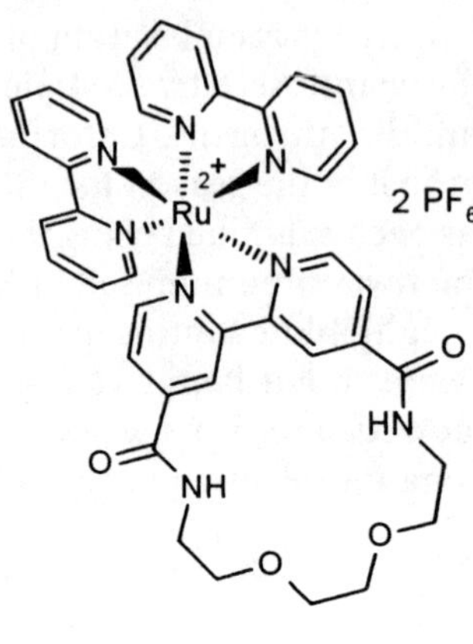

58

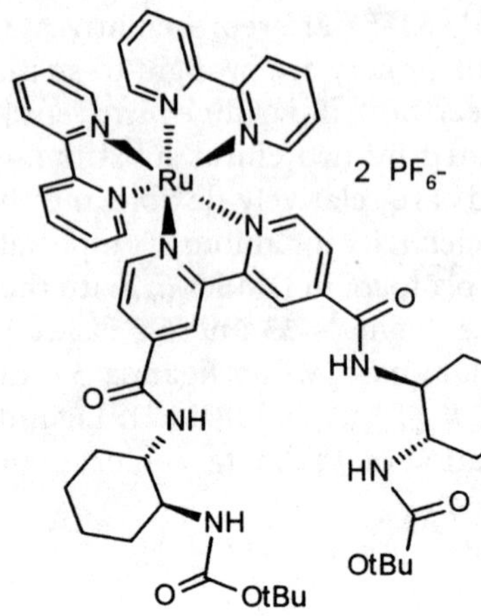

59

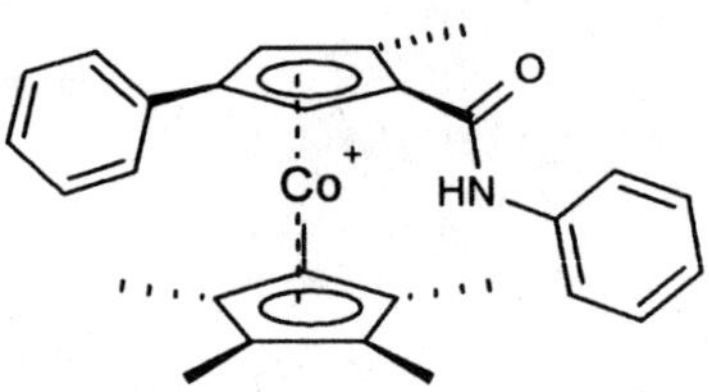

60

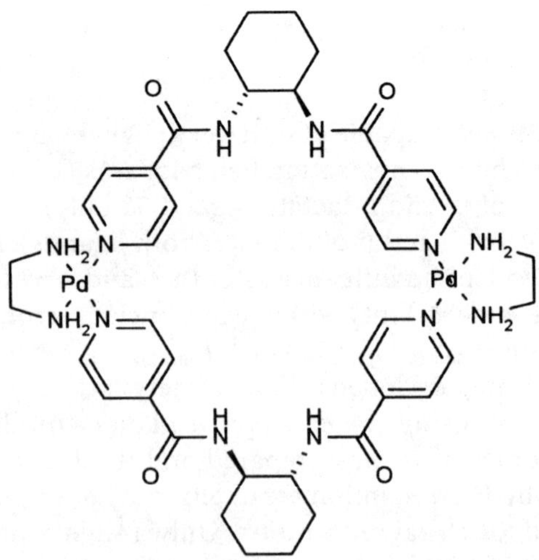

61

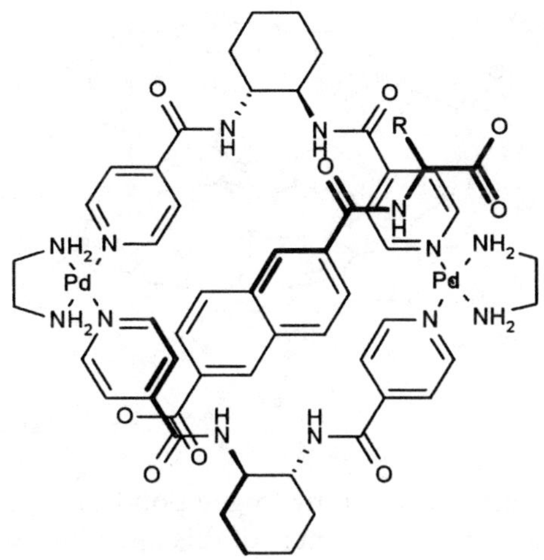

**Fig. 4** Complex of **61** and aromatic carboxylate

Coordination of flexible amidic structures to the palladium ion resulted in the formation of a rigid hydrophobic cavity of **61** [82]. This self-assembly takes place in water, and aromatic carboxylates are recognized in it. It was found using NMR titration in $D_2O$ (pD=8.5, borate buffer). The structure of the complex formed is shown schematically in Fig. 4. It is apparent that cooperative binding of both carboxylates is essential for the stability of the complex formed; nevertheless, the enantioselection of carboxylates is only very weak (Table 5).

Ligand **62** [83] was prepared for the enantioselective recognition of amino acids. Chiral carboxylates are bound by cooperative binding by electrostatic

**Table 5** Stability of complexes formed by ligand **61** with chiral carboxylates

| | K($M^{-1}$) | |
|---|---|---|
| 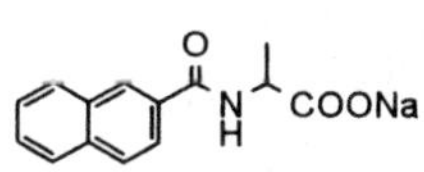 | L | 5 |
| | D | 8 |
| 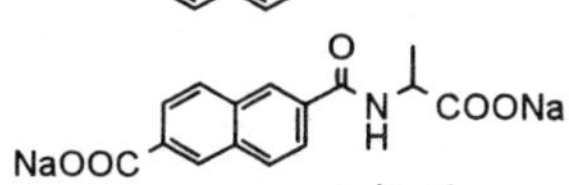 | L | 146 |
| | D | 134 |
| 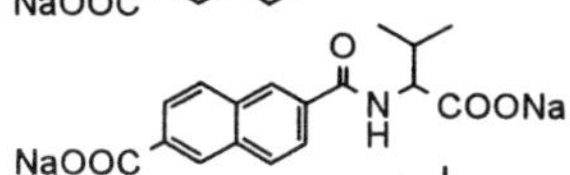 | L | 120 |
| | D | 155 |
| 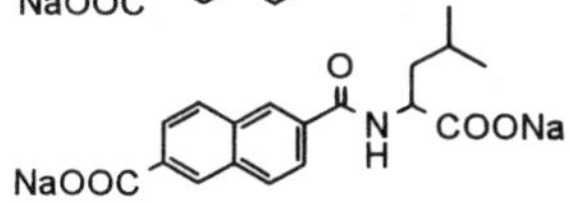 | L | 210 |
| | D | 125 |

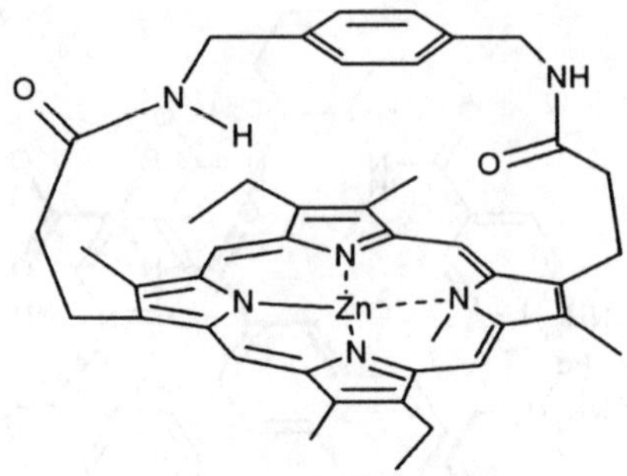

**62**

interaction, steric constraints, and hydrogen bonds; the complex is schematically shown in Fig. 5. Both enantiomers of this ligand were separated by chromatography on chiral stationary phases and used in complexation studies. The efficiency of these ligands is excellent. Chiral carboxylates were extracted from aqueous solution of the appropriate sodium salt into a chloroform solution of (+)-**62**. The ratio of diastereomeric complexes was determined by $^1$H NMR (signal of *N*-methyl group) using standards prepared from enantiomerically pure salts and (+)-**62**. In general, (+)-**62** binds L-enantiomers of amino acids with good to excellent preference, and obviously (–)-**62** behaves analogously as it binds D-amino acids with the same preference. Among the amino acids examined, the highest enantioselectivity was observed for *N*-(3,5-dinitrobenzoyl)phenylglycinate (Table 6). As for the effect of the N-protecting group of the

**Table 6** Single extraction of the sodium salts of racemic N-protected amino acids with racemic **62**, showing the ratio of major and minor diastereomeric complexes

| $Na^+$ salt of amino acid | Ratio |
|---|---|
| *N*-Cbz-alaninate | 91/9 |
| *N*-Cbz-*N*-methylalaninate | 50/50 |
| *N*-Cbz-valinate | 96/4 |
| *N*-Boc-valinate | 95/5 |
| *N*-acetylvalinate | 85/15 |
| *N*-Cbz-norvalinate | 91/9 |
| *N*-Cbz-leucinate | 85/15 |
| *N*-Cbz-norleucinate | 88/12 |
| *N*-Cbz-prolinate | 50/50 |
| *N*-Cbz-methioninate | 91/9 |
| *N*-Cbz-serinate | 23/77 |
| *N*-Cbz-fphenylglycinate | 91/9 |
| *N*-(3,5-dinitrobenzoyl)fphenylglycinate | 96/4 |
| *N*-Cbz-fphenylalaninate | 84/16 |
| *N*-Cbz-tryptophanate | 89/11 |

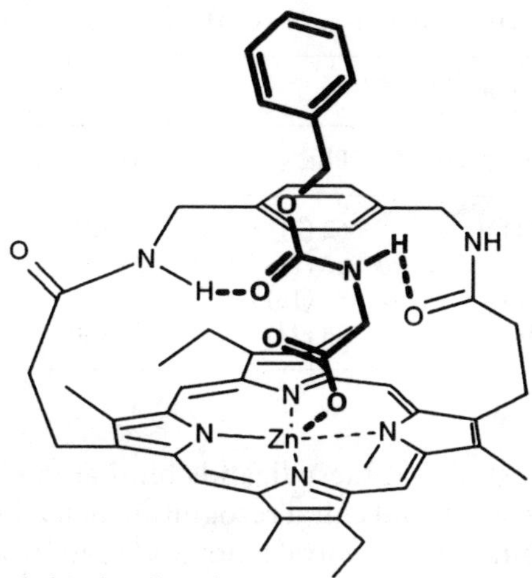


**Fig. 5** Complex of **62** and chiral carboxylate

amino acid salts, the enantioselectivity for the *N*-Boc group was comparable with that of *N*-Cbz, while the *N*-Ac analog was less enantioselectively recognized. In sharp contrast, no enantioselectivity took place for *N*-Cbz-Pro and *N*-Cbz-methylalaninate, having no NH–CO functionalities. The high degree of chiral recognition is considered a result of multiple interactions between the chiral ligand and the substrates. Evidence for hydrogen bond formation between the NH–CO moieties of the ligand and substrates was provided by IR and $^1$H NMR studies coupled with two-dimensional DQF-COSY and NOESY measurements. Also a reluctance for deuterium exchange was used to suggest the mode of binding shown in Fig. 5.

Lipophilic lanthanide complexes of fluorinated β-diketonate ligands were demonstrated to bind unprotected amino acids under neutral conditions. It is not clear whether amino acids are bound as anions or zwitterions. Chiral ligands **63–66** have been prepared and tested for extraction of amino acids from water into dichloromethane [84] (Table 7). NMR and CD spectroscopic

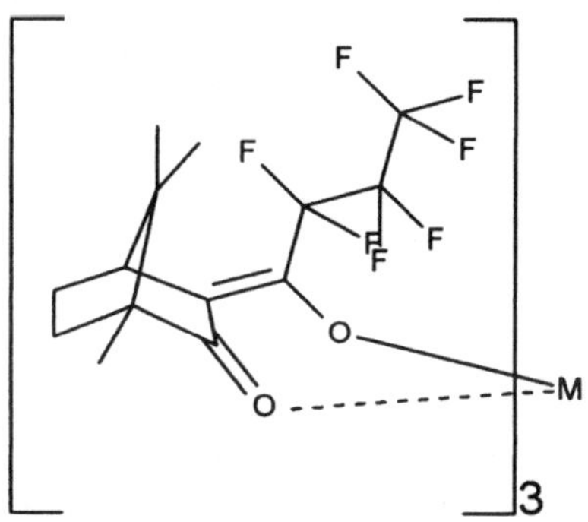


**63** M = Pr
**64** M = Er
**65** M = Eu
**66** M = Yb

**Table 7** Amounts and enantiomeric excesses of amino acids extracted by **63–66**

| Ligand % extracted and (ee) | Amino acid | | | |
|---|---|---|---|---|
| | Phenylglycine | Phe | Trp | Leu |
| **63** | 31 (11) | 62 (2) | 53 (3) | 45 (5) |
| **64** | 43 (13) | 62 (7) | 52 (4) | 61 (7) |
| **65** | 33 (19) | 55 (15) | 39 (23) | 45 (17) |
| **66** | 14 (49) | 36 (24) | 24 (30) | 26 (17) |

studies further suggested that these ligands bind amino acids at two points. Their extraction, transport, and chiral recognition behaviors were significantly controlled by a combination of central lanthanide cation and coordinating ligand. The chiral ytterbium complex offered good enantioselectivity in the extraction of unprotected amino acids, and the related praseodymium complex provided their efficient membrane transport. Chiral lanthanide complexes in aqueous media is the topic of an excellent recent tutorial review [85].

Parker reported the very promising behavior of two ligands **67** and **68** [86]. Their Eu and Tb complexes are reported to form complexes with (*S*)-lactate with up to 4.5 orders of magnitude higher stability than with (*R*)-lactate in aqueous collidine/HCl buffer, pH 7.4. Recently, structural and NMR investigation of the ternary adducts of 20 α-amino acids and selected peptides with the Yb complex of ligand **69** [87] have been studied.

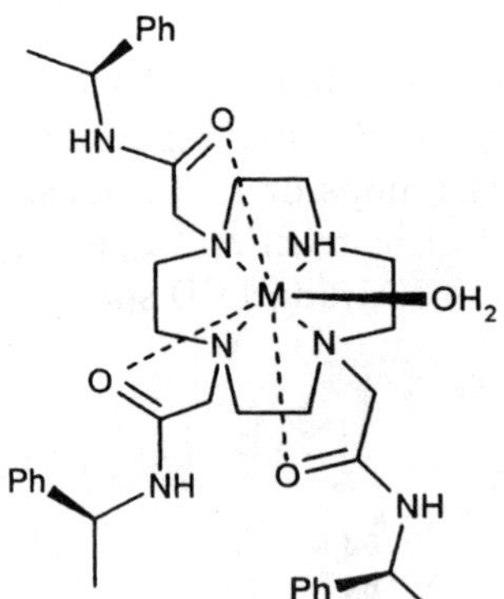

M = Eu **67**
M = Yb **69**

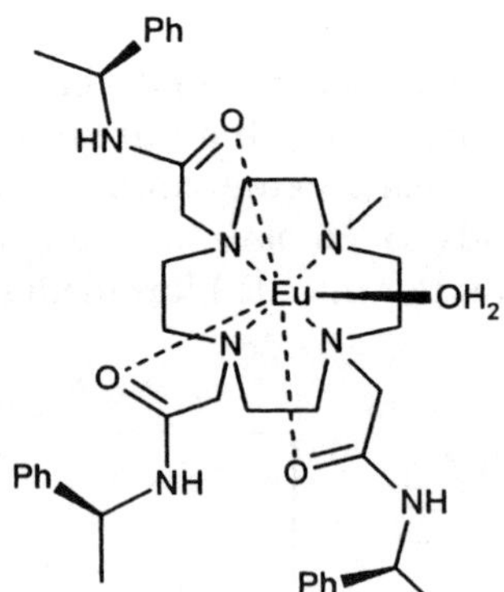

**68**

Enantioselection performed by the Yb complex of (*R*,*R*,*R*)-**69** has been examined with several racemic amino acids. Preferential binding of (*S*)-amino acid was observed. The degree of enantioselectivity measured after addition of

20 equivalents of racemic amino acid to chiral ligand was dependent on the amino acid side chain – namely, Ser (4:1), Thr (4:1), Asp (2:1), Asn (3:1), His (2:1).

Very recently Anslyn and Zhu reported the method of facile quantification of enantiomeric excess and concentration using an indicator-displacement assay exemplified by analysis of 2-hydroxyacids [88].

## References

1. Pirkle WH, Pochapsky TC (1989) Chem Rev 89:347; Ward TJ, Hamburg D-M (2004) Anal Chem 76:4635
2. Schmidtchen FP, Berger M (1997) Chem Rev 97:1609
3. Gale PA (2000) Coord Chem Rev 199:181
4. Gale PA (2001) Coord Chem Rev 231:79
5. Gale PA (2003) Coord Chem Rev 240:191
6. Beer PD, Gale PA (2001) Angew Chem Int Ed 40:486
7. Fitzmaurice RJ, Kyne GM, Douheret D, Kilburn JD (2002) J Chem Soc Perkin Trans I, 841
8. Choi K, Hamilton AD (2003) Coord Chem Rev 240:101
9. Suksai Ch, Tuntulani T (2003) Chem Soc Rev 32:192
10. Martínez-Máñez R, Sancenón F (2003) Chem Rev 103:4419
11. Ilioudis CA, Steed JV (2003) J Supramol Chem (in press)
12. Bondy CR, Loeb SJ (2003) Coord Chem Rev 240:77
13. Best MD, Tobey SL, Anslyn EV (2003) Coord Chem Rev 240:3
14. Tsukube H (1996) Coord Chem Rev 148:1
15. Tsukube H, Shinoda S (2002) Chem Rev 102:2389
16. Kano K, Kato Y, Kodera M (1996) J Chem Soc Perkin Trans II, 1211
17. Kitae T, Nakayama T, Kano K (1998) J Chem Soc Perkin Trans II 207
18. Kitae T, Takashima H, Kano K (1999) J Incl Phenom Macrocyclic Chem 33:345
19. Kano K, Kamo H, Negi S, Kitae T, Takaoka R, Yamaguchi M, Okubo H, Hirama M (1999) J Chem Soc Perkin Trans II, 15
20. Rekharsky MV, Inoue Y (2001) J Am Chem Soc 124:813
21. Li S, Purdy WC (1992) Anal Chem 64:1405
22. Lipkowitz KB, Raghothama S, Yang J (1992) J Am Chem Soc 114:1554
23. Liu Y, Li B, Han BH, Li YM, Chen RT (1997) J Chem Soc Perkin Trans II, 1275
24. Liu Y, Zhang YM, Sun SX, Li YM, Chen RT (1997) J Chem Soc Perkin Trans II, 1609
25. Tabushi I, Kuroda Y, Mizutani T (1986) J Am Chem Soc 108:4514
26. Liu Y, Zhang YM, Qi AD, Chen RT, Yamamoto K, Wada T, Inoue Y (1997) J Org Chem 62:1826
27. Corradini R, Dossena A, Impellizzeri G, Maccarrone G, Marchelli R, Rizzarelli E, Sartor G, Vecchio G (1994) J Am Chem Soc 116:10267
28. Pagliari S, Corradini R, Galaverna G, Sforza S, Dossena A, Marchelli R (2000) Tetrahedron Lett 41:3691
29. Pagliari S, Corradini R, Galaverna G, Sforza S, Dossena A, Montalti M, Prodi L, Zaccheroni N, Marchelli R (2004) Chem Eur J 10:2749
30. Godoy-Alcántar C, Nelen MI, Eliseev AV, Yatsimirsky AK (1999) J Chem Soc Perkin Trans II, 353
31. Lara KO, Godoy-Alcántar C, Eliseev AV, Yatsimirsky AK (2004) Org Biomol Chem 2:1712
32. Hancock RD, Bhavan R, Wade PW, Boeyens JCA, Dobson SM (1989) Inorg Chem 28:187
33. Alfonso I, Rebolledo F, Gotor V (2000) Chem Eur J 6:3331
34. Alfonso I, Dietrich B, Gotor V, Lehn JM (2001) Helv Chim Acta 84:280

35. Howarth J, Al-Hashimy NA (2001) Tetrahedron Lett 42:5777
36. Echavarren AM, Galán A, de Mendoza J, Lehn JM (1989) J Am Chem Soc 111:4994
37. Galán A, Pueyo E, Salmerón A, de Mendoza J, Tetrahedron Lett (1991) 32:1827
38. Galán A, Andreu D, Echavarren AM, Prados P, de Mendoza J (1992) J Am Chem Soc 114: 1511
39. Metzger A, Gloe K, Stephan H, Schmidtchen FP (1996) J Org Chem 61:2051
40. Breccia P, Van Gool M, Pérez-Fernández R, Martín-Santamaría S, Gago F, Prados P, de Mendoza J (2003) J Am Chem Soc 125:8270
41. Liu F, Lu GY, He WJ, Wang ZS, Zhu LG (2001) Chin J Chem 19:317
42. Lawless LJ, Davis AP (1999) J Chem Soc Chem Commun 9
43. Lawless LJ, Blackburn AG, Ayling AJ, Pérez-Payán MN, Davis AP (2001) J Chem Soc Perkin Trans I, 1329
44. Schmuck C (2000) Chem Eur J 6:709
45. Segura M, Alcázar V, Prados P, de Mendoza J (1997) Tetrahedron 53:13119
46. Alcázar V, Segura M, Prados P, de Mendoza J (1998) Tetrahedron Lett 39:1033
47. Chinchilla R, Nájera C, Sánchez-Agulló P (1994) Tetrahedron Asymmetry 5:1393
48. Davis AP, Dempsey KJ (1995) Tetrahedron Asymmetry 6:2829
49. Montero VA, Tomlinson L, Houk KN, Diederich F (1991) Tetrahedron Lett 32:5309
50. Alcazar V, Diederich F (1992) Angew Chem Int Ed Engl 31:1521
51. Hernández JV, Oliva AI, Simón L, Muniz FM, Mateos AA, Morán JR (2003) J Org Chem 68:7513
52. Pieters RJ, Cuntze J, Bonnet M, Diederich F (1997) J Chem Soc Perkin Trans II, 1891
53. Waymark CP, Kilburn JD, Gillies I (1995) Tetrahedron Lett 36:3051
54. Henley PD, Waymark CP, Gillies I, Kilburn JD (2000) J Chem Soc Perkin Trans I, 1021
55. Flack S, Kilburn JD (1995) Tetrahedron Lett 36:3409
56. Dowden J, Edwards PD, Flack SS, Kilburn JD (1999) Chem Eur J 5:79
57. Lustenberg P, Martinborough E, Denti MT, Diederich F (1998) J Chem Soc Perkin Trans II, 747
58. Hong JI, Namgoong SK, Bernardi A, Still WC (1991) J Am Chem Soc 113:5111
59. Almaraz M, Raposo C, Martin M, Caballero MC, Morán JR (1998) J Am Chem Soc 120: 3516
60. Hernández JV, Almaraz M, Raposo C, Martín M, Lithgow A, Crego M, Caballero C, Morán JR (1998) Tetrahedron Lett 39:7401
61. He Y, Xiao Y, Meng L, Zeng Z, Wu X, Wu CT (2002) Tetrahedron Lett 43:6249
62. Goswami S, Ghosh K, Dasgupta S (2000) J Org Chem 65:1907
63. Casnati A, Sansone F, Ungaro R (2003) Acc Chem Res 36:246
64. Voyer N, Coté S, Biron E, Beaumont M, Chaput M, Levac S (2001) J Supramol Chem 1:1
65. Sansone F, Baldini L, Casnati A, Lazzarotto M, Ugozzoli F, Ungaro R (2002) Proc Natl Acad Sci USA 99:4842
66. Sdira SB, Felix CP, Giudicelli MBA, Seigle-Ferrand PF, Perrin M, Lamartine RJ (2003) J Org Chem 68:6632
67. Murakami Y, Hayashida O, Nagai Y (1994) J Am Chem Soc 116:2611
68. Kyne GM, Light ME, Hursthouse MB, de Mendoza J, Kilburn JD (2001) J Chem Soc Perkin Trans I, 1258
69. Pernia GJ, Kilburn JD, Essex JW, Mortishire-Smith RJ, Rowley M (1996) J Am Chem Soc 118:10220
70. Jullian V, Shepherd E, Gelbrich T, Hursthouse MB, Kilburn J (2000) Tetrahedron Lett 41:3963
71. Tejeda A, Oliva AI, Simón L, Grande M, Caballero MC, Morán JR (2000) Tetrahedron Lett 41:4563

72. Sansone F, Chierici E, Casnati A, Ungaro R (2003) Org Biomol Chem 1:1802
73. Ballistreri FP, Notti A, Pappalardo S, Parisi MF, Pisagatti I (2003) Org Lett 7:1071
74. Stibor I, Holakovský R, Mustafina AR, Lhoták P (2004) Collect Czech Chem Commun 69:365
75. Sessler JL, Davis JM (2001) Acc Chem Res 34:989
76. Sessler JL, Davis JM, Král V, Kimbrough T, Lynch V (2003) Org Biomol Chem 1:4113
77. Sessler JL, Andrievsky A, Král V, Lynch V (1997) J Am Chem Soc 119:9385
78. Sessler JL, Andrievsky A (1998) Chem Eur J 4:159
79. Uppadine LH, Keene FR, Beer PD (2001) J Chem Soc Dalton Trans 2188
80. Uno M, Komatsuzaki N, Shirai K, Takahashi S (1993) J Organomet Chem 343
81. Komatsuzaki N, Uno M, Shirai K, Takai Y, Tanaka T, Sawada M, Takahashi S (1996) Bull Chem Soc Jpn 69:17
82. Kim HJ, Lim CW, Hong JI (2001) Mater Sci Eng C 15:265
83. Konishi K, Yhara K, Toshishige H, Aida T, Inoue S (1994) J Am Chem Soc 116:1337
84. Tsukube H, Shinoda S, Uenishi J, Kanatani T, Itoh H, Shiode M, Iwachido T, Yonemitsu O (1998) Inorg Chem 37:1585
85. Parker D (2004) Chem Soc Rev 33:156
86. Bruce JI, Dickins RS, Govenlock LJ, Gunnlaugsson T, Lopinski S, Lowe MP, Parker D, Peacock RD, Perry JJB, Aime S, Botta M (2000) J Am Chem Soc 122:9674
87. Dickins RS, Basanov AS, Howard JAK, Parker D, Puschmann H, Salamano S (2004) J Chem Soc Dalton Trans 70
88. Zhu L, Anslyn EV (2004) J Am Chem Soc 126:3676

Top Curr Chem (2005) 255: 65–95
DOI 10.1007/b101162


# Anion Receptors Based on Calixarenes

Pavel Lhoták (✉)

Department of Organic Chemistry, Institute of Chemical Technology, Technická 5,
Prague 6, Czech Republic
*lhotakp@vscht.cz*



**Abstract** Calix[*n*]arenes represent a well-known family of macrocyclic molecules with a broad range of potential applications in many branches of supramolecular chemistry. Because of their preorganisation, calix[*n*]arenes are frequently used as building blocks and molecular scaffolds in the construction of more elaborate systems, such as artificial enzyme biomimetics and receptors. This review is focused on the recent development of calixarene-based anion receptors.

## 1 Introduction

Calix[*n*]arenes [1] **1** are a well-known family of macrocyclic compounds possessing interesting three-dimensional structures with many unique properties. They are notorious for their ability to create host–guest-type complexes with many ions and/or neutral molecules (depending on the substitution pattern). To efficiently form given complexes, calix[*n*]arenes usually utilise the combination of several non-covalent interactions; among them hydrogen-bonding

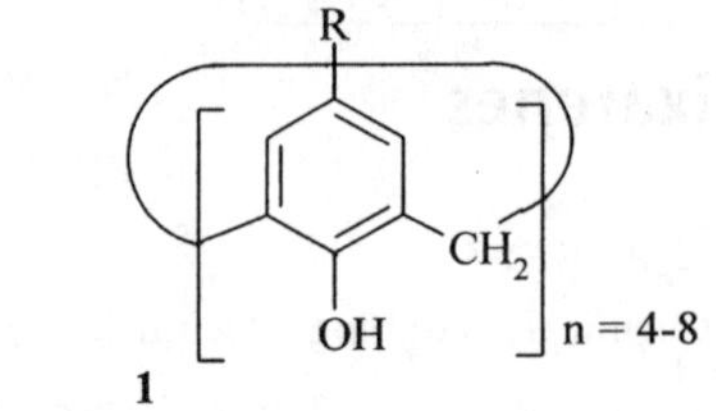

1

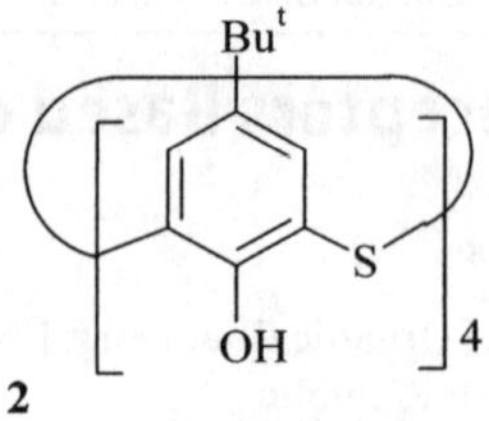

2

interactions, electrostatic interactions, cation–π interactions or hydrophobic effects could be cited as the most significant examples. Because of their easy one-pot preparation and simple derivatisation, they are frequently used in supramolecular chemistry as building blocks in the construction of more elaborate systems.

The exclusive "shaping" and "tuning" possibilities of calix[4]arene derivatives makes these compounds ideal candidates for the design of novel receptors. Thus, simple alkylation on the calixarene lower rim (phenolic hydroxyls) leads to four different conformers which are isolable and infinitely stable if the substituents are bulky enough (Fig. 1). Any of these four conformations, called *cone*, *partial cone*, *1,3-alternate* and *1,2-alternate*, represents a unique structure with inherent characteristic properties including complexation abilities. Consequently, the well-established chemistry of calix[4]arenes gives us a tool to arrange deliberately the functional groups in space to achieve the best interaction mode of the complexation process.

While the selective interactions of functionalised calixarenes with cations have been studied broadly for almost three decades, the application of calixarene-based receptors for anion recognition is a relatively new research topic [2]. This review is focused on recent developments in the design and synthesis of calixarene-based anion receptors. Although the name calixarene was originally designated only for phenol–formaldehyde derivatives **1**, recently many structural variations and mutations have been formed. Some of them, such as calixpyrroles [3], are widely used for anion recognition; nevertheless, this review is restricted only to "classical" calixarenes **1** and newly discovered thiacalixarenes **2** [4].

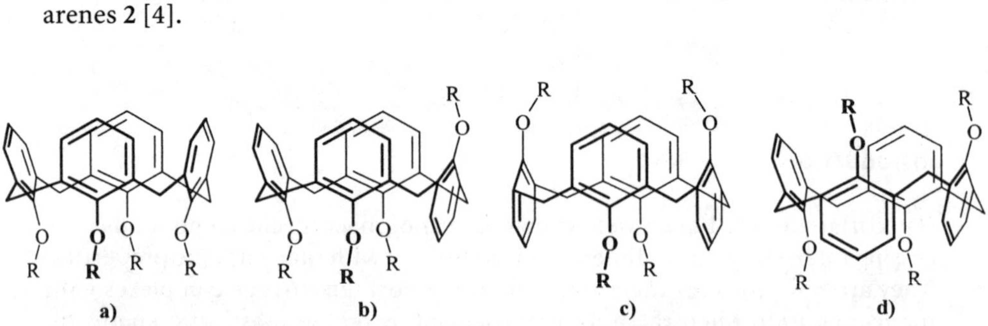

**Fig. 1** Four basic conformations of calix[4]arene: **a** *cone*, **b** *partial cone*, **c** *1,3-alternate* and **d** *1,2-alternate*

# 2
# Receptors Using Metals

## 2.1
## Metallocene-Based Systems

As mentioned above, the chemistry of calixarenes offers us a unique possibility of exact spatial arrangements of appropriate functional groups, which can be used for anion recognition. Among such functions, the cobaltocenium and ferrocene moieties are frequently used for anion recognition [5]. Both systems, albeit based on seemingly identical metallocene structures, possess some basic differences. Thus, cobaltocenium represents an inherently positively charged organometallic function that serves as a pH-independent redox-active moiety. These properties make this unit very promising for anion sensing in polar (potentially aqueous) solvents, where it can use electrostatic interactions with anions. On the other hand, it seems that only pure electrostatic interactions are insufficient for effective complexation of anions [6, 7]. To support the complexation process, combination with another interaction is needed. As the hydrogen bonding phenomenon is generally used for this purpose, it is not surprising that the appending of amidic or urea functions to metallocenes leads to very efficient systems [8].

Consequently, the application of calixarene as a molecular scaffold allows the design of receptors utilising the combination of spatially oriented hydrogen bonds (HBs) and synchronous coulombic interactions. Of course, both factors favourably influence the overall complexation process. The role of preorganisation can be illustrated by the comparison [9] of calixarene **3** with acyclic analogue **4**.

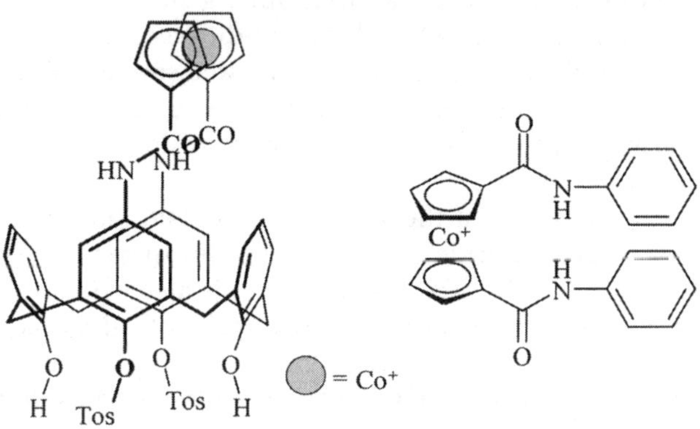


**3** **4**

As proven by $^1$H NMR titration experiments in DMSO-$d_6$, the upper rim bridged derivative **3** exhibits very good binding abilities towards many selected anions (in the form of tetrabutylammonium salts). On the other hand, recep-

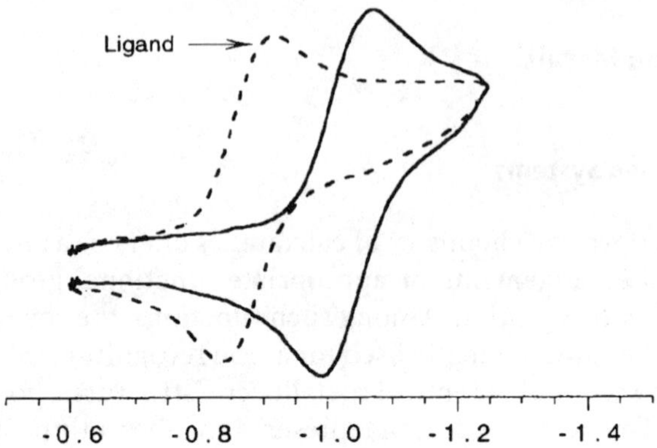


**Fig. 2** Cyclic voltammograms of **3** in MeCN; (- - -) absence and (—) presence of 3 equiv. of acetate

tor **4** shows much lower stability constants as a consequence of missing macrocyclic preorganisation. Thus, if we compare the association constants for $H_2PO_4^-$ ($K^3$=6,380 $M^{-1}$ vs $K^4$=320 $M^{-1}$) or the binding of acetates ($K^3$=41,500 $M^{-1}$ vs $K^4$=1,560 $M^{-1}$), it is evident that there is at least one order of magnitude difference in the strength of interactions. Calixarene **3** shows particularly high affinity for carboxylates, albeit without pronounced selectivity ($K^3$ [benzoate]= 38,400 $M^{-1}$, $K^3$ [phenylacetate]=22,270 $M^{-1}$).

The electrochemical behaviour of **3** was studied [10] using cyclic voltammetry in acetonitrile with $Bu_4NBF_4$ as supporting electrolyte. As shown in Fig. 2, anion recognition can be demonstrated on a substantial cathodic shift of the corresponding reversible cobaltocenium/cobaltocene redox couple. The addition of 3 equivalents of $Bu_4NAc$ led to cathodic perturbation up to 155 mV; other carboxylates or dihydrogen phosphate showed similar values. The above-described electrochemical properties of cobaltocene derivatives indicate that metallocene-derived calixarenes could be very promising as amperometric sensors for anions.

Crystallographic analysis of the **3**·$Cl^-$ complex (Fig. 3a,b) verified the proposed structure where chloride anion is held almost symmetrically by the amidic functions (Cl...HN=2.45 and 2.52 Å). It is also evident that there are several close contacts with the aromatic hydrogens of calixarene (ArH...Cl= 3.05, 3.15 Å) and metallocene (CH...Cl=2.75, 2.95 Å) moieties. As the whole structure creates a dimeric form of head-to-tail type, the chloride anion is held by additional HBs of the metallocene unit from the second molecule (CH...Cl= 2.77, 2.93 Å).

Contrary to derivative **3**, calixarene **5** was found to create a 1:2 complex with chloride anion in the solid state. Both amidic functions introduced into the upper rim of calixarene operated independently of each other and held one chloride anion with a Cl...HN distance of approx. 2.37 Å and two CH...Cl con-

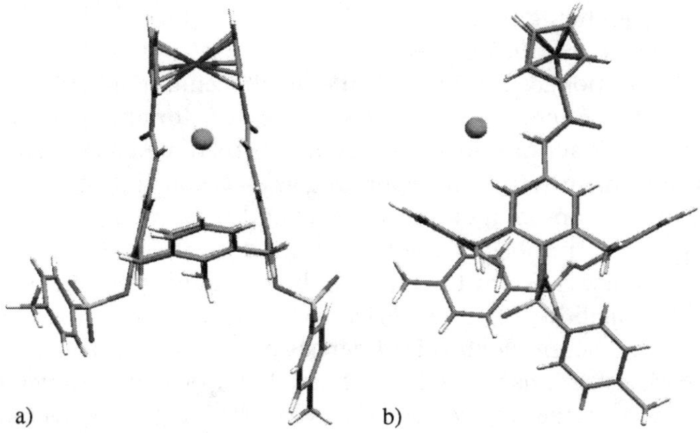

**Fig. 3** Crystallographic structure of complex **3**·$Cl^-$: **a** front view, **b** side view (solvent molecule removed for better clarity)

tacts (2.66, 2.80 Å). The solution experiments (DMSO-$d_6$) revealed that both isomeric structures **5** and **6** coordinate anions with 1:1 stoichiometry. Surprisingly, the substitution pattern on the lower rim affects fairly dramatically the effectiveness of complexation. With only one exception ($Cl^-$), derivative **5** possesses always much higher complexation constants towards anions (carboxylates, dihydrogen phosphate) when compared with those of isomeric structure **6**. Molecular modelling calculations proposed that this phenomenon can be explained by different conformations adopted by compounds in solution. While **5** prefers the *pinched cone* conformation, where cobaltocenium moieties on the upper rim are held in close proximity thus enabling the cooperative effect of both units during complexation, compound **6** adopts the *pinched cone* structure

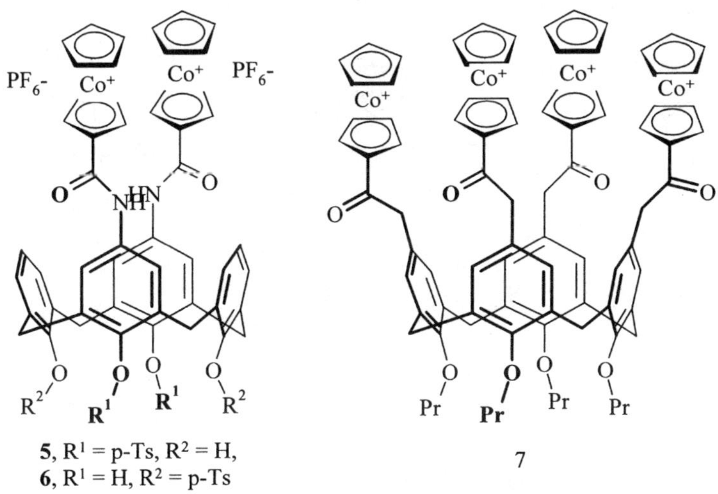


**5**, $R^1$ = p-Ts, $R^2$ = H,
**6**, $R^1$ = H, $R^2$ = p-Ts

7

with metallocenes being spaced apart. This assumption was also corroborated by X-ray structures of free ligands.

As already mentioned, in contrast to the cobaltocenium moiety, the ferrocene derivatives represent neutral redox-active receptors for anions. As we are losing coulombic interactions, the complexation is mediated solely via hydrogen bonding. Consequently, the corresponding association constants can be evaluated only in HB non-competitive solvents and they are usually much lower when compared with those of cobaltocenium derivatives. This apparent disadvantage is compensated by the much higher synthetic potential of ferrocene (commercial availability of many derivatives, much easier synthetic handling) and the excellent electrochemical properties of this compound.

The $^{1}H$ NMR titrations ($CD_3CN$) revealed that upper rim bridged ferrocene derivatives **8a–c** interact with carboxylates in solution [11, 12], as indicated by significant downfield shifts of NH protons upon addition of anions. Thus, the complexation-induced shift in the **8c**/acetate system is more than 3.5 ppm (7.3 to 11 ppm) leading to a complexation constant of 1,200 $M^{-1}$. Derivatives bearing methyl groups on the lower rim are conformationally unstable and represent a thermodynamic equilibrium mixture of two conformations: *cone* **8a,b** and *partial cone* **9a,b**. The complexation abilities of both conformers are comparable, which indicates that most likely only two amide groups are responsible for the binding of anions. The Job plot analysis confirmed the 1:1 binding mode for all receptors **8** and **9** with pronounced complexation of bidentate hydrogen bonding anions (carboxylates, $H_2PO_4^-$). Interestingly, with exception of chloride, halides are not complexed by these receptors. The electrochemical properties of **8a–c** and the lower rim bridged derivative **10** [13] were studied by cyclic voltammetry (CV) and square-wave voltammetry (SWV). It was shown that the complexation of anion induced a large cathodic perturbation of the ferrocene/ferrocenium redox couple during CV, which potentially enables the application of these compounds as electrochemical anion receptors. In the

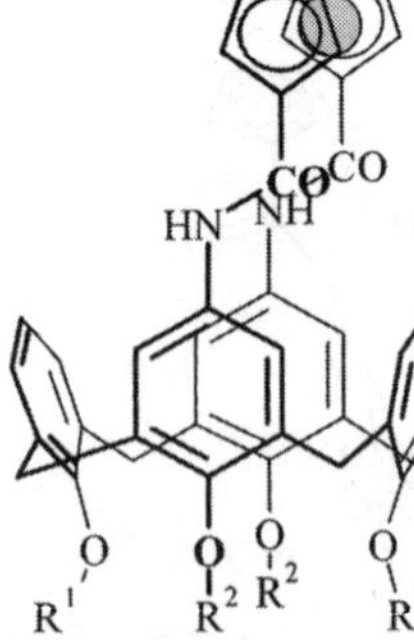


**8a**, $R^1 = R^2 = CH_3$,
**8b**, $R^1 = R^2 =$ -$CH_2COOEt$
**8c**, $R^1 = CH_3$, $R^2 =$ -$CH_2COOEt$

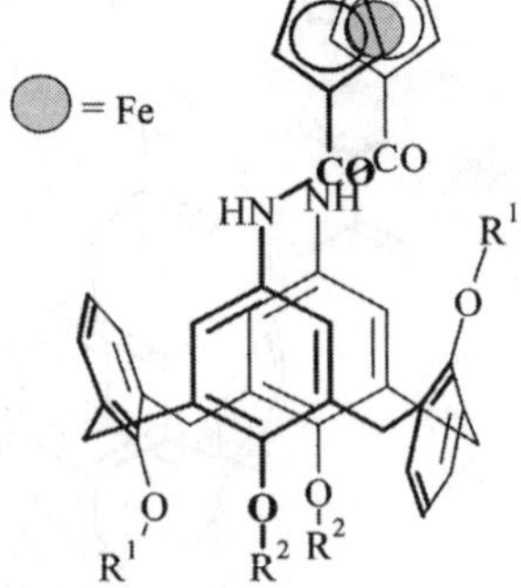


**9**

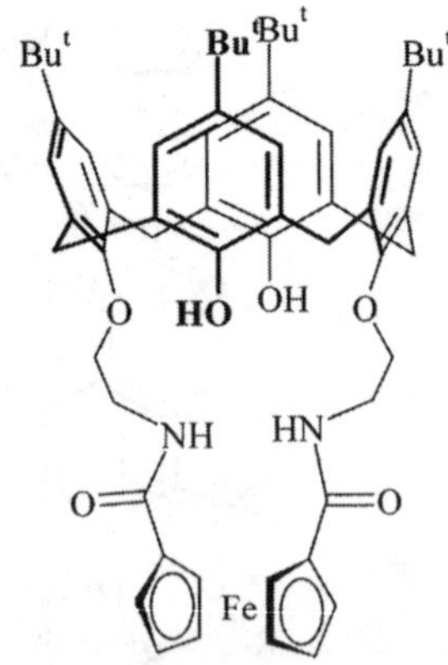


**10**

case of SWV, the addition of benzoate or acetate to **8c** gradually results in quasi-reversible voltammograms, and finally leads to a completely irreversible response. As the authors stated, the ferrocenium moiety presumably enhances [12] the anion binding by electrostatic interactions, thereupon the anion inhibits the electron transfer back to the ferrocene group. Similar behaviour was also observed in the case of calixarenes **11** and **12** bearing four ferrocene units on the upper rim [14].

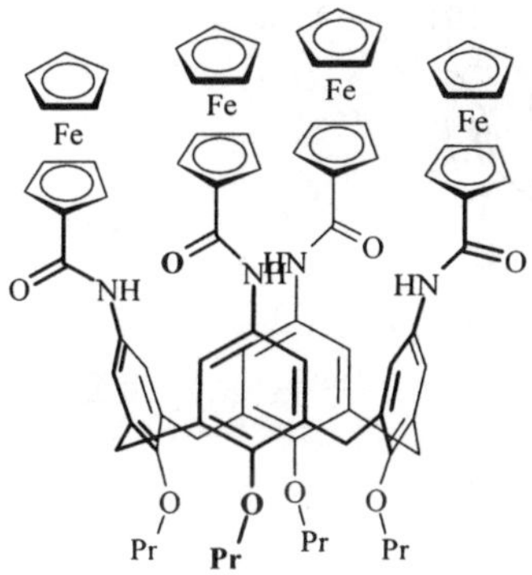

**11**

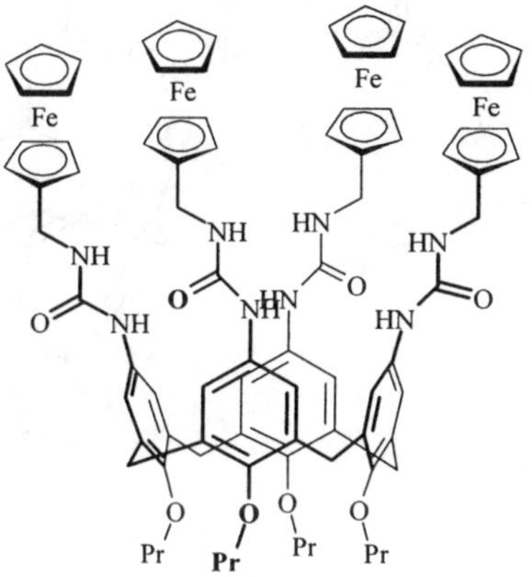

**12**

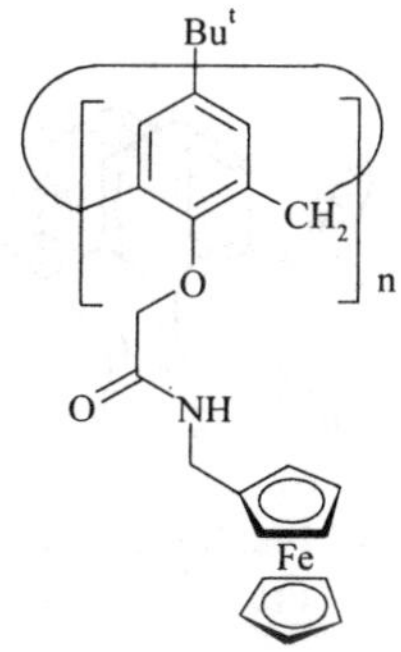

**13a**, n = 4
**13b**, n = 5

The comparison of lower rim tetrasubstituted calix[4]arene **13a** with the corresponding pentasubstituted calix[5]arene **13b** (both compounds immobilised in the *cone* conformation) revealed that the higher number of ferrocene units in **13b** does not positively influence the complexation ability of the receptor. The stability constants of **13a** and **13b** with chloride anion were found to be 55 and 15 $M^{-1}$, respectively. As indicated by cyclic voltammetry, both receptors are selective towards the dihydrogen phosphate anion (cathodic shift ≈160 mV).

Very interesting solvent-dependent anion complexation [15] was observed for receptor **14a**. The NMR titrations surprisingly revealed that the association constants are dramatically influenced by the solvent used. Thus, the *K* values for chloride anion measured in $CD_2Cl_2$, $CD_3CN$ or acetone-$d_6$ are 40, 70 and 5,200 $M^{-1}$, respectively. Acetate exhibits a similar general trend: 26, 120 and 6,000 $M^{-1}$, respectively. Interestingly, benzoate does not change the *K* value so dramatically, which leads to the change of overall selectivity depending on the solvent. While the selectivity factor $K_{(\text{acetate})}/K_{(\text{benzoate})}$ is approx. 0.22 in $CD_2Cl_2$, the same value in acetone-$d_6$ is 6.40. These results cannot be simply explained in terms of the differences in bulk solvent properties such as relative permittivity ($\varepsilon$) or dipole moment ($\mu$). The authors suggested an explanation based on the Gutmann acceptor number (GAN) of the solvent, which gives a quantitative measure of HB donor ability of the solvent. As the GAN decreases, the magnitude of binding increases and the overall selectivity shifts towards smaller,

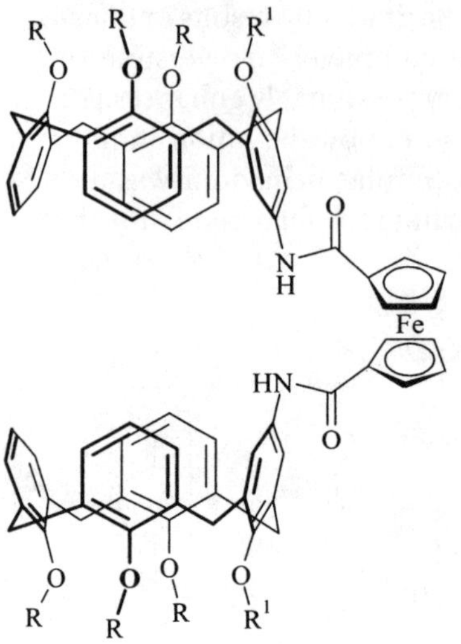

**14a**, R = $CH_3$, $R^1$ = H
**14b**, R = $R^1$ = -$CH_2COOEt$

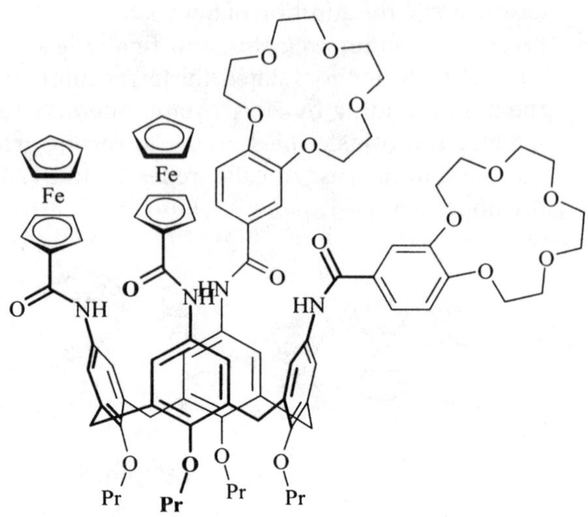

**15**

harder anions (chlorides, acetates). It is very surprising, in this context, that similar solvent-dependent behaviour was not reported [16] for analogous system **14b** possessing four ethyl acetate arms on the lower rim. As both compounds differ only in the substitution of the lower rim, it indicates that there are some additional factors governing the above solvent-dependent complexation of anions. Because of the presence of methyl groups, calixarene **14a** is obviously conformationally unstable. Hence, this compound exhibits thermodynamic equilibrium between several conformations, the distribution of which could be heavily influenced by the solvent used. This effect cannot be observed in the case of **14b** that is immobilised in the *cone* conformation.

Receptor **14b** was designed for ion pair recognition as it also contains a cavity suitable for cation complexation (ester groups on the lower rim). Similar recognition ability [17] was found for derivative **15** bearing additional crown ether units. In both cases the relationship between the cation present and the binding of anions was studied.

## 2.2 Bipyridine Complex-Based Systems

Another interesting and potentially very useful group of calixarene-based anion receptors is represented by systems with appended transition metal complexes of 2,2-bipyridine units. Technically, these systems utilise classical hydrogen bonding interactions of amidic/urea functions; hence, from this point of view, they do not differ from many other receptors. On the other hand, the covalent attachment of bipyridine complexes of ruthenium(II) or rhenium(I) metals imparts to the molecules many novel profitable features. As the 2,2-bipyridine complexes are well known as excellent luminophores, the introduction of these moieties near the complexation place in the host molecule enables sensing of the anion with the aid of fluorescence/luminescence techniques.

This type of receptor is represented by compounds **16a,b** bearing ruthenium(II) bipyridine moieties. Both calixarenes [18] exhibit 1:1 binding of chloride and bromide anions (DMSO-$d_6$), and they are especially suitable for the complexation of $H_2PO_4^-$ ($K^{16a}$=2.8·$10^4$ $M^{-1}$; $K^{16b}$=5.2·$10^3$ $M^{-1}$). On the other hand, if we compare these results with those for similar non-calixarene receptors, where the bipyridine unit is substituted by alkyl, aryl or ethylene glycol substituents, the introduction of calixarene does not bring any substantially new features into the complexation abilities of these derivatives. As shown by X-ray analysis, the anion is encapsulated within the cavity formed by amidic functions with the contributions of CH...anion interactions from the bipyridine unit.

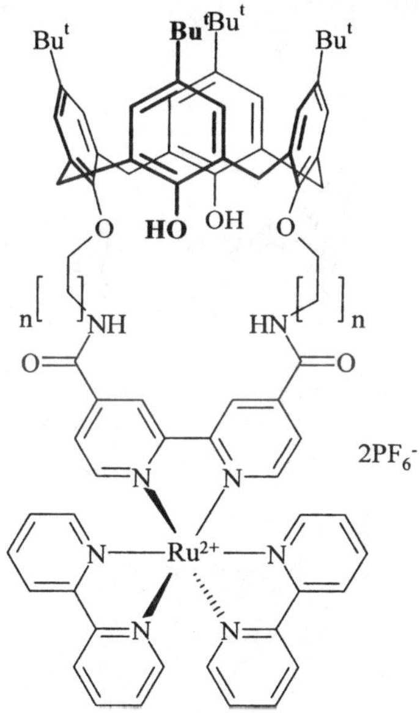


**16a**, n = 1
**16b**, n = 2

Several calixarenes **17–20** bridged on the lower rim with ruthenium(II) and rhenium(I) bipyridyl complexes have been prepared and studied for anion recognition [19]. When compared to acyclic receptor **21** having a similar active structure (ruthenium complex), all receptors show significantly better complexation of $AcO^-$ and chloride in DMSO-$d_6$. Thus, calixarene **17** exhibits an almost 30 times higher association constant for $AcO^-$ ($K^{17}$=9,990 $M^{-1}$ vs $K^{21}$=

350 M$^{-1}$), proving the importance of preorganisation imposed by the calixarene skeleton. In contrast, the situation is just reversed for $H_2PO_4^-$ where the acyclic receptor forms a stronger complex ($K^{17}$=215 M$^{-1}$ vs $K^{21}$=1,300 M$^{-1}$). It is noteworthy that in all cases the receptors based on ruthenium complexes possess significantly higher complexation constants than the corresponding rhenium analogues. The plausible explanation could be based on electrostatic effects. While the rhenium complexes are neutral, the ruthenium receptors are positively charged which leads to stronger binding of anions.

The luminescence spectra of all receptors in $CH_3CN$ were found to be dramatically affected by the addition of acetate or chloride. While compound **19** exhibits an emission decrease, the other receptors **17**,**18** and **20** show a remarkable intensity increase (up to 500%) with a slight concomitant blue shift of the emission maximum (660 nm for **17**). The anion-induced enhancement of luminescence intensity in the case of **17** is clearly due to the decrease of the electron transfer between the ruthenium(II) bipyridyl centre and the quinone moieties. Alternatively, receptors bearing ruthenium or rhenium complexes on the upper rim were also described [20].

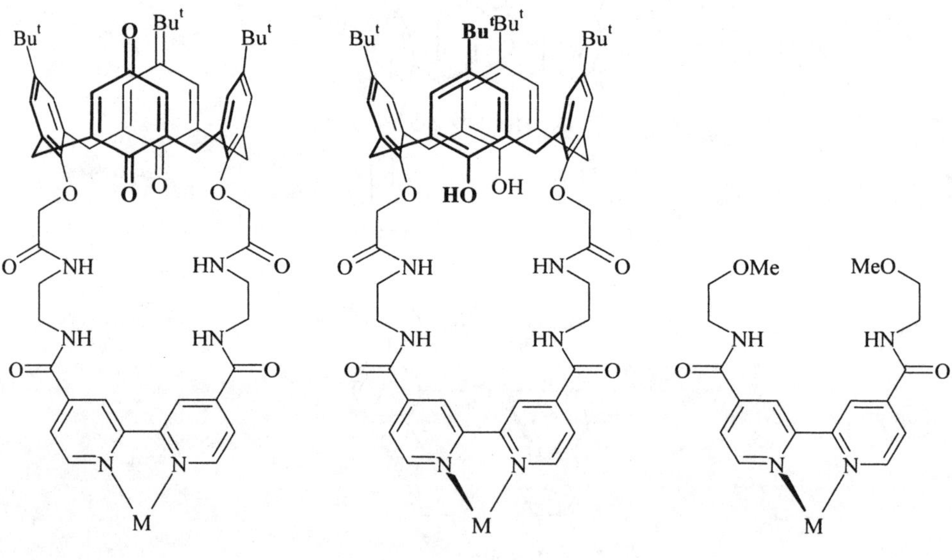

**17**, M = $Ru(Bipy)_2^{2+} (PF_6)_2$
**18**, M = $Re(CO)_3Cl$

**19**, M = $Ru(Bipy)_2^{2+} (PF_6)_2$
**20**, M = $Re(CO)_3Cl$

**21**, M = $Ru(Bipy)_2^{2+} (PF_6)_2$

Rhenium complexes **22a,b** were designed as ditopic receptors for the cooperative complexation of ion pairs [21]. As proved by $^1$H NMR titration experiments in $CD_3CN$, the receptors form complexes of 2:1 stoichiometry with alkali metal cations (interaction with both ethyl acetate cavities on the lower rim) and exhibit [16] a positive allosteric effect for iodide complexation; compare

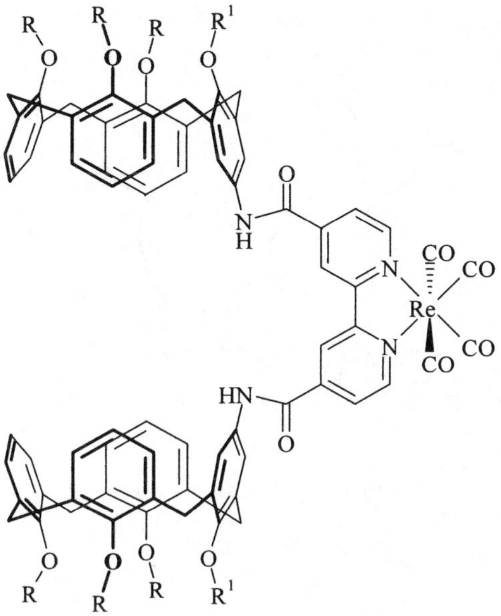

**22a**, R = $CH_3$, $R^1$ = H
**22b**, R = $R^1$ = $-CH_2COOEt$

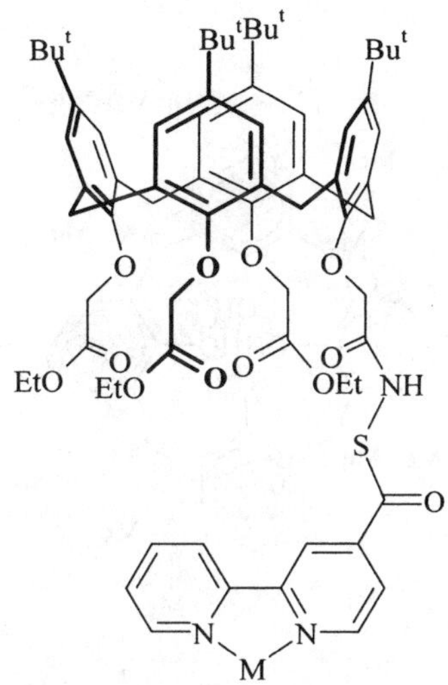

**23a**, M = $Re(CO)_3Cl$; S = $(CH_2)_2$
**23b**, M = $Re(CO)_3Cl$; S = $(CH_2)_3$
**23c**, M = $Re(CO)_3Cl$; S = $(CH_2)_4$
**23d**, M = $Re(CO)_3Cl$; S = o-phenylene
**23e**, M = $(bipy)_2Ru^{2+}(PF_6)_2$; S = $(CH_2)_2$
**23f**, M = $(bipy)_2Ru^{2+}(PF_6)_2$; S = $(CH_2)_3$
**23g**, M = $(bipy)_2Ru^{2+}(PF_6)_2$; S = $(CH_2)_4$
**23h**, M = $(bipy)_2Ru^{2+}(PF_6)_2$; S = o-phenylene

$K^{22b}$=40 $M^{-1}$ in the absence and $K^{22b}$=305 $M^{-1}$ in the presence of two equivalents of $Li^+$. Recently, receptors **23** for ion pair complexation based on the same strategy were reported [22]. As in the previous example, a positive allosteric effect of cation binding for anion recognition was observed depending on the spacer used. The highest enhancement of binding constants was observed for rhenium analogue **23d** bearing the most rigid spacer. For instance, the binding of bromide was enhanced 60 times in the presence of lithium cation ($K^{23d}$=40 $M^{-1}$ vs $K^{23d}$=2,400 $M^{-1}$ in $CD_3CN$).

## 2.3
## Other Metal-Based Systems

Another class of metal-employing anion receptors is represented by structure **24** [23]. Its function is based on the incorporation of positively charged transition metal complexes directly into the calixarene skeleton. Such calixarenes with enhanced electron deficiency of the aromatic walls provide well-preorganised cavities suitable for anion inclusion. The corresponding rhenium [24], ruthenium, rhodium or iridium complexes of this type were prepared and studied for anion recognition [25, 26].

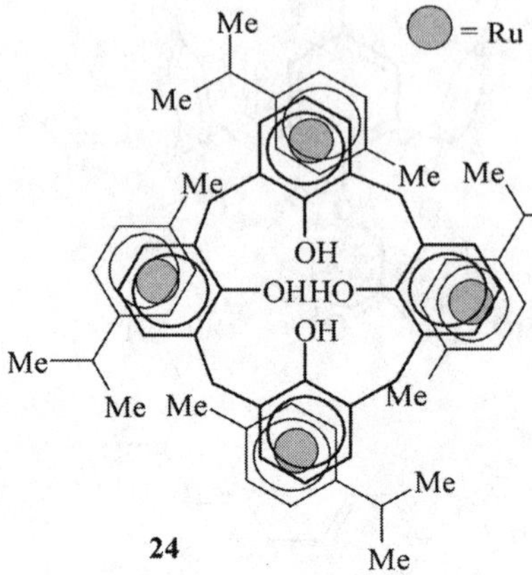

24

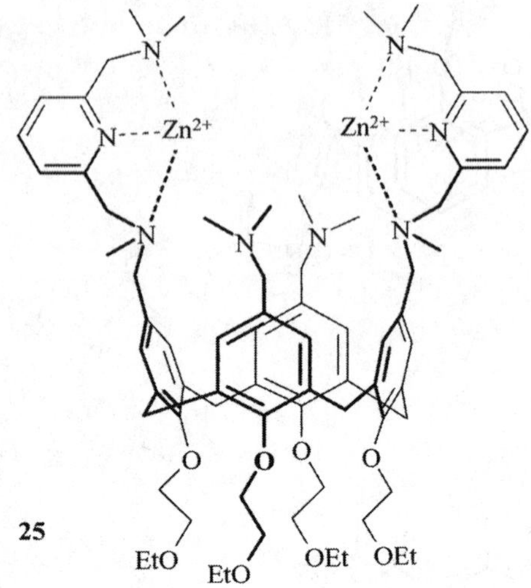

25

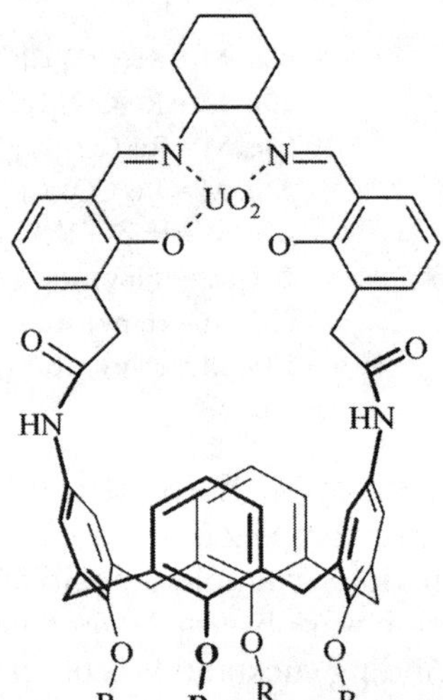

**26a,** R = -Pr
**26b,** R = $-CH_2COOEt$

Many artificial systems have been designed recently to imitate the function and behaviour of native enzymes – biomimetic chemistry [27]. Among them, calixarene-based receptors bearing one, two or three Zn(II) complexes on the upper rim were prepared as a model for phosphoesterases [28–31]. Dinuclear receptor **25** was reported to enhance the rate of transesterification of the RNA model substrate 2-hydroxypropyl-*p*-nitrophenyl phosphate more than 20,000 times compared with the non-catalysed reaction. The complexation mode for the phosphate anion can be described as cascade complexation where the anion is coordinated within the cavity formed by two zinc cations.

Macrocyclic ligand **26a** was shown to complex anions in solution via the synchronous functioning of the central uranyl ion and hydrogen bonding interactions of appended amidic functions. The operation mode of this type of receptor is based on the donor–acceptor interactions of the anion with the central Lewis acid unit (orbital overlap with $UO_2$) [32–34]. A similar type of receptor immobilised in a *1,3-alternate* conformation was also used for simultaneous transport of an ionic pair through a supported liquid membrane [35].

## 3
## Receptors Based on Amidic or Urea Functions

As demonstrated by many examples in the literature [36], the amidic and urea/thiourea functions could be very useful in the design of receptors for anion recognition. Contrary to the electrostatic interactions, the hydrogen bonds are usually inactive or almost diminished when used in more polar and hence HB-competitive solvents. On the other hand, the application of electroneutral host molecules does not raise the problems with appropriate counter-anions (such as competitive binding), which are frequently met when positively charged receptors are used. As the above-mentioned functional groups are capable of highly directional hydrogen bonding interactions, their incorporation into the well-preorganised macrocyclic skeleton can lead to the design of receptors with many interesting properties.

Systems based solely on amidic functions are still relatively rare in the calixarene literature. One of the earliest examples is based on chlorosulphonation of calix[4]arene and subsequent reaction with alkylamines, which leads to the corresponding sulphonamides **27a–d**.

It was demonstrated [37] using NMR titration experiments ($CDCl_3$) that these compounds can interact with $H_2PO_4^-$, $HSO_4^-$, $Cl^-$, $NO_3^-$ and $ClO_4^-$ with peak selec-

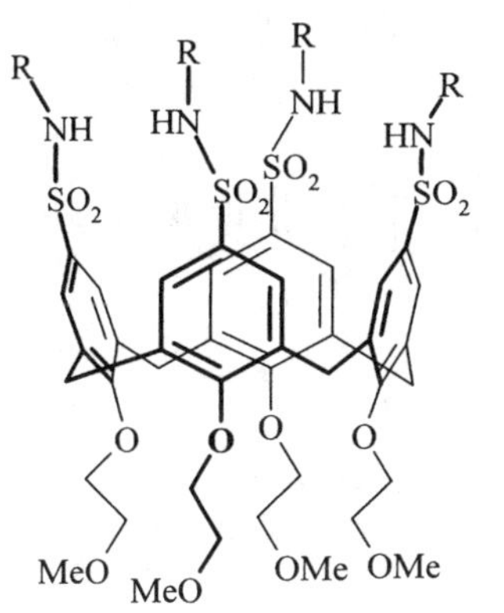


**27a**, R = H
**27b**, R = *n*-Pr
**27c**, R = *t*-Bu
**27d**, R = -$CH_2CH_2NHC(O)CH_3$

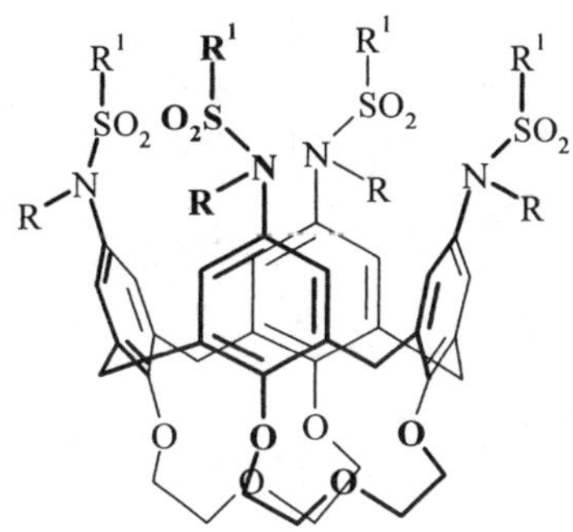


**28a**, R[1] = Bz, R = H
**28b**, R[1] = Bz, R = $CH_3$

tivity for $HSO_4^-$ in the case of **27d** ($K^{27d}$=$10^5$ $M^{-1}$). This result can be explained by the synchronous action of both sulphonamide and carboxamide functions during anion complexation which is enabled only in **27d**. Similar structure **28a** was designed as a heteroditopic receptor for the recognition of ion pairs [38] and it was found that the complexation of the corresponding tetramethylammonium salts strongly depends on the anion used. Derivative **28b** bearing a methyl group on the amidic function does not show any complexation ability towards anions, thus proving the crucial role of NH bonds in the recognition process.

It seems that the introduction of electron-withdrawing substituents into the amide moiety should lead to enhanced acidity of the N–H bond and consequently, to higher complexation constants of the binding process. Surprisingly, the situation is not so simple and there are some findings which question this assumption. Thus, while the chloromethyl derivative **29a** shows only weak interactions with the benzoate anion in $CDCl_3$ ($K^{29a}$=107 $M^{-1}$) [39], the corresponding dichloromethyl group in **29b** induces much stronger complexation ($K^{29b}$=5,160 $M^{-1}$). Contrary to expectations, trichloromethyl amide **29c** does not show any complexation ability. Moreover, the same trend in the binding constants was observed for simple benzamides bearing $-CH_2Cl$, $-CHCl_2$ and $-CCl_3$ moieties. This unexpected phenomenon was ascribed by the authors to the steric crowding of the $-CCl_3$ group.

The series of simple aromatic amides **30a–d** were tested for their complexation ability towards anions and significant differences were found [40]. The presence of the perfluorophenyl substituent in **30b** enhances the binding ability for the bromide anion almost ten times ($K^{30a}$=107 $M^{-1}$ vs $K^{30c}$=1,320 $M^{-1}$), and the value of the stability constant strongly depends on the position of this activating group ($K^{30b}$=370 $M^{-1}$). Amazingly, derivative **30d** with two electron-

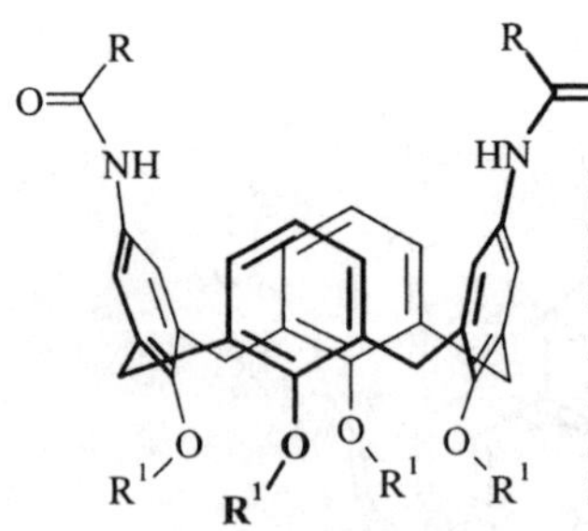

**29a**, R = $CH_2Cl$
**29b**, R = $CHCl_2$
**29c**, R = $CCl_3$
**29d**, R = -NH-Ph
**29e**, R = -Ph
**29f**, R = 2-pyridyl
**29g**, R = $-C_6F_5$

**29a-e**, $R^1$ = n-Pr
**29f,g**, $R^1$ = $-CH_2COOEt$

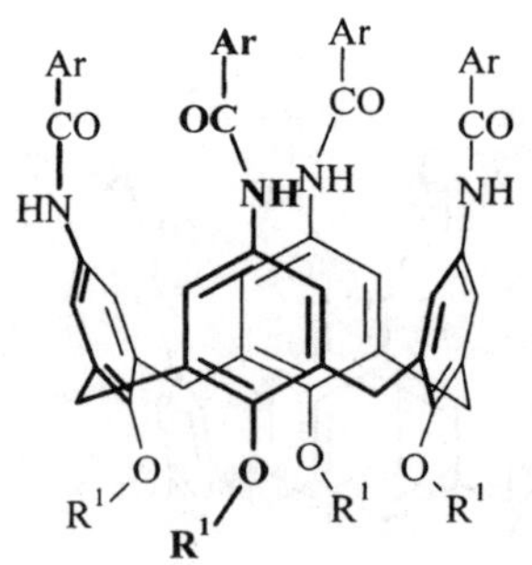

**31a**, R = $-C_6F_5$
**31b**, R = 3-pyridyl
**31c**, R = 4-pyridyl
**31d**, R = $-C_6F_5$
**31e**, R = 2-pyridyl
**31f**, R = 3-pyridyl
**31g**, R = 4-pyridyl

**31a-c**, $R^1$ = n-Pr
**31d-g**, $R^1$ = $-CH_2COOEt$

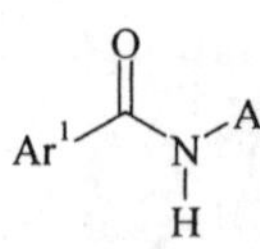

**30a**, $Ar^1$ = $Ar^2$ = -Ph
**30b**, $Ar^1$ = Ph, $Ar^2$ = $-C_6F_5$
**30c**, $Ar^1$ = $-C_6F_5$, $Ar^2$ = -Ph
**30d**, $Ar^1$ = $Ar^2$ = $-C_6F_5$

withdrawing units is inactive and does not bind bromide at all. These results indicate that the presence of at least one hydrogen atom in an α or *ortho* position to the –NH–CO– moiety is of crucial importance for the binding of the anion. As supported by theoretical studies [41, 42], such hydrogen atoms can interact with the anion via non-classical hydrogen bonding interactions which contribute significantly to the overall binding process. As confirmed by the X-ray analysis [10, 18], this type of CH...anion interaction is probably a general phenomenon in anion binding, as it was found in the solid-state structures of several amidic receptors (see e.g. Fig. 3).

Upper rim-substituted tetraamides of type **31** were prepared [40] and tested for anion recognition. All these receptors form 1:1 complexes with $Cl^-$, $HSO_4^-$, $BF_4^-$, $H_2PO_4^-$, $Cl^-$, $NO_3^-$ or squarate anions with a solvent-dependent selectivity. Thus, receptor **31a** exhibits the highest complexation constant for $HSO_4^-$ ($K^{30a}$= 860 $M^{-1}$ in $CD_3CN$, $K^{30a}$=4,450 $M^{-1}$ in acetone-$d_6$), while squarate shows the reversed selectivity ($K^{30a}$=2,150 $M^{-1}$ in $CD_3CN$, $K^{30d}$=650 $M^{-1}$ in acetone-$d_6$). The corresponding diamide **29g** showed good complexation for dicarboxylates with peak selectivity for terephthalate ($K^{29g}$=7,850 $M^{-1}$) and adipate ($K^{29g}$=6,500 $M^{-1}$) in acetone-$d_6$.

The macrocyclic receptors **32** [43] and **33** [44] combining two calix[4]arene motifs within the molecule were designed for anion recognition. While **32** creates 1:1 complexes with several anions (halides, $HSO_4^-$, $H_2PO_4^-$), compound **33** is too rigid to efficiently complex halides or benzoate. By contrast, similar cage molecule **34** with the ureido bridges showed complexation ability towards chloride or benzoate.

To compare the complexation ability of amide versus urea groups, the calixarenes **29e** and **29d** bearing two functions on the upper rim were prepared [45].

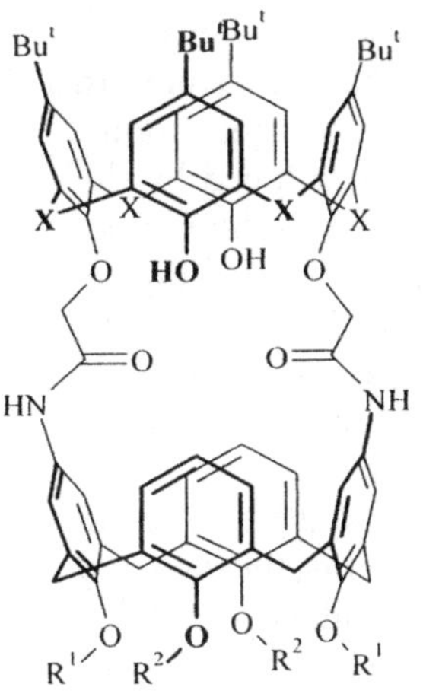

**32a**, R¹ = p-Ts, R² = H
**32b**, R¹ = p-Ts, R² = BzOC(O)-

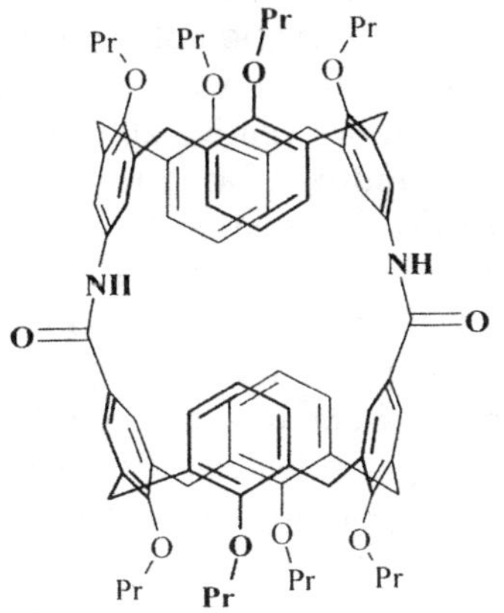

**33**

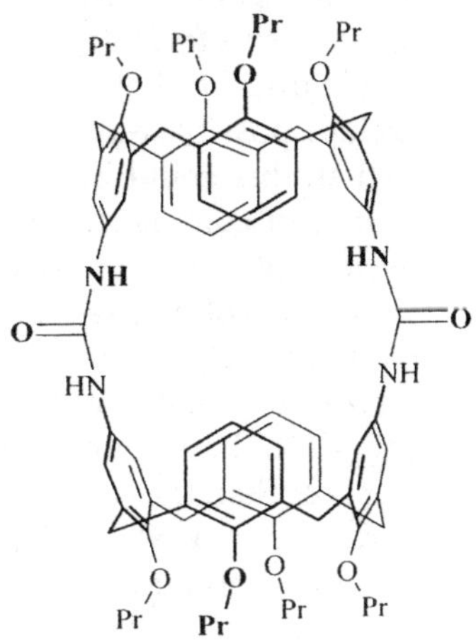

**34**

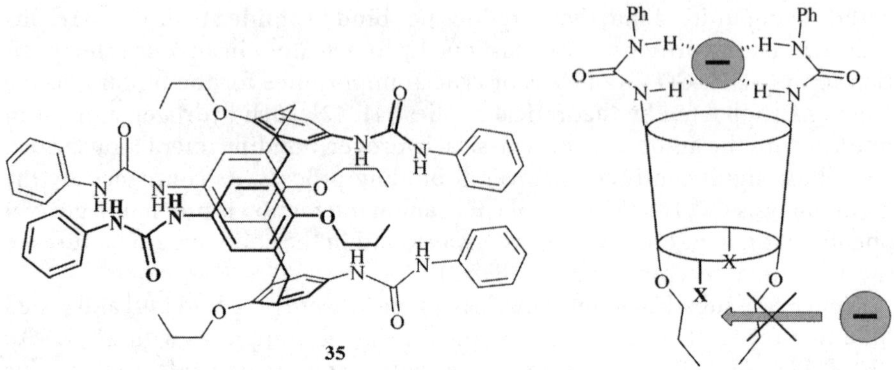

**Fig. 4** Negative allosteric effect in receptor **35**

The NMR titration experiments ($CDCl_3$:$CD_3CN$=4:1) clearly proved much stronger complexation of $H_2PO_4^-$ by ureido derivative **29d** ($K^{29d}$=2,300 $M^{-1}$) than by the amide receptor **29e** ($K^{29e}$=4 $M^{-1}$). No changes in the NMR spectrum of **29e** were observed upon addition of other anions ($Cl^-$, $Br^-$, acetate), showing dramatic reduction of complexation ability in the case of the amidic receptor. On the other hand, the ureido receptor **29d** binds strongly to many anions with a remarkable preference for benzoate ($K^{29d}$=161,000 $M^{-1}$), where additional π–π interactions with the aromatic part of the ureido functions was proposed. The Job plot analysis showed in all cases only 1:1 complexes indicating the synchronous operation of both ureido functions during the complexation.

The corresponding tetraureido derivative **35** was designed [45] as a ditopic anion receptor with two ureido functions preorganised at both sites of the *1,3-alternate* conformation. Surprisingly, the Job plot unambiguously confirmed the exclusive formation of only 1:1 complexes with halides, acetate or benzoate. As the distance between both complexation places is more than 11 Å, one can exclude possible repulsive interactions between two negatively charged anions. Rather, this could be ascribed to a negative allosteric effect induced by the change of conformation after binding of the first anion, which disables effective complexation of the second anion on the other site of the molecule (Fig. 4).

A series of neutral anion receptors based on the bisurea or bisthiourea derivatives of calix[4]arene (*cone* conformation) **36–38** was prepared. Due to the increased acidity of NH protons in thiourea ($pK_a$≈21) compared to urea ($pK_a$≈27), the thiourea derivative **36c** was expected [46] to be a better anion receptor than the corresponding ureido derivative **36a.** Surprisingly, as one can find by comparison of the corresponding complexation constants, the contrary is the case, and in all cases the ureido derivative **36a** creates stronger complexes with selected anions (halides, dihydrogen phosphate, acetate). The $^1$H NMR titration ($CDCl_3$) revealed the unexpected fact that the introduction of methyl groups into the lower rim of calixarene (compound **36b**) induces dramatic enhancement of the complexation abilities ($Cl^-$: $K^{36a}$=150 $M^{-1}$, $K^{36b}$=3,920 $M^{-1}$,

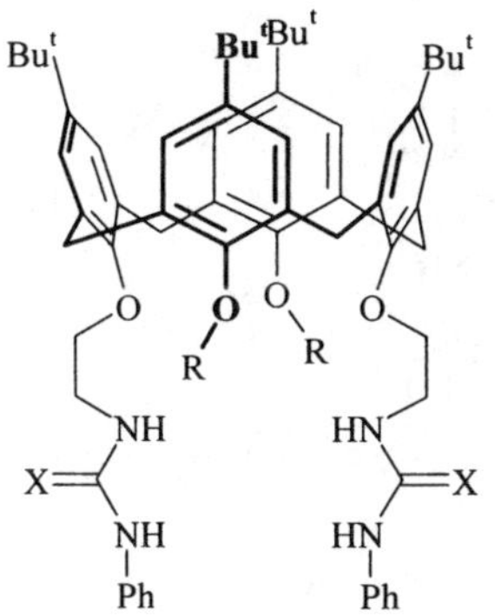
36a, X = O, R = H
36b, X = O, R = Me
36c, X = S, R = H

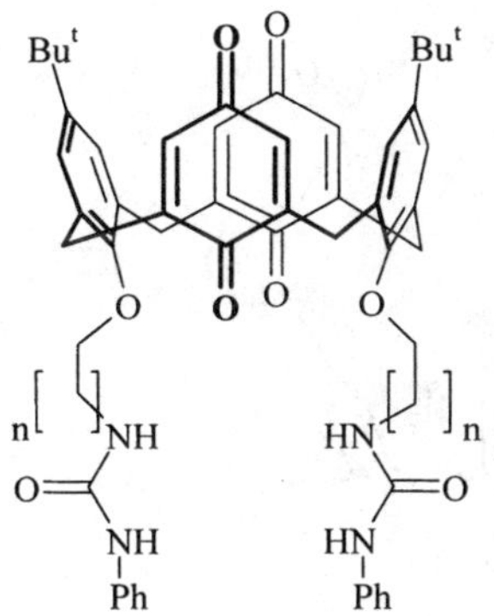
37a, n = 1
37b, n = 3

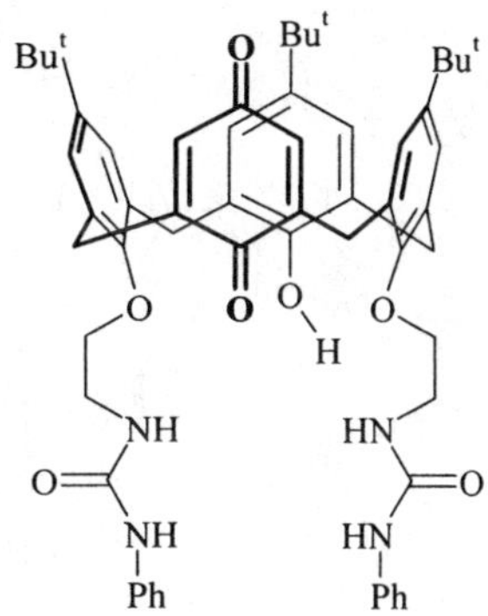
38

$K^{36c}$=60 $M^{-1}$; acetate: $K^{36a}$=500 $M^{-1}$, $K^{36b}$=1,150 $M^{-1}$, $K^{36c}$=50 $M^{-1}$). This phenomenon could be ascribed to possible hydrogen bonding of free OH groups with ureido functions inhibiting the anion binding.

Similar structures **37** and **38**, bearing ureido functions on the lower rim of calix[4]diquinone or monoquinone moieties, were used for anion recognition with the aid of electrochemistry (cyclic voltammetry) [47, 48]. Some of these compounds were found [49] to bind selectively $HSO_4^-$ or $H_2PO_4^-$ anions.

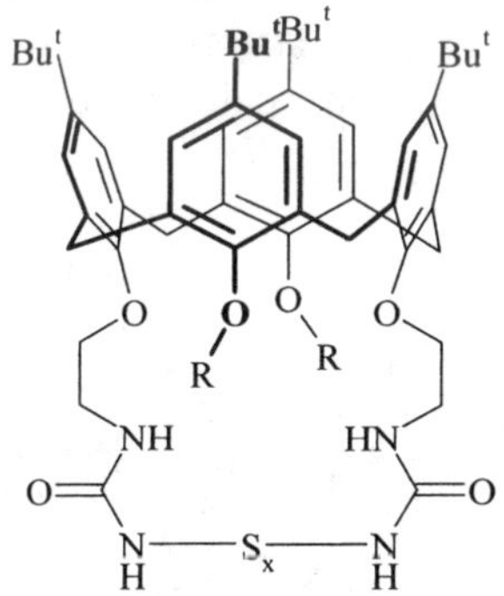
39a, $S_1$
39b, $S_2$

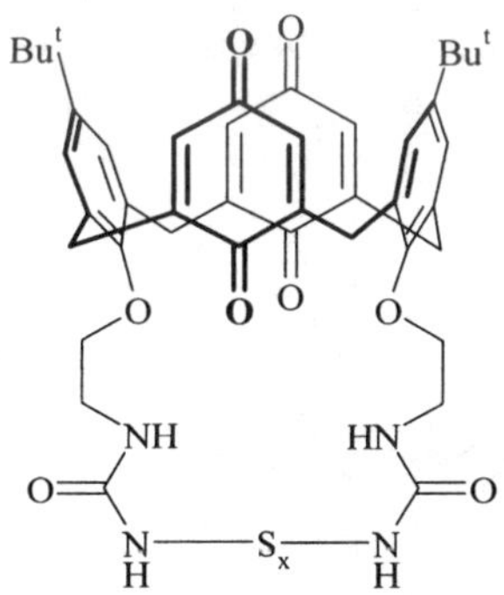
39c, $S_1$
39d, $S_2$

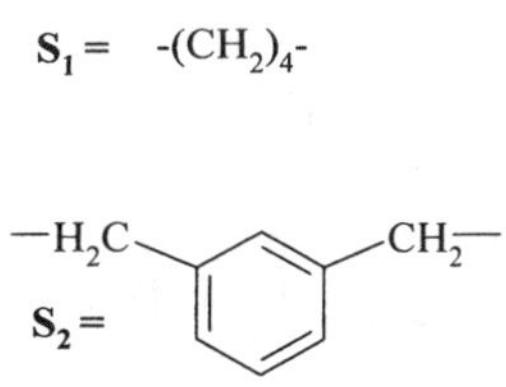

Very recently, interesting macrocyclic diureido derivatives **39a,b** lower rim bridged by a spacer unit were reported [50]. Their oxidation with $(CF_3COO)_3Tl$ led to the corresponding quinones **39c,d**, potentially useful as electrochemical anion receptors. A similar upper rim bridged system **40** was shown to exhibit good shape-selective recognition abilities towards various aromatic dicarboxylates in DMSO [51]. The molecular mechanics force field calculations indicate

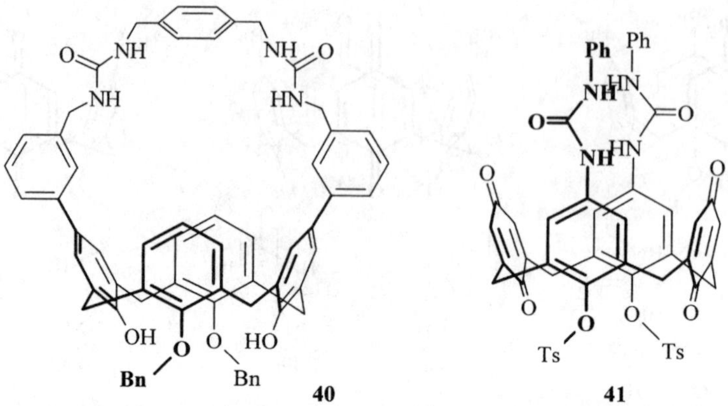

**40** **41**

that, contrary to the shape of the carboxylate, the additional non-covalent forces between guest and calixarene cavity (van der Waals, CH/π, π–π interactions) are not important in complex formation.

Two ureido functions introduced into the upper rim of calix[4]arene [52] or the corresponding calix[4]diquinone derivative **41** exhibit good complexation abilities for various anions. Even in highly competitive solvents such as DMSO-$d_6$, quinone **41** shows [53] very strong complexation for $H_2PO_4^-$ ($K^{41}$= 13,900 $M^{-1}$) or for benzoate ($K^{41}$=2,430 $M^{-1}$). The complexation process can be observed using electrochemical methods (cyclic voltammetry) where the addition of anions generates substantial cathodic shifts.

Several calix[4]arene [54, 55] or calix[6]arene [56, 57] derivatives **42–44** bearing ureido or thioureido functions on the lower rim have also been evaluated as anion receptors. As indicated by NMR spectroscopy, calixarene **42**

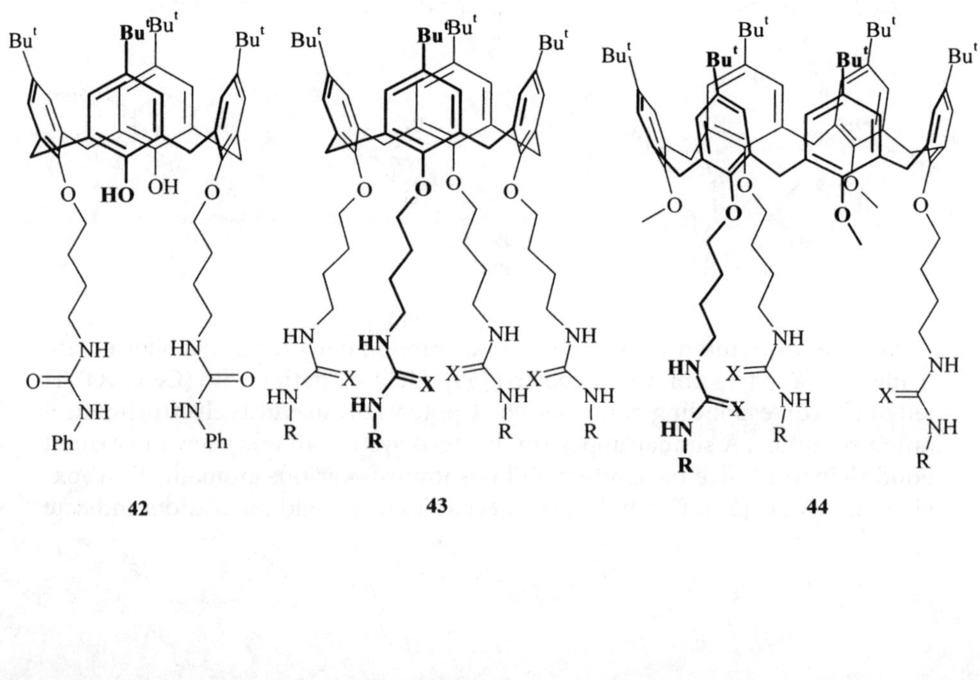

**42** **43** **44**

exhibits [58] pronounced selectivity towards $HSO_4^-$ ($K^{42}$=2,990 $M^{-1}$) while $H_2PO_4^-$ or acetate are bound much more weakly ($K^{42}$=410 $M^{-1}$). Triureido receptor **44** (X=S) [56] having $C_3$ symmetry shows very strong 1:1 binding ($CDCl_3$) with the shape-complementary 1,3,5-benzenetricarboxylate anion ($K^{44}$=290,000 $M^{-1}$), while the 1,3-benzenedicarboxylate (with lower symmetry) is complexed significantly more weakly ($K^{44}$=6,400 $M^{-1}$).

Recently we reported [59] the synthesis of novel calix[4]arene–porphyrin conjugates **45–48** connected to the upper rim via ureido functions. As the interaction with anions affects the mutual position of appended porphyrin moieties, the complexation process can be simply measured by the changes in UV-Vis spectra (Fig. 5). All conjugates create 1:1 complexes with anions and exhibit pronounced complexation ability towards halides and nitrate. The comparison of diureido derivatives **45–47** with the corresponding monourea **48** clearly shows the importance of a hydrogen bonding cooperativity for effective anion complexation. The role of solvent in the complexation process is demonstrated in Fig. 6. While in $CH_2Cl_2$ receptor **45a** forms a very strong complex with $Cl^-$, the binding in acetonitrile is much weaker, and the use of methanol leads to the disappearance of anion-binding abilities. This phenomenon clearly shows that HB-competitive solvent (methanol) disables anion complexation because of the competition with urea units of the receptor molecule. The introduction of the

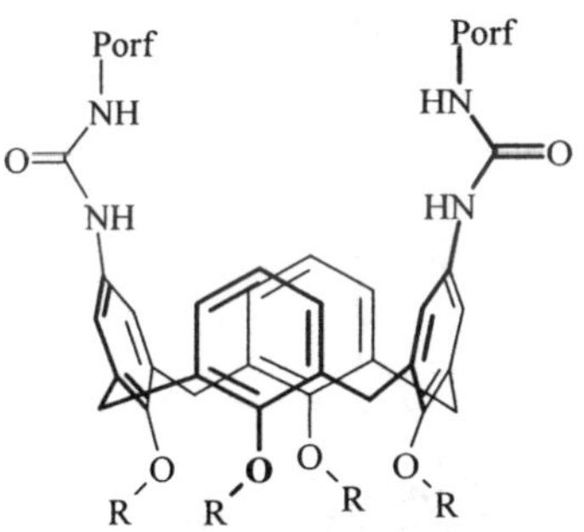


**45a**, R = *n*-Pr
**45b**, R = -$CH_2$COOEt

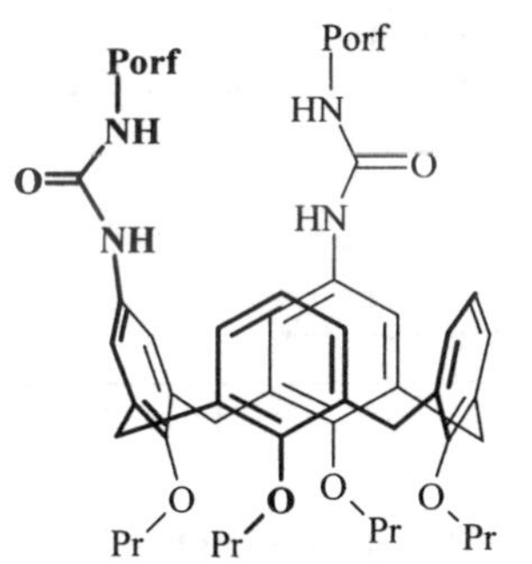


**46**

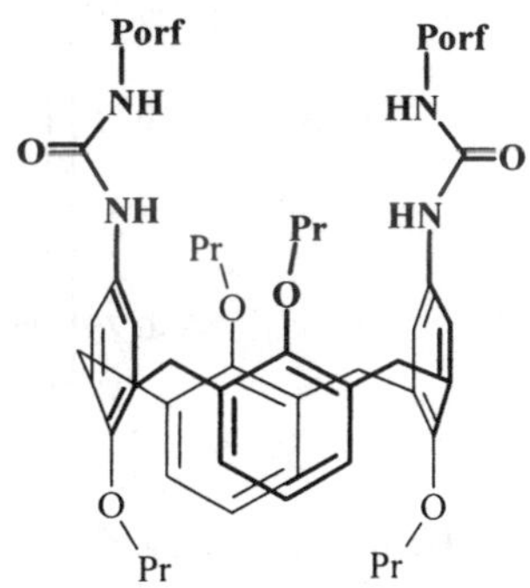


**47**

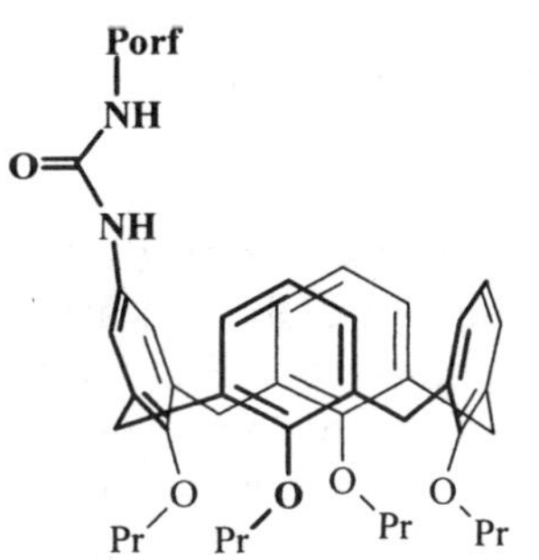


**48**

Porf =

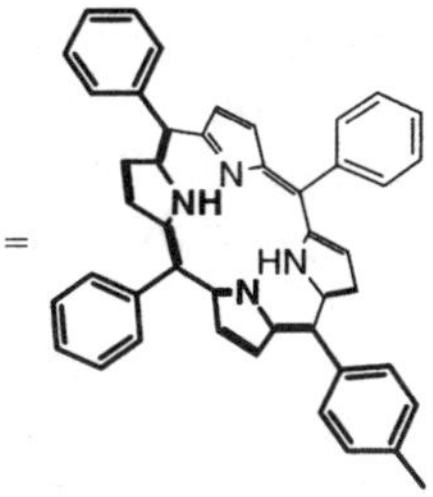

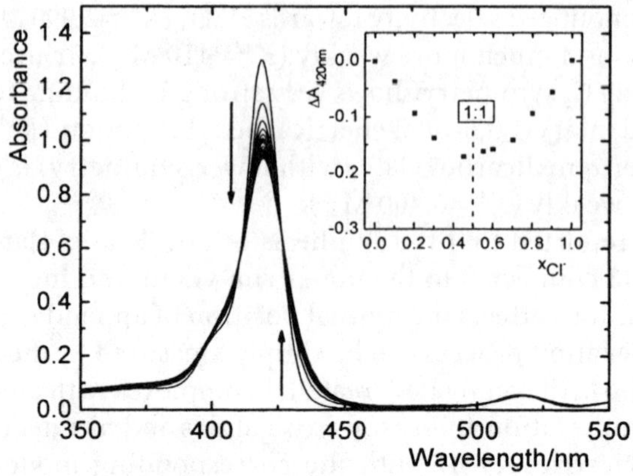


**Fig. 5** UV/Vis titration of **45a** with $Bu_4N^+Cl^-$ in $CH_2Cl_2$. The *arrows* show changes due to increasing concentration of $Cl^-$. *Inset*: Job plot

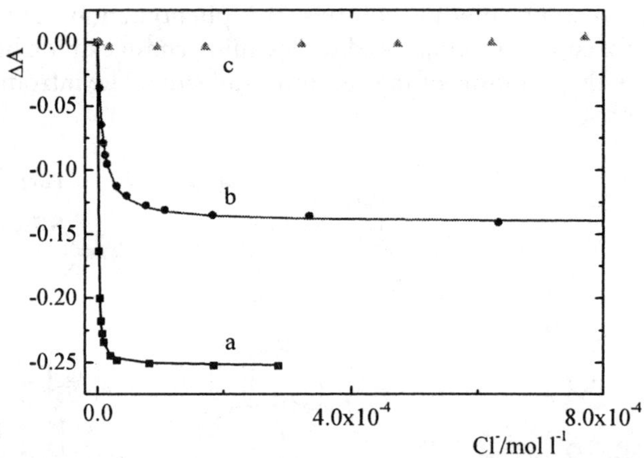


**Fig. 6** Absorbance changes of **45a** at the Soret maxima upon addition of $Cl^-$ in $CH_2Cl_2$ (**a**), MeCN (**b**) and MeOH (**c**)

acetate groups into the lower rim of calixarene **45b** forms the additional binding site for cations.

Very recently the first anion receptors **49** based on thiacalixarene units were reported [60] and their complexation abilities towards $Cl^-$ anion were studied by $^1H$ NMR titrations in $CDCl_3$. The Job plot analysis clearly confirmed the formation of the 1:1 complexes, whilst the corresponding complexation constants depend on the length of alkylene chains on the lower rim. The lower the rigidity of the system (the longer chain), the weaker is the binding (compare $K_c$ for **49a** (3,480 $M^{-1}$), **49b** (1,900 $M^{-1}$), **49c** (1,790 $M^{-1}$)). The comparison of **49a** and **49d** shows that the thioureido derivative is a slightly worse chloride receptor

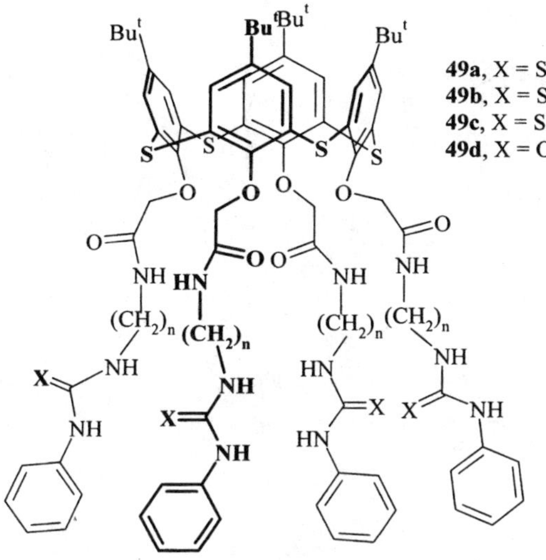


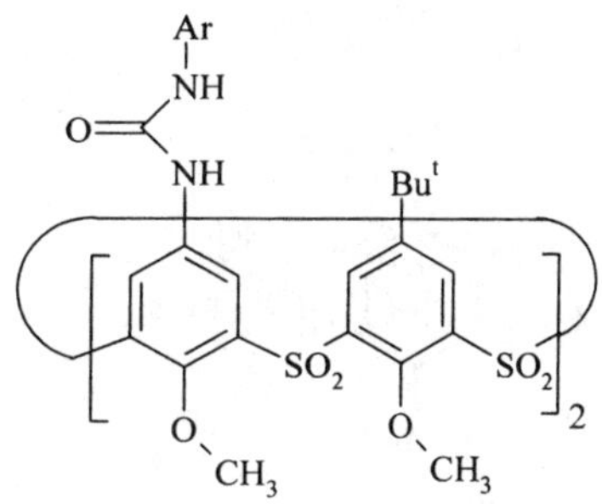

**50**

when compared with the corresponding ureido functions. On the one hand, both thiacalix[4]arene receptors **49a** and **49d** exhibit higher complexation abilities than similar classical calix[4]arene derivatives. On the other hand, classical calix[4]arene-based structures [61] bearing two thioureido functions were found to complex selectively α,ω-dicarboxylates, while other anions (halogens, acetate, dihydrogen phosphate) remain untouched. Tetrasulphone **50**, another thiacalixarene-based receptor [62], creates very strong 1:1 complexes with many anions in $CDCl_3$:$CD_3CN$ (1:1) mixture, e.g. $Cl^-$ ($K^{50}=10^4$ $M^{-1}$), benzoate ($K^{50}=5\cdot10^4$ $M^{-1}$), acetate ($K^{50}=1.2\cdot10^4$ $M^{-1}$), $HSO_4^-$ ($K^{50}=4\cdot10^4$ $M^{-1}$).

## 4 Receptors Based on Electrostatic Interactions

It is well known that anions can be bound by positively charged receptor molecules [63] using electrostatic interactions. Such receptors can be easily prepared by the protonation of suitable basic compounds, usually polyaza crowns and various ammonium compounds. Derivative **51** provides a nice example of this type of receptor [64]. The tripodal azacrown moiety appended to the lower rim of calixarene interacts with various anions in methanol ($I^-$, $Br^-$, $NO_3^-$) after the full protonation (four stepwise protonation constants were measured). As the compound also exhibits ethylene glycol moieties (cation binding site), the receptor can serve for ion-pair recognition or it can be switched between anion or cation complexation depending on the pH values [65]. Another example of an electrostatic anion receptor is represented by structure **52**. The comparison of **52a** (protonated form) with **52b** (methylated ammonium salt) [66] indicated that, in addition to the electrostatic forces, hydrogen bonding plays a key role

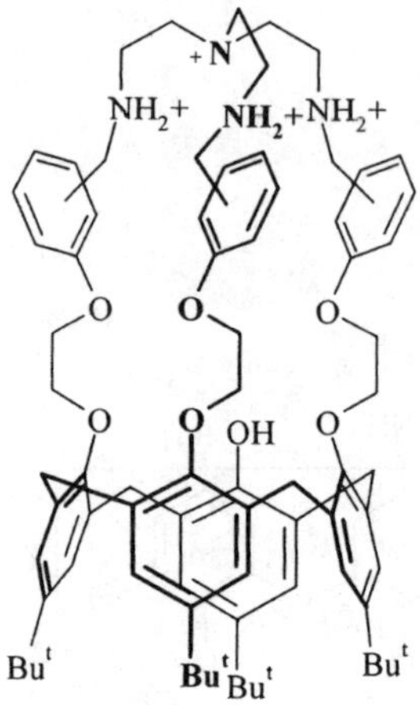

51

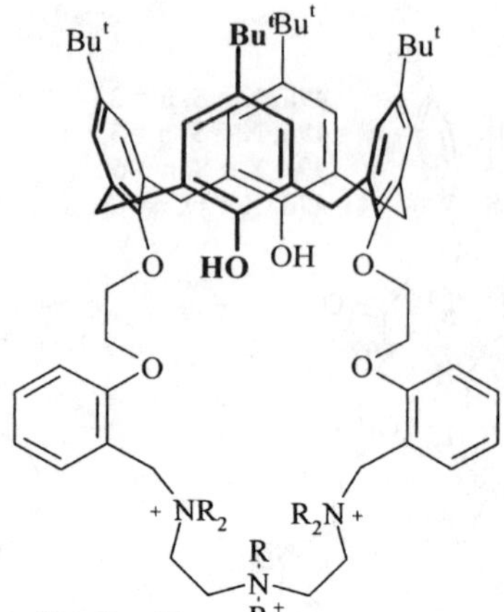

**52a**, R = H
**52b**, R = Me

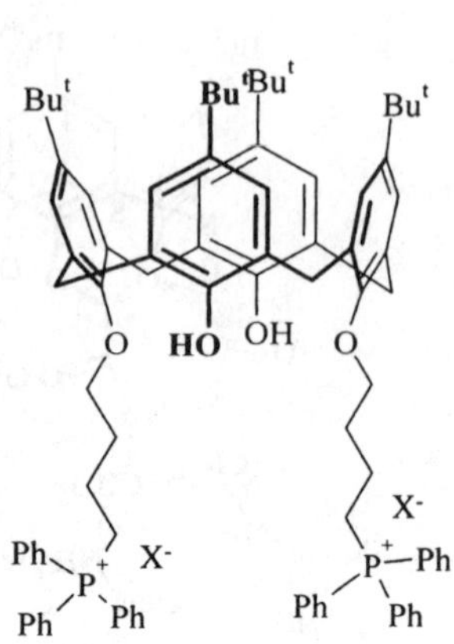

53

in the complexation, as only **52a** interacts with anions ($Cl^-$, $NO_3^-$, $AsO_2^-$, $CO_3^{2-}$). Recently, novel anionic receptor **53** consisting of calix[4]arene bearing two triphenylphosphonium moieties on the lower rim was reported [67]. As the authors proposed, the anions are bound by the synchronous action of electrostatic and π-anion forces.

A specific type of electrostatic anion receptor constitutes the compounds capable of chromate ($CrO_4^{2-}$) or dichromate ($Cr_2O_7^{2-}$) recognition. These anions are particularly important because of their toxicity and the presence of these pollutants in the environment (waste water and soils). The corresponding receptors are based on the protonated amines [68, 69] (such as **54**), protonated forms of similar compounds [70, 71] or on the quaternary ammonium salts [72] of type **55**. Thus, ammonium salt **54** (in the form of chloride) can transfer chro-

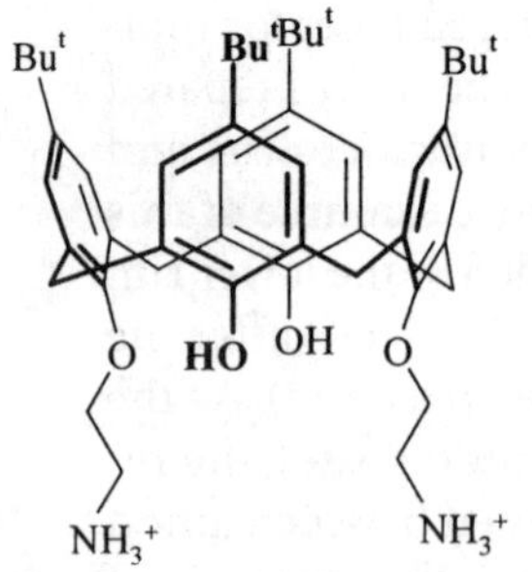

54

55

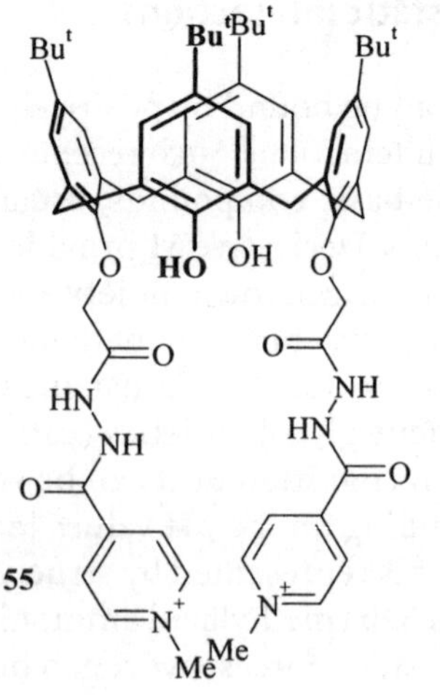

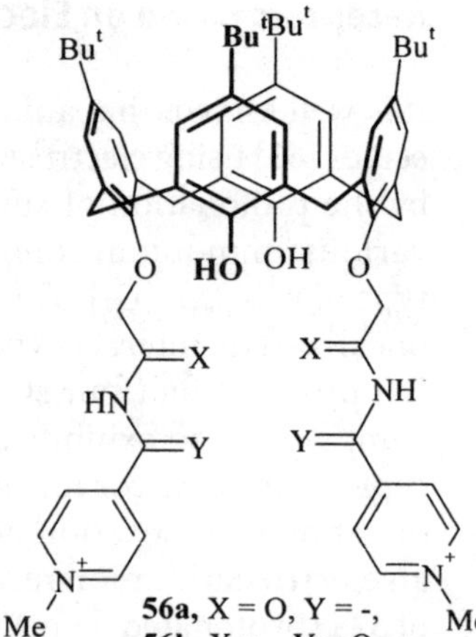

**56a**, X = O, Y = -,
**56b**, X = -, Y = O,

mate or dichromate anions from water to a chloroform layer. Similarly, preliminary studies showed that **55** can be efficiently used as an extractant for the selective extraction of $Cr_2O_7^{2-}$ from aqueous solution to the organic phase ($CH_2Cl_2$). Some of these calixarenes were anchored [73] to a polymer support using bromoethylated polystyrene and used for dichromate extraction at low pH values (pH=1.5). Similar types of compounds **56a,b** were shown [74] to bind a variety of anionic guest species (halides, $HSO_4^-$, $H_2PO_4^-$) in DMSO-$d_6$, some of them with surprisingly high complexation constants ($H_2PO_4^-$, $K^{56a}$=4.5·10$^4$ M$^{-1}$).

The synthesis of molecular capsules using non-covalent interactions is a very attractive area of supramolecular chemistry. Recently, novel systems based on the dimerisation of calixarene building blocks were described [75]. The introduction of sulphonic acid into the upper rim of calixarene gives negatively charged calixarene **58**, while tetraamidinium derivative **59** represents the charge-complementary unit (Fig 7). The formation of capsule-like dimers is a result of the electrostatic interactions between both building blocks. The capsules are stable in polar solvents like methanol and their formation was unambiguously proven by MS and NMR experiments. The combination of carboxylate **57** with

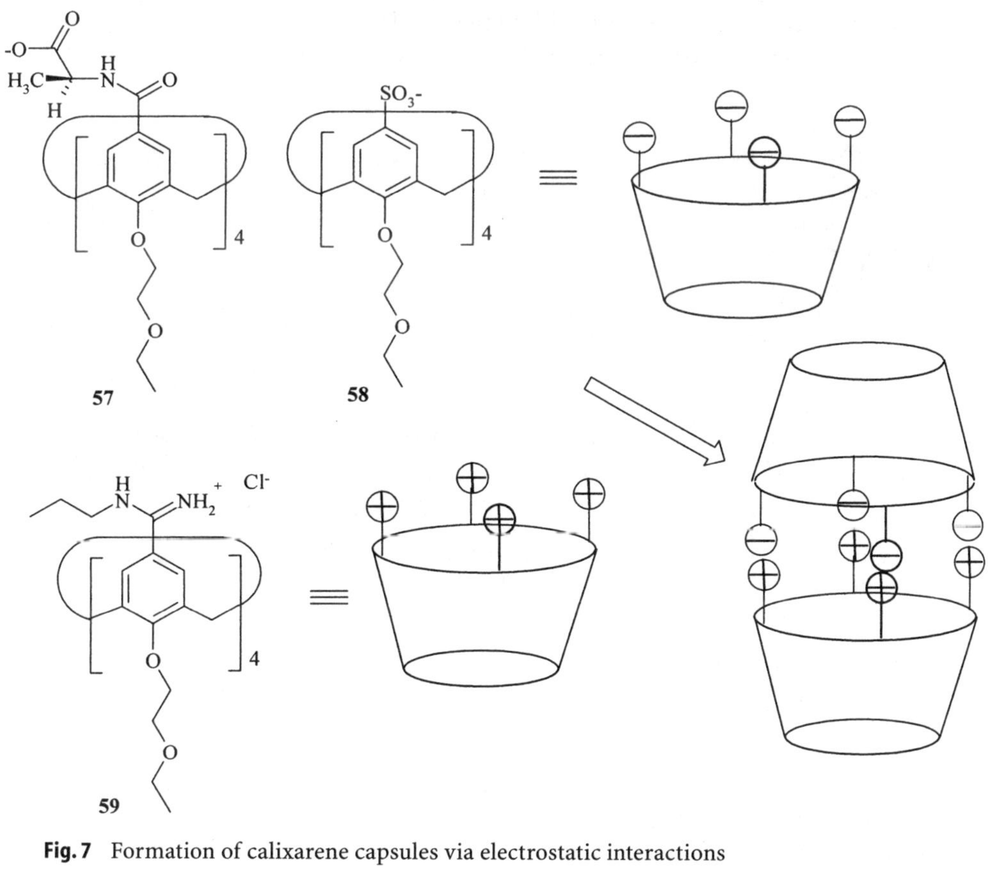

**Fig. 7** Formation of calixarene capsules via electrostatic interactions

**59** led to water-soluble dimers [76] that can encapsulate various guest molecules in aqueous solutions. Very recently, similar ionic capsules based on calix[4]arenes or calix[6]arenes [77] were prepared using the same strategy as described above.

## 5 Peptidocalixarenes and Glycocalixarenes

Another interesting family of anion receptors is represented by peptido- or glycocalixarenes. These compounds, albeit strictly speaking they usually belong to some of the above-described receptor types (amides or ureas), create a more or less independent group because of their properties. As they contain α-amino acid, peptide or carbohydrate moieties, they are chiral and can be potentially used as biomimetic receptors for binding of suitable guest molecules in polar solvents and/or aqueous solutions.

The lower rim C-linked *N*-tosyl peptidocalix[4]arenes [78] of type **60** bearing L- or D-amino acid residues were prepared and studied for their anion recognition ability. The $^{1}$H NMR titrations proved that in all cases only the 1:1 complexes are formed in $CDCl_3$ with high complexation constants (e.g. **60**, D,L-Ala, $n$=4, R=Pr; $K^{60}$ ($HSO_4^-$)=3,850 $M^{-1}$, $K^{60}$ (*N*-tosyl-L-alaninate)=6,900 $M^{-1}$).

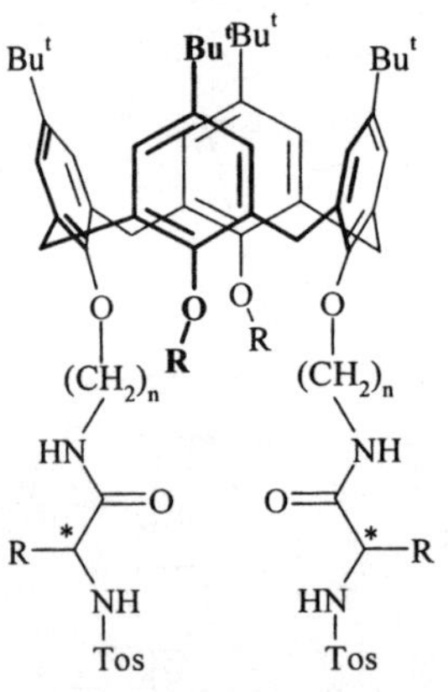

**60**, n = 2-4, R = H or Pr
D-amino acids: Ala,Val, Phe, Leu
L-amino acids: Ala,Val, Phe, Leu, Ile

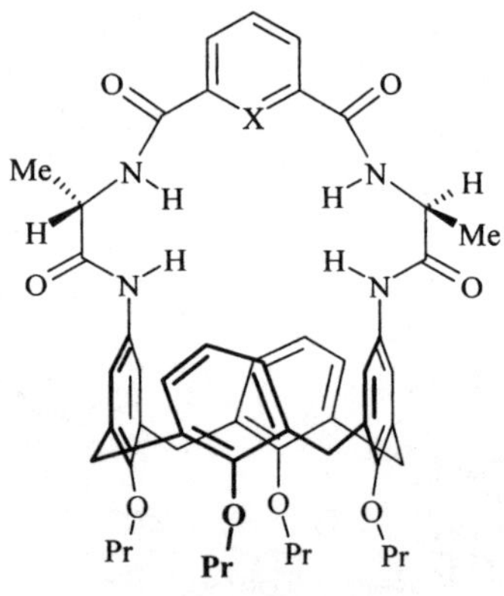

**61a**, X = CH
**61b**, X = N

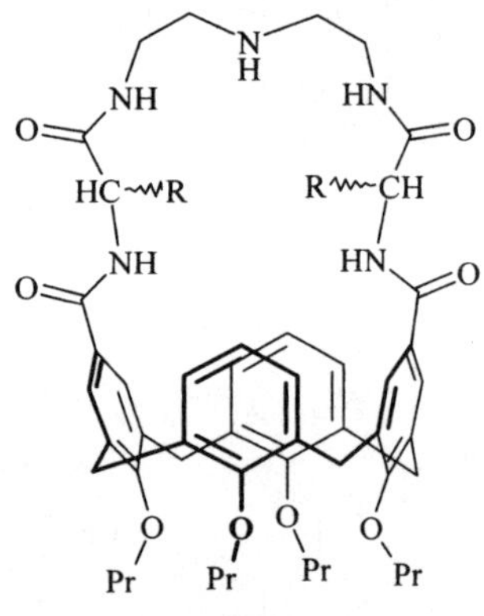

**62**

Upper rim bridged macrocyclic receptors **61a,b** possessing two L-alanine units show [79] very strong binding of various carboxylates in acetone-$d_6$ with pronounced selectivity towards benzoate ($K^{60a}$=4.4·$10^4$ $M^{-1}$, $K^{60b}$=4.0·$10^4$ $M^{-1}$) while simple inorganic anions are bound only weakly ($Cl^-$, $K^{60b}$=2.8·$10^3$ $M^{-1}$, $NO_3^-$, $K^{60b}$=200 $M^{-1}$). The enhanced binding of benzoate can be explained by the additional π–π interactions of anion with aromatic bridging unit (benzene or pyridine rings). A similar macrocyclic system **62** with L-alanine spacers binds *N*-Ac-L-Ala-L-Ala dipeptide [80]. The NMR measurements indicate that the hydrogen transfer from the guest to the host NH function results in electrostatic interactions that substantially contribute to the overall binding in $CDCl_3$. Very recently, novel C-linked peptidocalixarene **63** bearing dansyl groups on the upper rim was reported [81] as a highly selective fluorescence chemosensor for fluoride ions (no complexation of $Cl^-$, $Br^-$ or $I^-$).

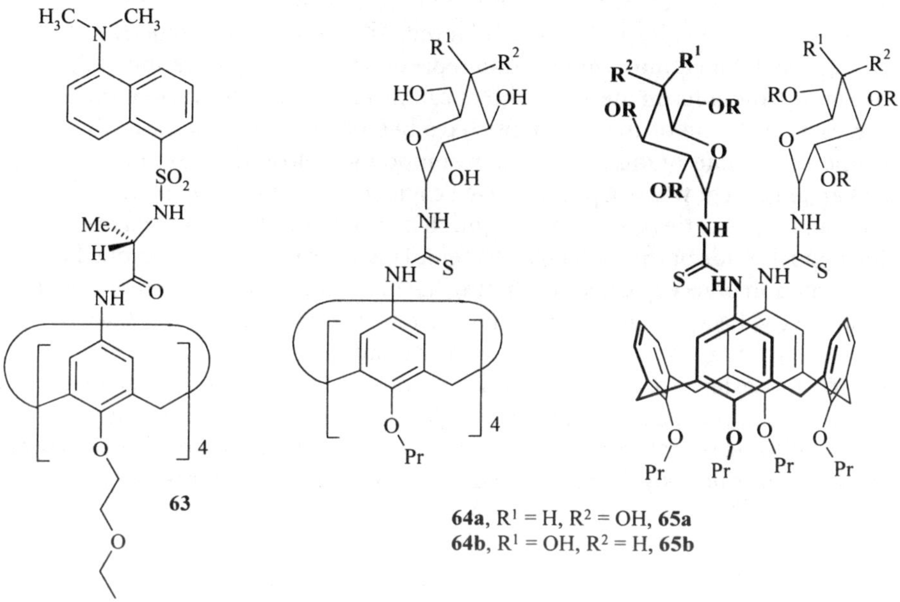

**63**

**64a**, $R^1$ = H, $R^2$ = OH, **65a**
**64b**, $R^1$ = OH, $R^2$ = H, **65b**

The introduction of sugar moieties into the calixarene skeleton results in the formation of glycocalixarenes (calixsugars) [82] which exhibit reasonable water solubility. These compounds, originally designed for the study of intercellular communication and the immune response mechanism (multivalent binding), were also reported as anion receptors [83]. Reaction of aminocalixarenes with peracetylated β-glucosyl or β-galactosyl isothiocyanates and subsequent deprotection resulted in the formation [84] of bis or tetrakis derivatives **64** and **65**. These compounds can bind various anions in DMSO-$d_6$ with selectivity towards benzylphosphonate ($K^{65a}$=170 $M^{-1}$) and benzoate ($K^{65a}$=103 $M^{-1}$).

## 6 Ion-Pair Complexation

Recently, ion-pair recognition has been attracting a lot of interest within the supramolecular community. This kind of molecular recognition is enabled by ditopic receptors [85] that simultaneously bind both cationic and anionic guest molecules. To achieve this goal the receptor must combine the corresponding binding places within its molecule. It means that the host contains any of the above-described systems designed for anion recognition together with functional groups and moieties intended for cation recognition. The chemistry of calixarenes offers a number of basic strategies for constructing the ditopic receptor.

Using calixarene as a molecular scaffold we can take advantage of many well-known principles for cation binding. Thus, one can introduce ethylene glycol units, crown ethers, cryptands or similar moieties into the calixarene. Another basic approach for cation complexation represents the preorganisation of the ester or amide functions on the lower rim of calix[4]arene *cone* conformation to form the cavities. Some of these systems have been already discussed in the corresponding sections of this review. Thus, compounds **14b** and **31** exemplify the typical arrangement where a preorganised cavity on the lower rim is supplemented with amidic moieties on the upper rim. Hence, both binding places are placed on the opposite side of the molecular system. The same strategy was applied for the construction of receptor **22b** combining the bipyridine rhenium complex with an ester cavity. Receptor **26b** and similar receptors immobilised in a *1,3-alternate* conformation are other examples of a two-sides combination of Lewis acidic centre and ester/crown cavities. Both compounds were used for the study of simultaneous ion-pair transport through a supported liquid membrane.

The ditopic receptors **66–68** containing crown ether and urea/thiourea functional groups have been synthesised. Compound **66** immobilised in the *1,3-al-*

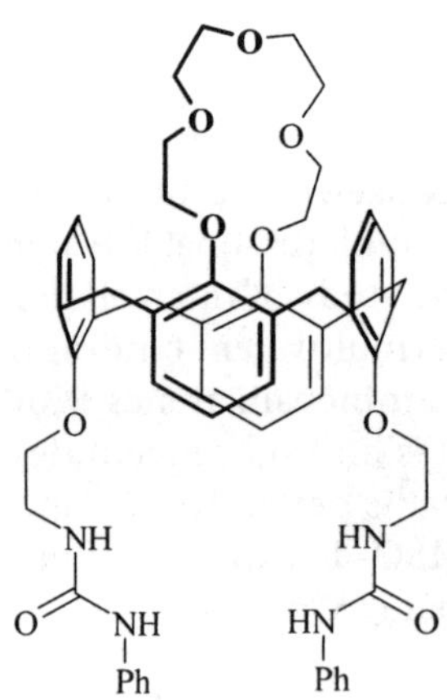

**66**

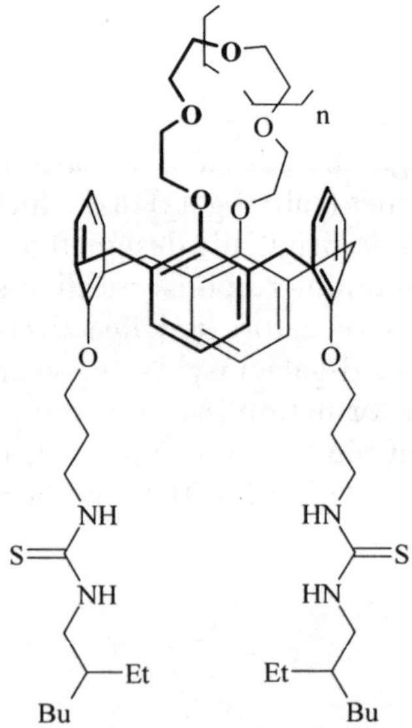

**67a**, n = 1
**67b**, n = 2

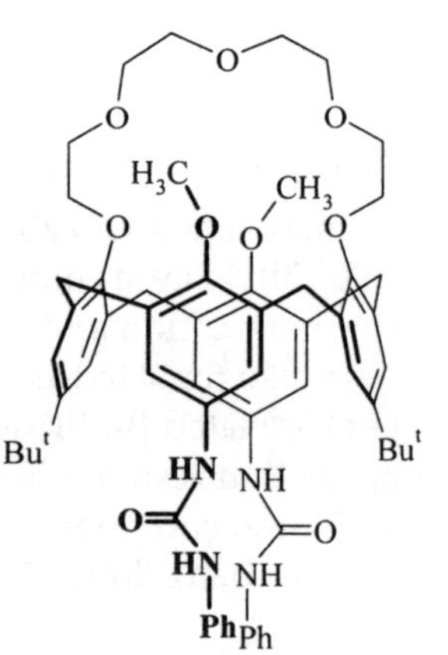

**68**

*ternate* conformation shows binding enhancement [86] towards $Cl^-$, $Br^-$ and $I^-$ in the presence of alkali metal cations. The corresponding complexation constants increase more than fivefold if potassium cation is added first; compare $K^{66}$=1,050 $M^{-1}$ for chloride ($CDCl_3$) and the same $K^{66}$=5,420 $M^{-1}$ in the presence of $K^+$. Related thiourea-based analogues **67a,b** [87] bearing lipophilic aliphatic substituents facilitate the transport of KCl and CsCl across the membrane phase where they operate as ion-pair carriers. It was shown that the transport efficiency is much higher if compared with the corresponding monotopic cation and/or anion receptors. Derivative **68** with a crown-5 unit [88] shows good binding abilities for the dihydrogen phosphate anion in DMSO-$d_6$ ($K^{68}$=200 $M^{-1}$), which was further enhanced by addition of $Na^+$ equivalent ($K^{68}$=1,030 $M^{-1}$).

A very interesting urea-based ditopic receptor, exhibiting a strong allosteric effect [89] of the cation, is represented by formula **69.** Free receptor cannot bind chloride or bromide anions as the ureido functions are connected together by strong intramolecular hydrogen bonds. The whole molecule adopts the so-called *pinched cone* conformation with approximately $C_{2v}$ symmetry. The addition of sodium cation leads to the complexation (into the lower-rim cavity of calixarene), and consequently, to the change of conformation. The complexation induces a conformational change ($C_4$ *cone* conformation) resulting in the interruption of hydrogen bonds. Under these circumstances the receptor **69** can interact with anions in $CDCl_3$: $K^{69}=1 \cdot 10^4$ $M^{-1}$ and $K^{69}=1.3 \cdot 10^3$ $M^{-1}$ for $Cl^-$ and $Br^-$, respectively.

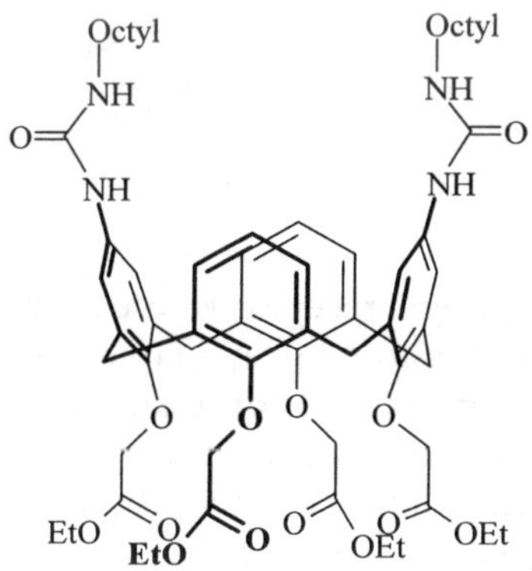

**69**

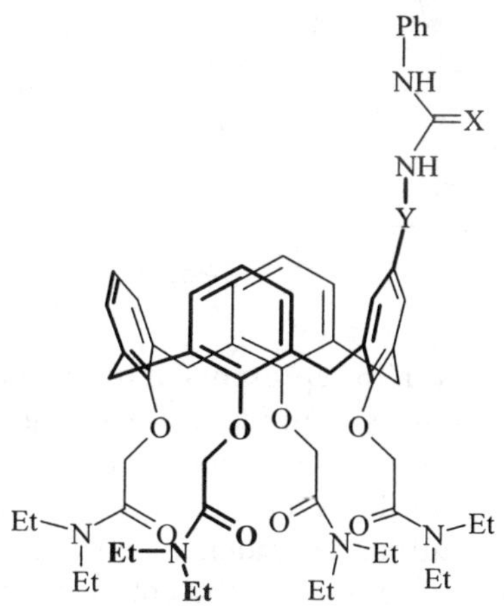

**70a**, X = S, Y = $CH_2$
**70b**, X = O, Y = $CH_2$
**70c**, X = S, Y = -

The $^1H$ NMR study of receptors **70a–c** revealed [90] that these compounds complex simultaneously sodium cation (lower rim amidic cavity) and anions (upper rim urea/thiourea), which markedly improves the solubility of the corresponding sodium salts in apolar solvents. If the urea unit is connected directly to calixarene (derivative **70c**) the addition of $Na^+$ increases the complexation constants towards anions while, at the same time, the selectivity decreases if compared with free host. The effect of cation binding on anion complexation in receptors **70a** and **70b** is not so pronounced.

A ditopic receptor **71**, comprising a semitubular calix[4]arene system [91] for cation recognition and urea moieties for anion recognition, was reported very recently. The complexation of sodium or potassium cations into the central ethylene glycol cavity invokes substantial enhancement of the binding strength (more than one order of magnitude) towards selected anions (halides, acetate).

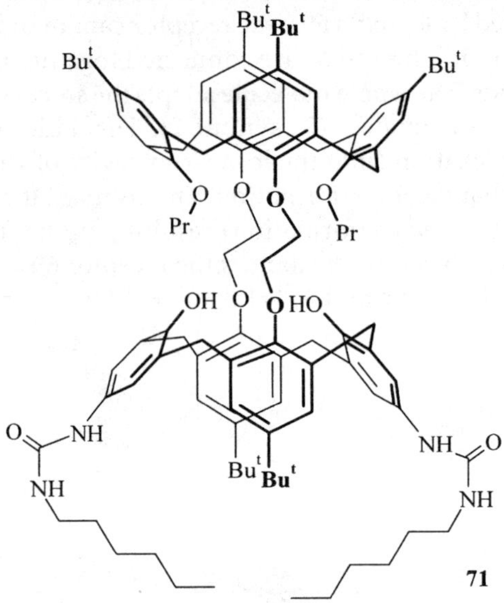

**71**

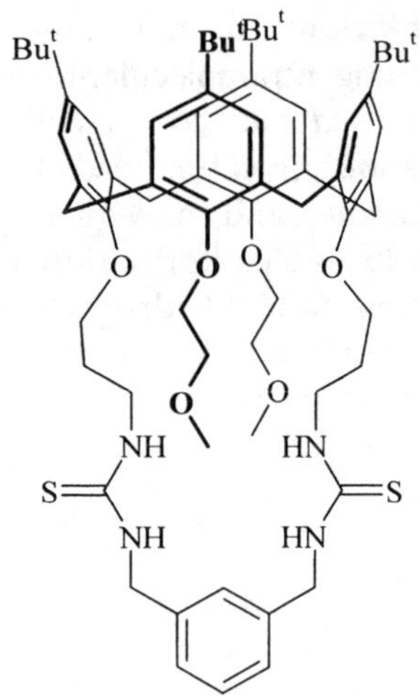

**72**

Receptor **15** represents another type of ditopic receptor where both cation (crown ethers) and anion (ferrocene amide) binding places are on the same side of the molecule. Such a configuration provides additional electrostatic interactions between oppositely charged counterions and usually exhibits apparent mutual dependence between the cation/anion complexation. The thiacalix[4]arene receptors **49** could, in principle, also bind ion pairs. Nevertheless, the cation/anion complexation processes were found to be competitive rather than synergistic. Macrocyclic ditopic receptor **72** binds selectively [92] acetate (in preference to diphenyl phosphate) in the absence of $Na^+$. The addition of sodium cation reverses the selectivity completely and the receptor prefers the $Ph_2PO_4^-$ over acetate. Very recently, the formation of water-stabilised dimer comprising calixarene–guanosine conjugate was reported [93]. The dimer serves as an ion-pair receptor and extracts alkali metal halides from water into organic solution.

## 7 Conclusions

This review has highlighted the contribution of calixarene chemistry to selective anion complexation. Using numerous recent examples we have demonstrated that calixarenes represent a very promising family of molecules with many potential applications in supramolecular chemistry, including anion recognition. Because of their unique topology and conformational behaviour, the employment of calix[4]arenes as building blocks/molecular scaffolds in the design of novel selective anion receptors seems to be very promising. The topic reviewed here indicates some novel interesting possibilities arising from the combination of the calixarene skeleton with functional groups/moieties designed for anion complexation. It is evident that the continuing progress of the chemistry of calixarenes and related molecules could bring many fascinating discoveries in the near future, with many possible applications in the field of anion recognition.

## References

1. For books on calixarenes see: (a) Asfari Z, Böhmer V, Harrowfield J, Vicens J (eds) (2001) Calixarenes. Kluwer , Dordrecht; (b) Mandolini L, Ungaro R (eds) (2000) Calixarenes in action. Imperial College Press, London; (c) Gutsche CD (1998) In: Stoddart JF (ed) Calixarenes revisited: monographs in supramolecular chemistry, vol 6. Royal Society of Chemistry, Cambridge; (d) Vicens J, Asfari Z, Harrowfield JM (eds) (1994) Calixarenes 50th anniversary: commemorative issue. Kluwer, Dordrecht; (e) Vicens J, Böhmer V (eds) (1991) Calixarenes: a versatile class of macrocyclic compounds. Kluwer, Dordrecht
2. Matthews SE, Beer PD (2001) In: Calixarenes, pp 421–439, see [1a]
3. Sessler JL, Camiolo S, Gale PA (2003) Coord Chem Rev 240:17
4. (a) Kumagai H, Hasegawa M, Miyanari S, Sugawa Y, Sato Y, Hori T, Ueda S, Kamiyama H, Miyano S (1997) Tetrahedron Lett 38:3971. For review see: (b) Iki N, Miyano S (2001) J Incl Phenom Macrocyc Chem 41:99; (c) Hosseini MW (2001) In: Calixarenes, p 110, see [1a]; (d) Lhoták P (2004) Eur J Org Chem 1675
5. Beer PD, Hayes EJ (2003) Coord Chem Rev 240:167
6. Beer PD, Keefe AD (1989) J Organomet Chem 375:C40
7. Beer PD (1998) Acc Chem Res 31:71
8. Beer PD, Cadman J (2000) Coord Chem Rev 205:131
9. Beer PD, Drew MGB, Hesek D, Nam KC (1997) Chem Commun 107
10. Beer PD, Hesek D, Nam KC (1999) Organometallics 18:3933
11. Tomapatanaget B, Tuntulani T (2001) Tetrahedron Lett 42:8105
12. Tomapatanaget B, Tuntulani T, Chailapakul O (2003) Org Lett 5:1539
13. Gale PA, Chen Z, Drew MBG, Heath JA, Beer PD (1998) Polyhedron 17:405
14. Evans AJ, Matthews SE, Cowley AR, Beer PD (2003) J Chem Soc Dalton Trans 4644
15. Beer PD, Shade M (1997) Chem Commun 2377
16. Cooper JB, Drew MGB, Beer PD (2000) J Chem Soc Dalton Trans 2721
17. Evans AJ, Beer PD (2003) J Chem Soc Dalton Trans 4451
18. Szemes F, Hesek D, Chen Z, Dent SW, Drew MGB, Goulden AJ, Graydon AR, Grieve A, Mortimer RJ, Wear T, Weightman JS, Beer PD (1996) Inorg Chem 35:5868

19. Beer PD, Timoshenko V, Maestri M, Passaniti P, Balzani V (1999) Chem Commun 1755
20. Beer PD, Drew MGB, Hesek D, Shade M, Szemes F (1996) Chem Commun 2161
21. Beer PD, Cooper JB (1998) Chem Commun 129
22. Cooper JB, Drew MGB, Beer PD (2001) J Chem Soc Dalton Trans 392
23. Atwood JL, Holman KT, Steed JW (1996) Chem Commun 1401
24. Steed JW, Juneja RK, Burkhalter RS, Atwood JL (1994) Chem Commun 2205
25. Steed JW, Juneja RK, Atwood JL (1994) Angew Chem Int Ed Engl 33:2456
26. Staffilani M, Hancock KSB, Steed JW, Holman KT, Atwood JL, Juneja RS, Burkhalter RS (1997) J Am Chem Soc 119:6324
27. Sansone F, Segura M, Ungaro R (2001) In: Calixarenes, p 496, see [1a]
28. Molenveld P, Kapsabelis S, Engbersen JFJ, Reinhoudt DN (1997) J Am Chem Soc 119:2948
29. Molenveld P, Engbersen JFJ, Kooijman H, Spek AL, Reinhoudt DN (1998) J Am Chem Soc 120:6726
30. Molenveld P, Engbersen JFJ, Reinhoudt DN (1999) Angew Chem Int Ed Engl 38:3189
31. Molenveld P, Stickvoort WMG, Kooijman H, Spek AL, Engbersen JFJ, Reinhoudt DN (1999) J Org Chem 64:3896
32. Rudkevich DM, Verboom W, Reinhoudt DN (1994) J Org Chem 59:3683
33. Rudkevich DM, Verboom W, Brzoska Z, Palys MJ, Stauthamer WPRV, VanHummel GJ, Franken SM, Harkema S, Engbersen JFJ, Reinhoudt DN (1994) J Am Chem Soc 116:4341
34. Rudkevich DM, Brzoska Z, Palys M, Visser HC, Verboom W, Reinhoudt DN (1994) Angew Chem Int Ed Engl 33:467
35. Rudkevich DM, Mercer-Chalmers JD, Verboom W, Ungaro R, DeJong F, Reinhoudt DN (1995) J Am Chem Soc 117:6124
36. Bianchi A, Bowman-James K, Garcia-Espana E (eds) (1997) Supramolecular chemistry of anions. Wiley-VCH, New York
37. Morzherin Y, Rudkevich DM, Verboom W, Reinhoudt DN (1993) J Org Chem 58:7602
38. Arduini A, Brindani E, Giorgi G, Pochini A, Secchi A (2002) J Org Chem 67:6188
39. Cameron BR, Loeb SJ (1997) Chem Commun 573
40. Stibor I, Hafeed DSM, Lhoták P, Hodačová J, Koča J, Čajan M (1997) Gazz Chim Ital 127:673
41. Čajan M, Stibor I, Koča J (1999) J Phys Chem A 103:3778
42. Čajan M, Damborský J, Stibor I, Koča J (2000) J Chem Inf Comput Sci 40:1151
43. Beer PD, Gale PA, Hesek D (1995) Tetrahedron Lett 36:767
44. Stastny V, Lhoták P, Michlová V, Stibor I, Sykora J (2002) Tetrahedron 58:7207
45. Budka J, Lhoták P, Michlová V, Stibor I (2001) Tetrahedron Lett 42:1583
46. Nam KC, Kang SO, Seung WK (1999) Bull Korean Chem Soc 20:953
47. Jeong H, Choi EM, Kang SO, Nam KC, Jeon S (2000) J Electroanal Chem 485:154
48. Nam KC, Kang SO, Jeong HS, Jeon S (1999) Tetrahedron Lett 40:7343
49. Jeong H, Choi EM, Kang SO, Nam KC, Jeon S (1999) Bull Korean Chem Soc 20:1232
50. Yang YS, Ko SW, Song IH, Ryu BJ, Nam KC (2003) Bull Korean Chem Soc 24:681
51. Haino T, Nakamura M, Kato N, Hiraoka M, Fukazawa Y (2004) Tetrahedron Lett 45:2281
52. Nam KC, Chun JC, Kang SO, Ko SW (1999) Bull Korean Chem Soc 20:1108
53. Cho EJ, Hwang SS, Oh JM, Kyoung LH, Jeon S, Nam KC (2001) Bull Korean Chem Soc 22:782
54. Scheerder J, Fochi M, Engbersen JF, Reinhoudt DN (1993) J Org Chem 59:7815
55. McDonald NA, Duffy EM, Jorgensen WL (1998) J Am Chem Soc 120:5104
56. Scheerder J, Engbersen JF, Casnati A, Ungaro R, Reinhoudt DN (1995) J Org Chem 60:6448
57. Nam KC, Kim DS, Yang YS (1998) Bull Korean Chem Soc 19:1133
58. Kang SO, Oh JM, Yang YS, Chun JC, Jeon S, Nam KC (2002) Bull Korean Chem Soc 23:145
59. Dudic M, Lhoták P, Stibor I, Lang K, Prosková P (2003) Org Lett 5:149

60. Zlatusková P, Stibor I, Tkadlecová M, Lhoták P (2004) Tetrahedron 60:11383
61. Liu SY, He YB, Wu JL, Wei LH, Qin HJ, Meng LZ, Hu L (2004) Org Biomol Chem 2:1582
62. Svoboda J, Stibor I, Lhoták P (2003) Thiacalix[4]arene-based receptors for anion recognition. 7th international conference on calixarenes, Vancouver, 13–16 August 2003
63. Schmidtchen FP, Berger M (1997) Chem Rev 97:1609
64. Tuntulani T, Poompradub S, Thavornyutikarn P, Jaiboon N, Ruangpornvisuti V, Chaichit N, Asfari Z, Vicens J (2001) Tetrahedron Lett 42:5541
65. Tuntulani T, Thavornyutikarn P, Poompradub S, Jaiboon N, Ruangpornvisuti V, Chaichit N, Asfari Z, Vicens J (2002) Tetrahedron 58:10277
66. Rojsajjakul T, Veravong S, Tumcharern G, Seangprasertkij-Magee R, Tuntulani T (1997) Tetrahedron 53:4669
67. Hamdi A, Nam KC, Ryu BJ, Kim JS, Vicens J (2004) Tetrahedron Lett 45:4689
68. Georgiev EM, Wolf N, Roundhill DM (1997) Polyhedron 16:1581
69. Wolf NJ, Georgiev EM, Yordanov AT, Whittlesey BR, Koch HF, Roundhill DM (1999) Polyhedron 18:885
70. Yilmaz A, Memon S, Yilmaz M (2002) Tetrahedron 58:7735
71. Memon S, Yilmaz M (2001) J Mol Struct 595:101
72. Tabakci M, Memon S, Yilmaz M, Roundhill DM (2003) J Incl Phenom Macrocyc Chem 45:265
73. Memon S, Akceylan E, Sap B, Tabakci M, Roundhill DM, Yilmaz M (2003) J Polym Environ 11:67
74. Beer PD, Drew MGB, Gradwell K (2000) J Chem Soc Perkin Trans II 511
75. Corbellini F, Fiammengo R, Timmerman P, Crego-Calama M, Versluis K, Heck AJR, Luyten I, Reinhoudt DN (2002) J Am Chem Soc 124:6569
76. Corbellini F, Di Costanzo L, Crego-Calama M, Geremia S, Reinhoudt DN (2003) J Am Chem Soc 125:9946
77. Zadmar R, Junkers M, Schrader T, Grawe T, Kraft A (2003) J Org Chem 68:6511
78. Sdira SB, Felix CP, Giudicelli MBA, Seigle-Ferrand PF, Perrin M, Lamartine RJ (2003) J Org Chem 68:6632
79. Sansone F, Baldini L, Casnati A, Lazzarotto M, Ugozzoli F, Ungaro R (2002) Proc Natl Acad Sci USA 99:4842
80. Casnati A, Sansone F, Ungaro R (2003) Acc Chem Res 36:246
81. Miao R, Zheng QY, Chen CF, Huang ZT (2004) Tetrahedron Lett (in press)
82. Fulton DA, Stoddart JF (2001) Bioconjug Chem 12:655
83. Dondoni A, Marra A, Scherrmann MC, Casnati A, Sansone F, Ungaro R (1997) Chem Eur J 3:1774
84. Sansone F, Chierici E, Casnati A, Ungaro R (2003) Org Biomol Chem 1:1802
85. Kirkovits GJ, Shriver JA, Gale PA, Sessler JL (2001) J Incl Phenom Macrocyc Chem 41:69
86. Kang SO, Nam KC (2002) Bull Korean Chem Soc 23:640
87. Chrisstoffels LAJ, de Jong F, Reinhoudt DN, Sivelli S, Gazzola L, Casnati A, Ungaro R (1999) J Am Chem Soc 121:10142
88. Tongraung P, Chantarasiri N, Tuntulani T (2003) Tetrahedron Lett 44:29
89. Scheerder J, van Duynhoven JPM, Engbersen JFJ, Reinhoudt DN (1996) Angew Chem Int Ed Engl 35:1090
90. Pelizzi N, Casnati A, Friggeri A, Ungaro R (1998) J Chem Soc Perkin Trans II 1307
91. Webber PRA, Beer PD (2003) J Chem Soc Dalton Trans 2249
92. Tumcharern G, Tuntulani T, Coles SJ, Hursthouse MB, Kilburn JD (2003) Org Lett 5:4971
93. Kotch FW, Sidorov V, Lam YF, Kayser KJ, Li H, Kaucher MS, Davis JT (2003) J Am Chem Soc 125:15140

Top Curr Chem (2005) 255: 97–124
DOI 10.1007/b101164


# The Construction and Operation of Anion Sensors: Current Status and Future Perspectives

Frank Davis (✉) · Stuart D. Collyer · Seamus P. J. Higson

Institute of Bioscience and Technology, Cranfield University at Silsoe, Silsoe, Beds., MK45 4DT, UK
*f.davis@cranfield.ac.uk, s.d.collyer@cranfield.ac.uk, s.p.j.higson@cranfield.ac.uk*



**Abstract** This chapter is intended to provide an insight into the construction and operation of anion sensing devices, rather than the chemistry of anion sensing per se, and therefore will concentrate more on the practical aspects of anion sensing rather than on the design of molecules or receptors that are used to specifically detect the presence of anions. Two main types of anion sensing, namely electrical and optical, will be discussed. We firstly discuss ion-selective electrodes, their construction and use, and progress through to more complex devices such as ISFETs and CHEMFETs.

This review will then consider optical sensors, together with details of their basic principles, construction and operation. Relative advantages and disadvantages compared to electronic sensors will also be discussed. Current applications of anion sensors within the field will be described together with a forward view towards some possible future technologies and applications.

**Keywords** Anion sensor · Electrode · Optrode · ISE · ISFET

**Abbreviations**

| | |
|---|---|
| ATP | Adenosine triphosphate |
| ATR | Attenuated total reflection |
| CHEMFET | Chemical field-effect transistor |
| CVD | Chemical vapour deposition |
| CWE | Coated-wire electrode |
| DMSO | Dimethyl sulphoxide |
| FET | Field-effect transistor |
| FTIR | Fourier-transform infra-red spectroscopy |
| HRP | Horseradish peroxidase |
| ISE | Ion-selective electrode |
| ISFET | Ion-selective field-effect transistor |
| MEMFET | Membrane field-effect transistor |
| PMMA | Poly(methyl methacrylate) |
| PVC | Poly(vinyl chloride) |
| REFET | Reference field-effect transistor |
| TCNQ | Tetracyanoquinodimethane |
| UV | Ultraviolet |

## 1 Introduction

The detection of anions (usually in aqueous solution) is a vital part of our maintenance and understanding of the environment. Many industrial and agricultural processes can lead to the release of anions to the environment and if unchecked these can have devastating effects. Much research is being focussed towards finding inexpensive, reliable and simple ways of detecting anions in solution, and discussion of some of these approaches form the bulk of this review.

Anion-selective receptor design is still less developed than that reported for its cation counterpart. This lack of development can be related to the various differences that exist between anions/cations and include:

- Ionic size – anions are generally greater in size and therefore require larger binding sites, e.g. $Cl^-$ has an ionic radius of 0.167 nm, larger than that of $K^+$ (0.133 nm, the same as the smallest anion $F^-$).
- There are a number of differing geometries of anion shape, e.g. chloride is spherical, cyanide linear, sulphate tetrahedral and nitrate trigonal planar.
- The pH windows of anion existence are narrower in comparison to cations, and ionisation state can be variable, e.g. carbonate vs bicarbonate.
- Anions undergo strong hydration in comparison to similar sized cations.

However, despite these obvious differences, various types of anion-selective sensor are being increasingly reported within the literature, with new aspects of development covering the transducer, receptor and membrane elements being described. Frant and Ross reported the mechanism of the first anion-

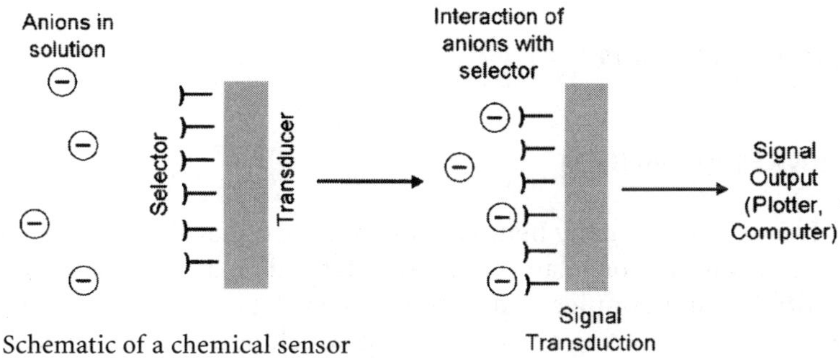


**Fig. 1** Schematic of a chemical sensor

selective sensor in 1966 [1], closely followed by Park and Simmons' work on synthetic inorganic anion receptors [2]; however, the development of anion-selective receptors and their devices has remained behind that reported for their cation counterparts. This does not mean that commercial anion-selective sensors have remained an unachievable goal, as new methods for their production are being described more and more regularly. Typical applications, for example, include measurement of nitrate in rivers or fluoride in drinking water.

There are many ways of detecting species of interest in solution, however, specific sensors often offer the least expensive and simplest approach. There are journals dedicated to reviewing the entire sensors field from a variety of publishers [3]. More specifically there is a large field of literature available concerned with sensing of species of chemical or biological interest, some of the most recent being books by Eggins [4], Diamond [5], Fraden [6] and Mulchandani and Sadik [7]. Reviews include those by Janata [8] and Bakker and co-workers [9, 10].

Most sensors consist of three elements, as shown in Fig. 1:

1. The first of these includes a chemical receptor capable of recognising the anion of interest specifically with a high degree of selectivity; this is usually concurrent with a binding event between receptor and analyte.
2. The second component that must be present is a transducer, where the binding event is converted into a measurable physical change such as a change in optical spectrum or a change in electronic state.
3. Thirdly there must be a method of measuring the change detected at the transducer and converting it into useful information.

Anion sensors come in two main types that can give rise to either (1) an electrical or (2) an optical signal. These will be considered separately.

# 2
# Electronic Anion Sensors

## 2.1
## Ion-Selective Electrodes (ISEs)

Electrical sensors offer many benefits due to their versatility and ease of use. Another important factor relates to the ease by which an electrical event may be monitored, using simpler and therefore less expensive approaches than those required for following an optical event. Much of this work on electronic sensors has been previously reviewed [11].

The earliest ion sensors were the ion-selective electrodes or ISEs. These initially were used for the measurement of protons, i.e. as a pH-sensitive electrode. Later ISEs were developed for the detection of other cations and then anions. The history of the ISE has been reviewed elsewhere [12], however, a brief introduction will be given here.

A schematic of a typical ISE is shown in Fig. 2. In this case the selector is an ion-selective membrane with the transducer being an Ag/AgCl electrode. The ion-selective membrane usually consists of the actual receptor molecule, which is a neutral or charged compound that is able to complex ions reversibly and to transfer them through the organic membrane in which they are dispersed. This compound is usually referred to as the ionophore or ion carrier. The membrane is usually polymeric in nature, with poly(vinyl chloride) (PVC) being the most common material used. Although other polymers like polystyrene, PMMA, polyamide or polyimide can be used as a membrane, PVC is the most widely used as its material properties lend themselves to simplicity of membrane preparation. An appropriate plasticiser is added to a membrane in order to obtain good chemical, physical and electrical properties. A typical polymeric membrane may consist of 65% plasticiser and 30% PVC, with the remaining 5% consisting of the ionophore. Such a membrane is quite similar to a liquid phase because diffusion coefficients for dissolved low molecular weight ionophores are of the order $10^{-7}$–$10^{-8}$ $cm^2/s$. Most commercial ISEs are in the form of a penlike probe, which can easily be immersed into the test solution.

The ISE is placed in the sample solution and the anions of interest pass through the membrane into the electrode. This results in a stable potential at the interface of the internal Ag/AgCl electrode which is different from that of the external reference Ag/AgCl electrode. An electromotive force is then measured between the ISE and the reference electrode. This can then, with suitable calibration, give the concentration of the anion of interest in the analyte.

Commercial ISEs are widely available for various ions. Usually the ion-selective layer is made from an insoluble salt of the ion in question. For example, a chloride-detecting electrode can be made where the selective layer is a pellet of AgCl. Because of the very low solubility of silver chloride, the pellet never reaches equilibrium with the solution. Instead a small amount of chloride dissolves in the sample, leaving a relative surplus of silver atoms at the pellet

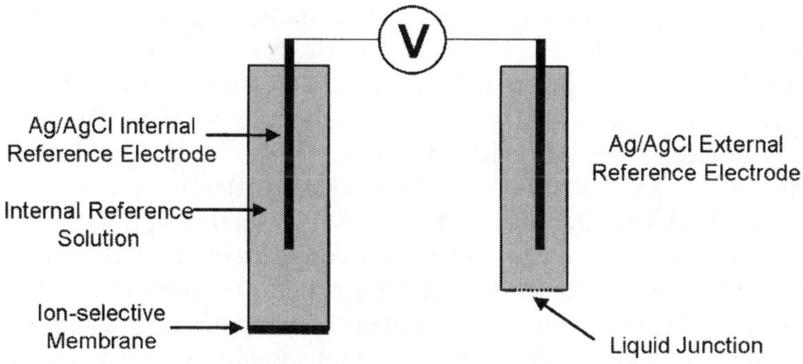


**Fig. 2** Schematic of an ion-selective electrode

surface. The resulting charge imbalance creates a measurable electrical potential and this is affected by the concentration of chloride ions, thus providing a detectable measurement of chloride concentration in the sample.

ISEs of the type shown in Fig. 2 are often expensive to fabricate as well as being relatively large in size. Conventional ISEs remain expensive to fabricate when transferred from the laboratory to mass-production scale and more work is being directed towards changing this trend. The high production cost is mainly due to the presence of an internal reference solution and therefore production normally requires large electrode dimensions. At present, most of the commercially available potentiostatic devices are based on conventional ISE technology.

Smaller, less expensive ISEs have been developed using coated-wire electrodes (CWEs) and also screen-printed ISEs. CWEs involve the direct printing of the ion-selective membrane onto a noble metal wire, such as for example, gold. However, the inherent design of CWEs can lead to erroneous results being obtained. The absence of an internal electrolyte can result in a thermodynamically ill-defined electrode-to-membrane interface. This problem has been overcome by, for example, the incorporation of a hydrogel (e.g. poly(vinyl alcohol) soaked with NaCl solution) between the electrode and membrane [13]. This method of production can also be readily applied to screen-printing techniques. Incorporation of a hydrogel has led to the production of disposable devices for selective anion measurement for environmentally important species such as nitrate [14].

Screen printing is an excellent method for fabrication of ISEs on an industrial scale. Ag/AgCl electrodes can be screen printed onto a suitable support such as Kapton polyimide. The selective membrane is then printed onto the ISE, often with an intervening layer of a hydrogel such as poly(vinyl alcohol) soaked with NaCl solution to act as the internal reference solution. The reductions of cost associated with the mass-production of these electrodes allow them to be sold as single-use disposable devices. Also their small size relative to conventional ISEs allows them to be assembled into sensor arrays.

There exist many other methods for miniaturising ISEs; however, these methods lie outside the scope of this article, but are extensively reviewed elsewhere [15]. A wide variety of ions have been studied, many of which are mentioned in reviews [11, 12, 15]; we include here only the most recent papers since a complete review is outside the scope of this article.

ISEs for all the halides have been developed and attention is concentrated on improving selectivity, stability and sensitivity. Metal porphyrins have been studied, since variation of the central atom appears to affect the selectivity of the electrode produced. Figure 3a shows the structure of some of the porphyrins [16], which are dispersed into a polyurethane matrix. Super-Nernstian responses can be obtained for electrodes containing these materials, with a central In(III) atom conferring good response for chloride, Ga(III) for fluoride and Co(III) for nitrate. An ISE for bromide was obtained by using a PVC membrane containing a xanthenium salt (Fig. 3b), and showed good ($10^{-5}$ M) sensitivity and high selectivity over other anions [17].

Iodide ISEs have been made using a carbon paste electrode containing a Schiff base complex of Fe(III) (Fig. 3c), and shows good ($10^{-6}$ M) sensitivity [18] and good selectivity over other ions. Nickel cyclams (Fig. 3d) have been incorporated into PVC membranes to give $10^{-5}$ M sensitivity to iodide with superior response times (<3 s) compared to other iodide-selective electrodes [19]. The best sensitivity is conferred by a mercurocarborand compound [20] which can detect iodide down to $2\times10^{-9}$ M, although the selectivity is not as good as that found for cation-selective electrodes.

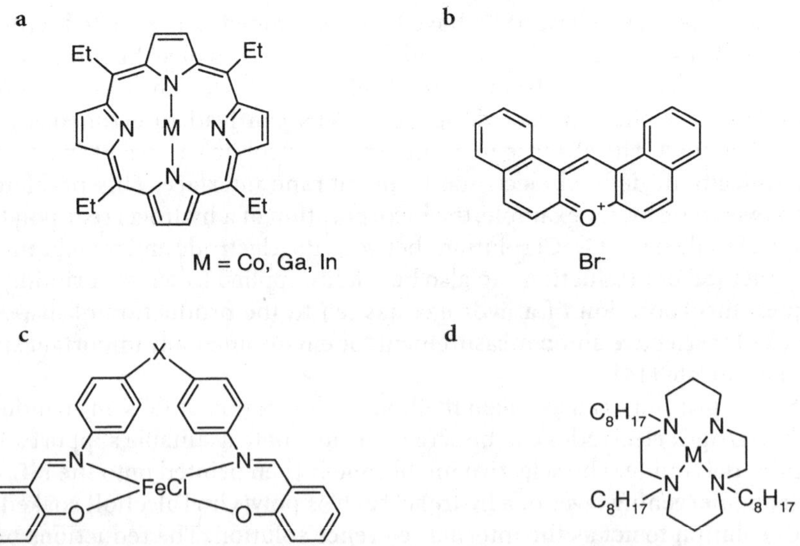


**Fig. 3** Compounds used in ISEs (see text)

Carbonate ISEs have been studied using molecular "tweezer" compounds [21] capable of being incorporated into PVC membranes to give electrodes that can measure physiological levels of carbonate in serum with reasonable selectivity over other interferent compounds likely to be present. This method has also been shown to give accurate measurements of oceanic carbon dioxide levels with real samples [22].

Nitrate ISEs have been formulated using amino acid betaines with the general formula $(CH_2{=}CHCH_2)_3N^+CHRCOOH{\cdot}Cl^-$ within styrene–butadiene membranes. These have been shown to give a Nernstian response from $10^{-1}$–$10^{-7}$ M with selectivity over chloride (selectivity coefficient 0.0034), representing a significant improvement on commercial sensors with response times <1 min [23].

At physiological pH, phosphate ions can exist in a variety of ionised states. Fibbioli et al. [24] made a zwitterionic bis-guanidinium-based sensor which, when incorporated into a PVC membrane, gives an ISE which displays noted selectivity for monophosphate ($H_2PO_4^-$) over diphosphate ($HPO_4^{2-}$) and other anions. Interestingly a close analogue of this compound displays a marked sensitivity ($10^{-6}$ M) and good selectivity for sulphate ion [24]. Other workers have demonstrated that careful choice of the internal electrolyte can greatly extend the stability and durability of the ISE [25].

ISEs for other ions have been recently developed. A manganese porphyrin, when incorporated into a PVC membrane for example, gave an ISE with good sensitivity to $BF_4^-$ ($10^{-1}$–$10^{-7}$ M), with a response time of 15 s and good selectivity [26]. Hassan et al. [27] studied a range of pyridine-based ionophores in PVC membranes and demonstrated sensitivity to thiocyanate ($10^{-2}$–$10^{-6}$ M) with good selectivity for the best materials. This electrode could also be used in conjunction with thiocyanate ions to titrate various metal ions.

A gold(I)/tri(methoxyphenyl) phosphine complex incorporated into PVC membranes led to formulation of ISEs with fast (14 s) Nernstian response ($10^{-2}$–$10^{-6}$ M) to perchlorate, with excellent selectivity even over chlorate in both water and urine [28]. Chromium porphyrins in PVC gave ISEs with near-Nernstian response to salicylate ($10^{-2}$–$10^{-6}$ M) with UV spectra confirming binding of salicylate to the chromium atom [29].

Studies have been made towards connecting low-selectivity anion-selective electrodes to chromatographic columns, using the column to separate the ions and the ISE to detect them. For instance [30], a CWE has been reported for the detection of species such as carboxylic acids, sugar phosphates and nucleotides from mixtures which had been passed through an anion-exchange column.

Direct binding of the ion-selective component to the electrode has also been studied. For example, graphite combined with an antimony compound has been screen printed and the resultant electrodes shown to give selective responses to sulphide ion in simulated wastewater samples (0.01–0.7 mM sulphide) with high stability to repeated testing and low interference from other compounds [31].

Besides the standard screen-printed or coated-wire electrodes, other architectures can be used. Examples include a rotating-disc electrode coated with a PVC or polyurethane membrane for the detection of polyions such as heparin [32].

## 2.2
## ISFETs, CHEMFETs and MEMFETs

Field-effect transistors (FETs) are often used to detect ions. The basic structure of an FET is shown in Fig. 4. The FET is made from p-type silicon, with two n-type regions, the source and the drain, being implanted within the bulk p-type silicon. This is then coated, usually by chemical vapour deposition (CVD), with a silicon oxide insulator gate layer. When a positive voltage is applied above a certain threshold voltage, the region between the source and drain becomes depleted of holes close to the surface. Electrons are attracted to this region and bridge the potential gap between the source and the drain. This leads to formation of a conducting channel between the source and the drain, near the silicon dioxide/silicon interface. The conductivity of this channel can be modulated by adjusting the strength of electrical field between the gate electrode and the silicon, perpendicular to the substrate surface. Applying a voltage between the drain and the source results in a drain current between the two n-type regions.

The electrical current flows from the source, via the channel, to the drain. However, the channel resistance depends on the electric field perpendicular to the direction of the current and the potential difference over the gate oxide. Should this surface be in contact with an aqueous solution, any interactions between the silicon oxide gate and ions in solution will affect the gate potential. Therefore, the source-drain current is influenced by the potential at the $SiO_2$/aqueous solution interface. This results in a change in electron density within the inversion layer and a measurable change in the drain current. This means we have an ion-selective FET (an ISFET), since the drain current can be related to ion concentration. Usually these are operated in feedback mode, so that the drain current is kept constant and the change of potential compared to a reference electrode is measured.

If $SiO_2$ is used as the insulator, we then have a surface which contains numerous SiOH groups. These are affected by interactions with the $H^+$ and $OH^-$ moieties present in any aqueous solution. In this way changes of pH will change the ionisation of the SiOH groups and therefore change the $SiO_2$ surface po-

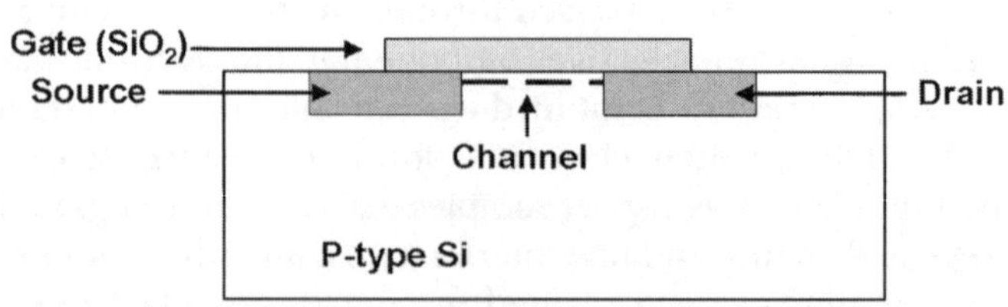


**Fig. 4** Schematic of a FET

tential. $SiO_2$ ISFETs have typical pH sensitivities of 37–40 mV/pH unit [33]. Other insulating materials can be deposited in place of $SiO_2$ such as silicon nitride and aluminium oxide. The ISFETs shown so far are inherently very unselective; however, they can be made much more so by either chemical modification of the SiOH groups or by deposition of an ion-selective membrane.

Early work on these systems studied their modification with a sensing membrane containing an ionophore which determined the response of the sensor [34]. MEMFETs (short for membrane FETs) are basically ISFETs where the gate oxide layer is covered by an ion-sensitive membrane [35]. The membrane is permeable to ions and therefore a potential is generated throughout the membrane, which is detected by the FET structure. The first ISFETs contained species such as valinomycin, as reported by Moss [36]. In this device a valinomycin/PVC solution was cast onto the gate oxide surface, leading to a $K^+$-sensitive FET. Deposition of AgBr membranes [37] led to further FETs which were sensors for $Ag^+$ or $Br^-$.

ISFETs and their derivatives (MEMFETs and CHEMFETs, chemical FETs) have been utilised in conjunction with solid-state or ion-exchange membranes to develop anion-selective microsensors, with the selectivity being established via ion-selective membranes placed on top of the transducer element. The selectivity of these membrane-based sensors has been established via the partition coefficients of lipophilic anions between solution and the membrane. This can be deduced from the Hofmeister series [38]. Positively charged membranes are most commonly used. This can, however, lead to limitations in sensor selectivity if strong interferents are present in test solutions in high quantities. This remains an inherent problem with all membrane-based selective sensor methods. Whilst these systems can be used for hydrophobic anions, hydrophilic anion-specific receptors (in the form of molecules, for example) need to be added to the membrane to specifically bind the target anions. The mechanisms of anion binding to receptors include Coulomb interactions with charged receptors (such as quaternary ammonium species), reversible covalent bond formations (for species such as $CO_3^{2-}$) and hydrogen-bonding/Lewis acid coordination of metal centres that are capable of binding anions with neutral receptors.

The solvent casting of PVC membranes containing ionophores and plasticisers has been widely used for the construction of MEMFETs [39, 40]. These are readily fabricated by casting PVC membranes with incorporated plasticiser and ionophore on the top of the ISFET gate oxide. Problems with this technique include poor adhesion of the membrane to the gate oxide, frequently leading to peeling. Also if exposed to solutions for prolonged periods, the electroactive components may leach out. Leaching can be reduced by using extremely hydrophobic receptors or by covalently linking the ionophore to the organic matrix [41].

Membrane attachment can be improved by chemically modifying the gate oxide to allow covalent anchoring of the membrane [42, 43]. The usual method is to react the SiOH groups at the surface with a silane, such as 3-(trimethoxysilyl)propyl methacrylate. This gives us a surface coated with active methacrylate groups which will readily react with other monomers. Often these reactions are initiated photochemically, which means that techniques such as photolithography can be used to mass-produce CHEMFETs. PVC is the commonest material used for the sensing membrane, however, the leaching and peeling problems have led to attempts to use alternative polymers such as polyurethane [44] and silicone rubber [42], as well as a number of other polymers.

ISFETs modified with plasticised PVC membranes lack a thermodynamically well-defined interface between the sensing membrane and the solid contact. This can cause a number of problems, since variables such as the concentration of protons and the gate oxide-membrane interfaces have large effects on the measurements attained. Therefore impurities capable of dissolving readily in the plasticised membrane, such as carbon dioxide [45], can lead to unreliable and erroneous readings. This can be a problem for some sensors, although often ISFETs modified with PVC membranes will perform acceptably due to reasonable amounts of water being present in the membrane, thus preventing large fluctuations in proton concentration. However, the development of a thermodynamically well-defined interface was widely studied.

A simple method of eliminating the $CO_2$ interference was to apply an intermediate Ag/AgCl layer on the gate-insulator surface [46]. One effect of this, however, is that the Ag/AgCl-membrane interface becomes critical, with its equilibrium state relying on the exchange of scarcely present $Cl^-$ ions in the membrane. Other approaches utilise the deposition of an additional layer between the polymer or alternatively depositing an Ag/AgCl layer and then coating it with a hydrogel such as poly(vinyl alcohol) – before applying the ionophoric membrane [47]. The hydrogel is often used as a host for a buffered salt solution, which tends to stabilise the equilibrium at the gate oxide surface and also eliminate the $CO_2$ interference [44, 47, 48].

Development of sensors for environmentally important anionic species is becoming increasingly reported. For example, nitrate-selective sensors have been described for both CWE- [49] and ISFET-based sensors [50]. An ISE has also been reported for nitrate detection based upon crystalline membranes of silver substituents, although this too suffers from interference from other anions in solution, such as $F^-$ and $SO_4^{2-}$. Another important anion of environmental importance that is being actively investigated, with a view to designing a selective disposable sensor system, is chloride ($Cl^-$). Methods for its determination recently reported include simple deposition of AgCl on a suitable surface, leading to the development of a $Cl^-$-selective ISFET [51]. This silver salt has also been doped into a polyfluorinated polymeric matrix to yield a $Cl^-$-selective ISFET [52]. Indeed, an AgCl precipitate has also found use within a highly sensitive $Cl^-$-sensing electrode [53]. $Cl^-$-selective sensors have also been produced using polypyrrole membranes [54]. More recently a screen-printed electrode, based upon a doped polypyrrole film, has been reported whose properties were comparable to those observed at a standard ISE, when used for chloride detection [55]. Indeed, the use of carbon as a screen-printed electrode surface for use within anion-selective sensor systems is becoming a highly investigated research area.

Lawrence et al. have recently reported the use of an *N,N′*-diphenyl-*p*-phenylenediamine mixture with carbon to produce a sensor selective for hydrosulphide [56]. An all-solid-state sensor for use within the beverage industry has been developed using various doped polymeric membranes deposited upon a carbon surface [57]. This new method of production of anion-selective electrode systems has also led to recent interest in the miniaturisation of selective electrode sensors. This is being driven by new trends in, for example, water quality analysis, or in disease monitoring. Here, there is a major interest in moving away from the use of centralised laboratory facilities and towards testing that can be undertaken, for example, in the doctor's surgery or even as continuous in vivo monitoring of a disease. Selective electrochemical sensors are of great interest because they can undergo miniaturisation and therefore be mass-produced to allow for the availability of low-cost, portable and direct-reading devices. Recent reviews covering modern developments in microsensor fabrication and production have also been published [58, 59], with a review linking ISFET technologies to microfabrication methods published by Cané et al. [60].

The majority of the devices mentioned thus far rely on the Hofmeister series for anion selectivity. However, for anions that deviate from this series, organometallic receptors can be utilised. The type of ligand or metal centre will influence the sensor selectivity due to the characteristics of the electron acceptance of the complex. An interesting development that is being explored here is the use of *calixarenes*. These have previously found use as cation-selective species, but with suitable substitution are now being incorporated within anion-selective devices. Compounds suitable as receptors for halides [61], benzoate [61] and acetate [62] have been developed. Reinhoudt and his co-workers have reported the production of a $H_2PO_4^-$-selective CHEMFET based on a uranyl cation immobilised within a salophene ligand (Fig. 5), which shows selectivity over more lipophilic anions such as $Br^-$ and $NO_3^-$ [63].

Metalloporphyrins, similar to those shown in Fig. 3a, offer an excellent method for exploring anion selectivity due to the various types of metal ion that can be incorporated into the complex. A metal centre of $Fe^{3+}$ has been shown to exhibit selectivity for $SCN^-$ [64], whereas $Sn^{4+}$ shows good selectivity for salicylate [65].

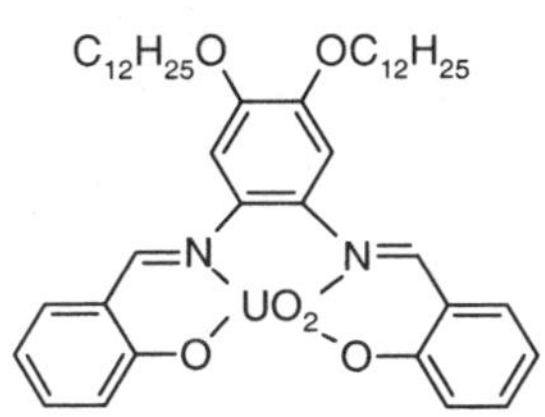


**Fig. 5** A salophene reagent which is phosphate sensitive

Although less developed than anion binding via Lewis acid centres, hydrogen-bonding interactions have shown some recent success with anionic moieties [66]. Covalent bonds have also found limited success for use as a method for inducing anion selectivity, with sensors for $CO_3^{2-}$ [67] and $SO_4^{2-}$ [68] being reported. A recent paper [69] varied the formulation of PVC membranes with various salts to give membranes with improved selectivity to nitrate; use of a Siloprene membrane led to slightly lower performance but much better membrane adhesion and stability.

As with the majority of ISEs, all of the aforementioned receptors are immobilised within close proximity to the transducer element. However, conducting polymers (electroactive conjugated polymers) are now emerging rapidly as one of the most promising classes of transducer for use within chemical sensors. Here, the receptor can be doped within the polymer matrix, i.e. within the transducer element itself. This will facilitate the production of reliable, cost-effective, miniaturised anion-selective sensors, as it will be possible to move away from plasticiser-based membranes, but allow for ion recognition sites in conjunction with all-solid-state ion-to-electron transducers.

Conducting polymers have already been well documented in conjunction with the classical ionophore-based solvent polymeric ion-selective membrane as an ion-to-electron transducer. This approach has been applied to both macro- and microelectrodes. However, with careful control of the optimisation process (i.e. ionic/electronic transport properties of the polymer), the doping of the polymer matrix with anion-recognition sites will ultimately allow selective anion recognition and ion-to-electron transduction to occur within the same molecule. This is obviously ideal and would allow for the production of durable microsensors, as conducting polymer-based electrodes, and due to the nature of their manufacture these are suited to miniaturisation. There are various examples of anion-selective sensors formed using this technique reported in the literature, some of which are listed below.

A wide range of materials can be electropolymerised at electrode surfaces to form conducting polymers such as polypyrrole, poly(3-methylthiophene) and polyaniline (Fig. 6). Polypyrrole has proved to be a popular choice as the conducting polymer. Initial work showed that when a polypyrrole-coated Pt electrode is placed in a flow-injection apparatus, concentrations of anions down to 1 μM can be detected due to their effect on the doping and conductivity of the polymer [70]. When pyrrole was electropolymerised from a solution containing 0.1 M LiCl, the resultant polymer electrode was found to give a Nernstian response to chloride ion, with a limit of detection of $3.5\times10^{-5}$ M of $Cl^-$ [71], although with some interference from other ions [72].

Cadogan et al. [73] deposited polypyrrole onto platinum from a variety of solutions and found the resultant films to be anion sensitive but very unselective. When polypyrrole was deposited from $NaNO_3$ solution, however, the resultant electrode gave near-Nernstian responses to nitrate ion in the range 0.5–0.0005 M with selectivities of about 20 over other lipophilic ions such as iodide and perchlorate [74]. Similar electrodes could be fabricated via a screen-

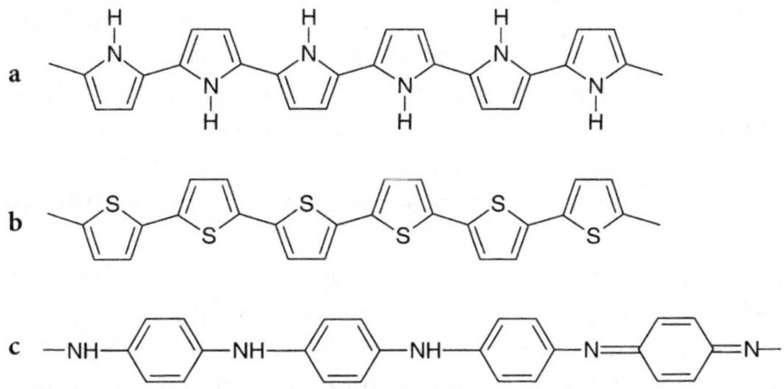


**Fig. 6** Structure of some conducting polymers: **a** polypyrrole, **b** poly(3-methylthiophene) and **c** polyaniline

printing process and covered with a PVC-based membrane, which improved the lifetime of the sensor [55]. The sensitivity of these types of electrodes has been greatly improved by coating the polypyrrole with a PVC-based ion-selective membrane [75].

Other early work in this field included the use of tetrakis(*p*-aminophenyl)-porphyrin (Fig. 7a), which was electrodeposited onto glassy carbon and showed a near-Nernstian response to iodide [76]. Electrodeposited methylthiophene–methylpyrrole copolymer was deposited and shown to give a near-Nernstian response to bromide [77]. Pyrrole-3-boronate (Fig. 7b) could be deposited to give films with a good response and marked selectivity to fluoride [78]. A cobalt aminophthalocyanine could also be electropolymerised to give a good sensor for nitrite [79] or sulphide ion [80].

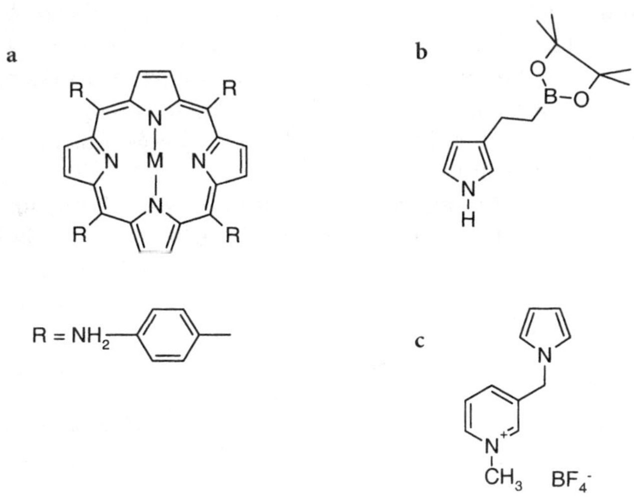


**Fig. 7** Structures of **a** tetrakis(*p*-aminophenyl)porphyrin, **b** pyrrole-3-boronic acid and **c** 1-methyl-3(1-pyrrole-1-ylmethyl)pyridinium tetrafluoroborate

A tetraruthenated porphyrin was electropolymerised onto glassy carbon and used to catalyse the oxidation of nitrite to nitrate, with the resultant current giving a selective measure of the concentration of nitrite ion [81]. As an alternative method, soluble poly(3-octyl thiophene) [82] was cast along with tridodecylmethylammonium chloride onto glassy carbon, to give electrodes with superior selectivity over PVC-based membranes to lipophilic ions such as bromide or nitrate.

A substituted pyrrole (Fig. 7c) was electropolymerised across the pores of an alumina membrane [83] and the resultant polymer film then acted as a permselective membrane across which anions could cross. This was then used to separate an analyte solution from an internal sensing solution containing nitrate reductase enzyme and methyl viologen.

Use of conventional reference electrodes is a limiting factor in reducing the size of the various CHEMFETs. This could be solved by incorporating the reference electrode into the CHEMFET chip. An example of this is the on-chip fabrication of an Ag/AgCl electrode containing a gel-filled cavity sealed with a porous silicon plug [84]. Unfortunately, sensor lifetime can be limited by leakage of the reference solution.

Another approach to the solution of this problem is the development of the REFET or reference FET (Fig. 8). Two ISFETs are manufactured on the same sensor chip and the response of both measured and compared. One of the ISFETs is chemically modified to render it less sensitive (or preferably totally insensitive) to ions present in the sample solution. The signal here is actually the difference between the two ISFETs, thereby meaning that effects due to temperature changes and other external effects tend to cancel out and be minimised.

The reference ISFET can be modified by chemically reacting the SiOH groups with trimethoxysilanes [7]. An alternative method involves deposition of a hydrophobic polymer on the surface, for example by thermal deposition of parylene [85]. These layers tend to be chemically bound to the SiOH surface, leading to enhanced stability and long lifetimes. Also the layers are very thin, which is essential because of decreased electrical sensitivity as the insulator thickness increases [9]. For ion-blocking layers, a stable attachment has been realized by plasma deposition [86, 87].

An alternative to ion-blocking layers is to use hydrophobic ionically conductive polymers. A wide variety of polymers have been studied, however, they

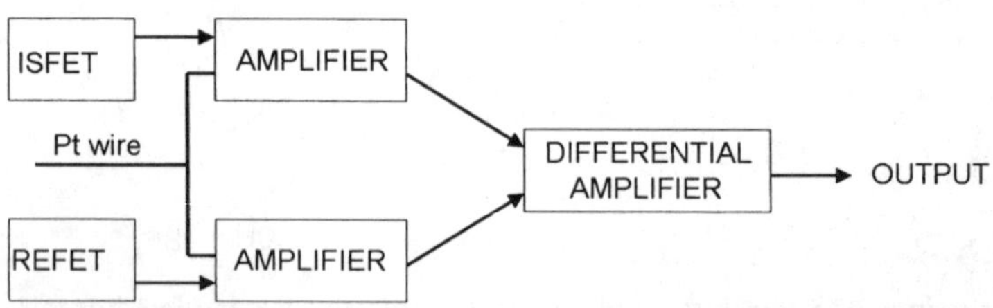


**Fig. 8** Schematic of a REFET

must be covalently linked to the gate oxide surface or else very poor stability and short lifetimes become problematic [88]. In this context, polyacrylate-based polymers tended to give the best results [88].

Urushi, a natural lacquer, has been used on the ISFET surface [89] to make, amongst other devices, a chloride sensor. The natural lacquer has a long curing time (10 days) but this can be shortened to 2–3 days with the use of formaldehyde as a crosslinking agent [90] and has been successfully used in the production of a nitrate sensor.

Sol-gel methods have been used to form membranes on sensor surfaces by the mixing of the ion-sensitive molecule with alkoxysilanes which then crosslink to form a hard, transparent coating on the sensor. For example, tridodecylmethylammonium chloride-doped films were used as a chloride sensor and showed Hofmeister-type selectivity [91]. Long cure times (several days), however, represent a disadvantage in this process [91].

Other siloxane polymers are of interest because their very low glass transition temperatures render plasticisers superfluous. However their low polarity and high hydrophobicity means that the simple commercial polydimethylsiloxanes have high membrane resistances [92], probably due to the low solubility of ions in such media. Plasticisers could aid this but would defeat the object of using polysiloxanes in the first place. Incorporation of more polar moieties as co-monomers in the polymer, such as cyanopropyl [93], lowered membrane resistance by a factor of up to 20, with the incorporation of trifluoropropyl groups leading to sensors for nitrate and cations [94]. Studies have also been made on using photopolymerisable crosslinkers, for example in the construction of a nitrate sensor [95].

Two types of REFET structures can be distinguished with respect to the penetration of ions into the polymeric layer, resulting in two different mechanisms of the REFET operation. In a non-ion-blocking REFET structure there is ion exchange between the solution and the polymer; consequently a thermodynamic equilibrium between ions in the solution and in the polymer is achieved and the electrical potential is a membrane potential. In an ion-blocking REFET structure, this ion exchange is negligible and in this case the electrical potential measured is a surface potential resulting from reversible ion-complexation reactions at the surface of the polymer.

This modern technology provides a possibility to design multi-ion sensors integrated with the reference cell (REFET). These sensors are very small, enjoy a high longevity and use only very small amounts of the necessary selective (and often relatively expensive) compounds for ion sensing, such as valinomycin.

## 2.3 Electroactive Sensors

Besides the sensors mentioned earlier, a series of sensors using redox-active compounds have been developed. Some anions such as ferricyanide, are redox active in themselves and can easily be detected by a variety of techniques such as cyclic voltammetry. However, most anions of interest lack this property and cannot be detected directly. Therefore various redox-active receptors have been synthesised that can selectively bind to anions, with the binding process modifying the redox behaviour of the receptor. This can be thought of as a soluble sensor and if a solution of this material is combined with the solution to be tested, electrochemical techniques such as cyclic voltammetry can be used to detect the anions present.

The first of this class of receptor reported was a cobalticenium-based macrocycle that complexed bromide ions, with shifts of up to 45 mV in the position of the redox peaks of the cobalt moieties [96]. Other species have been used as active receptors including calixarene-based compounds [97] and porphyrins [98].

Many of the receptors synthesised interact with a wide variety of anions; however, selectivity has been introduced to these systems. Ferrocene-based systems have been shown to detect aqueous phosphate [99] and sulphate [100] ions. Ferrocene boronic acid has been shown to associate with fluoride ion with much stronger binding than to chloride, bromide and other anions [101].

Although at present their use has been restricted to redox-active sensors in solution, it should be possible to immobilise these receptors at an electrode, and we may then have a simple redox-active electrode whose behaviour in solution is modified by the presence of ions. The high sensitivity of electrochemical techniques would then give us a sensitive and selective method of anion detection.

## 2.4 Biosensors

For the ability to selectively recognise anionic species in solution, chemistry comes second to nature. Biological materials have a very high degree of selectivity in their interactions for many molecules or ions, and this can be exploited in the manufacture of highly selective biosensors. For example, early work included exploitation of a light-driven chloride pump protein immobilised on the surface of an ISFET [102]. When illuminated with yellow light, the response was dependent on the level of chloride ion. Glassy carbon electrodes were modified by immobilising a nitrite reductase plus a maltose binding protein onto an electropolymerised pyrrole derivative [103], with the resultant composite being suitable for the electrochemical detection of nitrite down to $10^{-6}$ M concentrations.

Phosphate is another anion which is of biological significance as well as being biologically active. Numerous phosphate biosensors have been developed, with one based on a multi-enzyme system immobilised on a cellulose membrane on a Pt electrode being able to detect levels down to $10^{-8}$ M [104]. Simpler methods, however, use enzymes such as polyphenol oxidase combined with alkaline phosphatase bound within an electropolymerised layer based on a substituted pyrrole [105]. This was reported to give a sensor with a detection limit of $2\times10^{-6}$ M with good storage ability for a biosensor. Other workers immobilised pyruvate oxidase with a polyelectrolyte (DNA or a substituted dextran) onto carbon [106], with their detector giving linear responses between $5\times10^{-3}$–$10^{-5}$ M, with a detection limit of $5\times10^{-6}$ M and reasonable stability over 24 weeks.

Besides using the bioactive agent to detect the ion of interest, another approach can include monitoring an ion by its inhibitory effect upon enzymatic activity. For example, horseradish peroxidase (HRP) can be immobilised onto one gate of a REFET [107] allowing the presence of cyanide ion to be measured at concentrations of $10^{-3}$–$10^{-7}$ M. The approach used here is to monitor the inhibition of the enzymatic HRP effect, by the cyanide ion, on ascorbic acid. Even lower levels ($10^{-10}$ M) of detection can be obtained using a polyphenol oxidase/clay composite immobilised on carbon, with no interference from chloride, nitrate or bromide [108].

Biosensors are obviously of use in this field due to their high selectivity, however, the problems of storage stability (biological species such as enzymes can rapidly lose their activity) and possible poisoning effects from species such as heavy metals (that may be present in "real" samples) need to be addressed.

## 2.5 Microelectrodes

Most electrodes are of the planar type and can be affected by such factors as convectional mass transport within the sample in which they are immersed. Any type of turbulence will tend to increase the supply of ions to an electrode surface, above and beyond that of diffusional supply. This can have an effect on the magnitude of sensor response and give rise to erroneous results. Diffusion to a large planar electrode can be approximated to be perpendicular to the electrode surface. However, when electrodes become very small, the diffusion profile is hemispherical, mass transport is greatly increased (Fig. 9) and diffusion becomes much less of a limiting factor in the sensor response [109]. Thus, turbulence has a much smaller effect and this means that very small microelectrodes display stir-independent responses. Also the small size of micro-

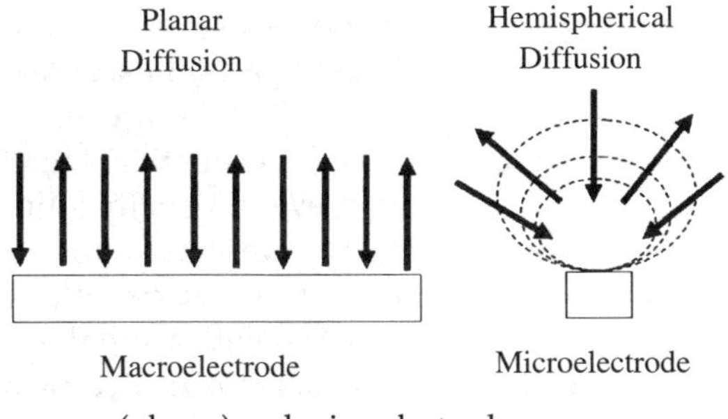


**Fig. 9** Diffusion at macro- (planar) and microelectrodes

electrodes makes them perfect for either analysing very small samples or for use with samples in vivo, e.g. inserting them into blood vessels or a single cell [110].

Anion detection at microelectrodes has not been studied widely. Amongst the first was the work of de Beer et al. [111] who manufactured a nitrite sensor with a tip just a few microns in diameter, which could detect nitrite ions down to 1 µM. This proved to be suitable for profiling the concentrations of nitrite anion within biofilms less than 1-mm thick inside water treatment plants. Other workers have found that use of an interdigitated microelectrode array [112] allows measurement of iodide via monitoring of its redox peak down to sub-micromole levels, making it a suitable technique for analysing mineral water. Carbon nanotubes coated onto Pt microdiscs have been utilised to make a nitrite sensor [113, 114] with detection levels of 0.1 µM. Sulphide has also been detected at nickel microdiscs (50 µm diameter) [115].

## 3 Optical Sensors

### 3.1 Fibre-Optic and Membrane Sensors

Optical sensors display several advantages over electronic sensors. Since photons rather than electrons carry the information, they are almost immune to electrical interference. Usually the optical components are made from glass chips or fibre-optic cable fibres, giving them excellent mechanical and thermal stability and often, moreover, a high chemical resistivity. Finally the use of glass, especially fibre-optic, fibres helps to minimise the size and weight of these sensors. Optical sensors, especially fibre-optic sensors have been the subject of several recent reviews [116–120].

A schematic of a fibre-optic sensor is shown in Fig. 10. The sensor follows the basic principles discussed earlier with three main components. Firstly, there is the chemical sensing element comprising an optical chip/fibre (known as the optrode). Secondly, the sensor contains an indicator with optical properties which depend on the presence/absence of an analyte. Usually the indicator has either a strong optical absorption, or is fluorescent and its absorption/fluorescence changes with the presence of the analyte. In most cases, it is necessary to use an indicator because the analyte does not give or exhibit changes of optical properties. This is the principle for the transduction. Finally, there has to be a method of following the transduction event. Usually a light source is used with a wavelength matched to the absorption maximum of the indicator, so ensuring that the best sensitivity of the sensor can be obtained. After passing through the optrode, the output is measured with a suitable detector – often a photodiode which converts the optical signal into an electrical one to allow this to be processed.

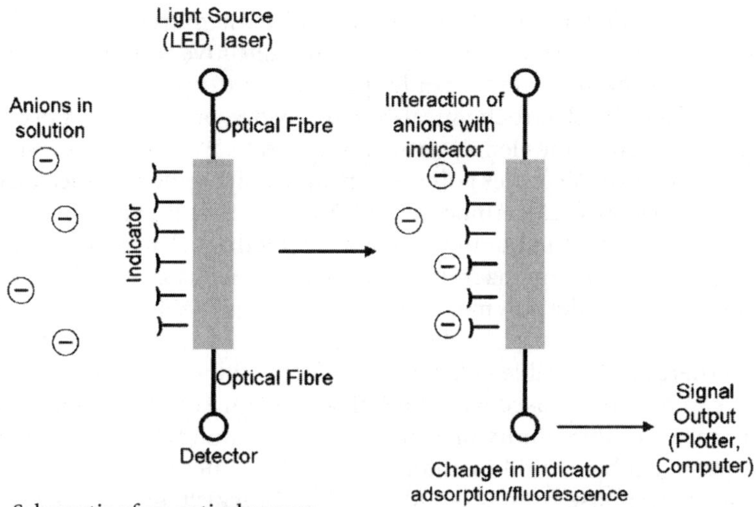


**Fig. 10** Schematic of an optical sensor

For example, an optical fibre may be encased in a polymer membrane which contains a pH-sensitive dye. This assembly is then placed into the test solution. If the dye has a different colour at different pH, then the membrane will change colour depending on the sample pH. Light of a certain wavelength, corresponding to the absorption maxima of either the protonated or unprotonated dye, is passed through the fibre with the resultant output light intensity being dependent on the solution pH. Measurement of the light output then gives us a signal which can be related to the pH of the solution. Sensors of this can be either single fibres, or often a two-fibre system. Only one of the fibres is coated with the membrane and the output from both fibres may be measured and compared. This has the advantage in that effects due to the solution temperature, colour and turbidity often serve to cancel each other out.

As earlier reported for electrochemical sensing, often the active chromophore will be dispersed in a polymeric matrix. For example, Mohr and Wolfbeis reported a nitrate sensor [121] where the active chromophore is a rhodamine B dye which had been modified with an octadecyl side chain to render it hydrophobic and prevent leaching. The dye was dispersed in a plasticised PVC membrane containing a hydrophobic anion carrier (tridodecylmethylammonium chloride). On exposure to nitrate, the fluorescence of the dye increased. This membrane, however, only displayed Hofmeister-type selectivity and was also affected by pH. Replacing the quaternary ammonium anion carrier with a palladium phospine chloride carrier led to selectivity for nitrite [122], probably due to a preferential interaction between Pd and nitrite ion.

A selective membrane for chloride was obtained by spin coating onto glass a PVC membrane containing a novel mercury-containing carborand, a compound which displayed strong Lewis acid activity [123]. This membrane, which also contained a pH-sensitive dye to give an optical response, exhibited strong selectivity for chloride over other anions. PVC membranes containing an indium porphyrin [124] also showed a strong, selective response to chloride, as measured by the shift in the visible spectrum of the porphyrin.

As was found for electrochemical-based sensors, problems can occur with peeling of the membrane element from its substrate. PVC is not the only membrane used for optrodes; other polymers include polyurethane, which displayed improved adhesion [125]. Ambrose and Meyerhoff [126] incorporated ion-selective species into thin liquid films of decyl methacrylate on glass and then photopolymerised the methacrylate, with simultaneous crosslinking and covalent attachment of the polymer to its substrate to give substantially improved lifetimes.

A commercially available chloride-quenchable fluorescent probe, lucigenin (Fig. 11) can be immobilised within a Nafion film and used as a sensing membrane to measure the salinity of seawater [127]. In solution lucigenin is completely quenched by 50 mM of chloride, but when immobilised gave a sensor with a linear response from 0.2–2 M chloride (brackish and seawater contain high levels of salt), a detection limit of 80 mM and no interference from ions commonly found in seawater. Lucigenin is particularly suitable because it can be excited by radiation of wavelength 451 nm, i.e. a blue LED. Incorporation of a second chloride-insensitive ruthenium-based fluorophore can be used to provide an internal reference whilst increasing the sensitivity of the device [128].

A polymeric hydrogel blended with a hydrophobic plasticiser gave a system in which phase separation led to formation of plasticiser droplets entrapped within the hydrogel matrix to allow for nitrate sensing [129]. Incorporation of a fluorescent dye within the hydrophobic regions gave an optical sensor displaying good reversibility, with fluorescence increasing linearly with nitrate concentration in the range 0.1–50 mM.

Incorporation of the optical dye within polymer microspheres has been developed as an alternative to membrane-based approaches. Crosslinked dodecyl acrylate microspheres containing a fluorescent dye (Nile Blue derivative) could be made by dispersion polymerisation [130]. These were plasticiser-free, thereby removing any leaching problem, and could easily be imaged in solution, so giving information on the anion activity within the test solution, with selectivity being of the Hofmeister type. Polystyrene beads coated with a phospholipid membrane containing lucigenin and a mercuro-carborand-based

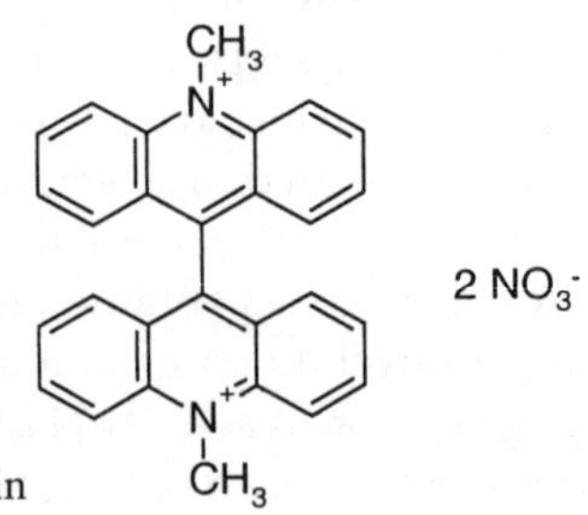


**Fig. 11** Structure of lucigenin

chloride ionophore could be used to measure chloride concentrations within biological fluids [131], with fluorescence quenching of a lipobead suspension being related to $Cl^-$ concentration.

There has been some development of optical biosensors. Nitrate reductase was immobilised within a sol-gel matrix, with binding of nitrate ion (down to a limit of $10^{-6}$ M) causing a characteristic change in the optical absorption [132]. It is notable that this sensor was reversible, allowed selective nitrate detection over other physiologically significant anions and did not lose activity even over six months. Phosphate-binding protein was immobilised on a fibre-optic detector and could be used to measure phosphate with a detection limit of about $10^{-6}$ M [133].

## 3.2 Colorimetric Sensors

These can be considered as a variant on optical sensors where the detection device is the human eye. Although it is unlikely that they would give a quantitative measurement of anion concentration, they can be used as simple tests to determine whether a particular anion is present above a certain concentration. If the dyes could be adsorbed onto paper and still remain active, an anion equivalent of pH indicator paper would be created.

The simple compound shown in Fig. 12a is used as a colorimetric sensor for fluoride; on addition of fluoride ions to a solution in dichloromethane or DMSO, a visible change from a yellow to purple solution becomes evident [134]. This is quenched, however, by addition of water.

Other methods have been used, for example complexing a dye such as Methyl Red with a receptor for nitrate [135]. Addition of nitrate ion causes formation of a nitrate-receptor complex, leading to the release of the chromophore and a large absorbance change. Similar methods have been developed [136] using complexes between calixpyrrole and 4-nitrophenolate ion which are colourless, but on addition of a suitable anion, release the intensely yellow free 4-nitrophenolate anion.

Other workers have used pyrilium cations [137] as sensors for anions such as ATP or sulphate. Charge-transfer complexes between calixpyrroles and chlo-

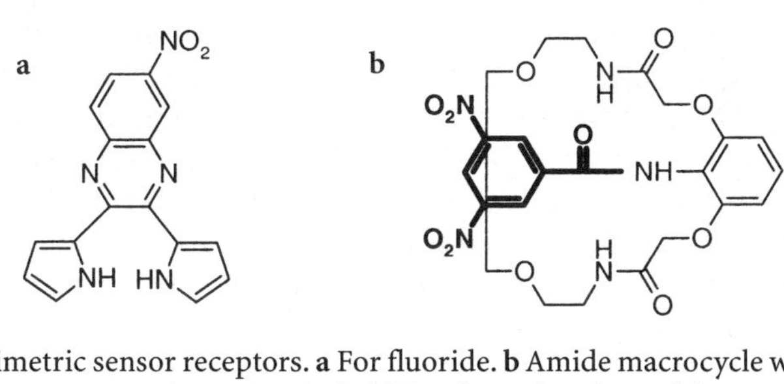


**Fig. 12** Colorimetric sensor receptors. **a** For fluoride. **b** Amide macrocycle which can detect fluoride, acetate or $H_2PO_4^-$; the moiety in *bold* lies above the plane of the rest of the molecule

ranil have been shown to respond selectively to either fluoride or $H_2PO_4^-$ [138]. An interesting amide macrocycle (Fig. 12b) can selectively show the presence of three different anions [139]. The macrocycle, which normally gives a clear solution, undergoes colour changes upon exposure to fluoride (green), acetate (yellow) or $H_2PO_4^-$ (magenta).

A dipyrrole compound when complexed with TCNQ shows a blue to orange transition upon exposure to $5\times10^{-5}$ M $PO_4^{3-}$ and a blue to purple transition upon exposure to $2.5\times10^{-4}$ M $CO_3^{2-}$ [140]. An alizarin-based compound, when adsorbed onto nanocrystalline $TiO_2$, gave a composite orange material which turned red when exposed to 1 mM aqueous fluoride or cyanide [141]. Other selective fluoride sensors include a polyhydroxybenzoxazole material which has a colourless to yellow transition [142] and a trialkylborane-porphyrin conjugate which displays a purple to green transition upon exposure to fluoride [143]. An indoaniline–thiourea system also gave selective blue to green transitions upon exposure to $2.5\times10^{-5}$ M of the tetrahedral anions $H_2PO_4^-$ and $HSO_4^-$ [144].

## 4 Other Types of Anion Sensors

A few other types of anion sensors have been mentioned recently in the literature. Tetrathiofulvalene microcrystals immobilised at a platinum electrode displayed electrochemical properties that were affected by the presence of anions in solution, with some selectivity for anions such as bromide [145]. A flow-injection analysis system using anion-exchange columns for separation and polyaniline electrodes as detectors could detect dichromate down to 0.004 ppb and could be used for seawater samples [146].

A microcantilever beam has been used in conjunction with a hydrogel to detect chromate ions, the presence of chromate causing the hydrogel to swell and deflect the microcantilever. Concentrations of chromate as low as $10^{-11}$ M are detectable with excellent selectivity over bromide, phosphate, nitrate, sulphate and carbonate [147].

A series of papers have concerned the incorporation of various sensors into lab-on-a-chip devices with, for example, conductivity measurements being combined with poly(methyl methacrylate) microfluidic devices to analyse mixtures of mono- and polyanionic molecules such as proteins [148].

An interesting technique recently published uses FTIR spectroscopy to detect anions [149]. Thin films of tetraalkylated ferrocenium salts were evaporated onto ATR crystals. Exposure to anion-containing solutions caused incorporation of the anions, and they could then be detected via their IR absorbances. Limits of detection for 10-min analyses varied from $10^{-6}$ to $10^{-8}$ M for a variety of anions; however, this technique cannot be used to detect anions that do not have IR absorbances, such as halides.

## 5 Applications of Anion Sensors

Anion sensors are beginning to find many applications. For example, chloride sensors are now being used for environmental monitoring. Due to the relative inertness of chloride ions with respect to rock or soil components, chloride is an effective tracer for pollution from chemicals moving from man-made sources into natural water bodies, or alternatively for monitoring salt water intrusion. Chloride ion measurements can aid the monitoring of landfills for leaks, tracing the movement of point- or non-point-source pollutants (for instance, storm water runoff) within a natural water body, monitoring estuarine waters for changes in salinity, and detection of salt water intrusion into drinking water supplies (ground or surface waters). NexSens, amongst other companies, manufactures ISEs for this application [150].

Another ion widely monitored within the environment is nitrate. Nitrate is the commonest biologically available form of nitrogen, an essential nutrient for all forms of life, including all levels of aquatic organisms. It can be found in both suspended solids and dissolved compounds in natural waters. Often, the amount of life that a natural system such as a lake can support is limited by the nitrate concentration. Small changes in biologically available nitrogen levels can dramatically affect the levels of microbiological, plant, and eventually, animal life. Low levels mean the biodiversity is limited by the nitrate level; conversely, high levels of accessible nitrogen are also damaging. High nitrate levels can effectively act as a fertiliser, leading to an overabundance of microorganisms, or a so-called algal bloom. Unfortunately this can lead to a depletion of dissolved oxygen, resulting in mortality to higher organisms such as fish and shrimp. The monitoring of nitrate levels is used to trace pollution from agriculture (nitrate fertilisers), for checking fish farms and other aquaculture for waste build-up and for surveying nutrient levels in natural water bodies. Companies such as Hydrolab (Hach) manufacture nitrate ISEs for this purpose [151].

Fluoridation of water is often used to help reduce tooth decay. However, whereas levels of 1 ppm are commonly used, excessive fluoride levels (>2 ppm) can be a problem and need to be corrected. Analytical Technology Inc. manufactures a fluoride-sensitive ISE for this purpose [152].

## 6 Future Perspectives

The future of anion sensors is a very promising one. Greater interest and legal requirements for environmental, food and water monitoring will require the detection of anionic species, often at lower and lower concentrations. Technical problems, however, need to be overcome. These include some of the considerations below:

- Sensor stability
  Problems such as peeling of the polymer films and leaching of the membranes need to be overcome to increase sensor lifetime. New membranes such as those based on siloxanes show possible increases in performance and stability.
- Cost
  The simplest way to deal with the problem of sensor lifetime is to circumvent it completely by using each sensor once. This requires the production of inexpensive, disposable sensor strips for a variety of anions, possibly by a method such as screen printing. An alternative method may be to further develop photoprocessable polymers, which can then be assembled into the final sensor by a process such as photolithography.
- Selectivity
  Often sensors only display Hofmeister selectivity or are affected by anions of a similar nature to that of interest. The development of sensors that will detect small concentrations of the anion of interest in the presence of strong interferents, possibly in higher concentrations, will be of great use.
- Multi-anion detectors
  Processes such as screen printing and lithography allow the reduction in sensor size and cost. Assembly of multiple sensors on a single chip, with each sensor being of good selectivity, would enable us to perform complex measurements in a single step. Pattern recognition technology could be used here to aid the determination of the analyte constitution.

Microelectrodes often display superior properties to those of larger planar electrodes. They experience hemispherical diffusion profiles which can confer stir independence on the results, which may be important if measuring turbulent systems such as in a river or in the bloodstream. They also allow measurements in high-resistance media, which may be experienced if electrolyte concentrations are very low.

## References

1. Frant MS, Ross JW (1966) Science 154:1553
2. Park CH, Simmons HE (1968) J Am Chem Soc 90:2437
3. Journals include Sensors (Goepel W, Hesse J, Zemel JN (eds)) Vols 1–9 and Sensors Update (Baltes HJ, Feddler GK, Goepel W, Hesse J, Korvink JG) Vols 1–13, Wiley, New York
4. Eggins BR (2002) Chemical sensors and biosensors. Wiley, New York
5. Diamond D (1998) Principles of chemical and biological sensors. Wiley, New York
6. Fraden J (2003) Handbook of modern sensors. Springer, Berlin Heidelberg New York
7. Mulchandani A, Sadik OA (2000) Chemical and biological sensors for environmental monitoring (ACS Symposium Series). ACS, Washington
8. Janata J (2001) Anal Chem 73:150A
9. Bakker E, Meyerhoff ME (2000) Anal Chim Acta 416:121
10. Bakker E, Telting-Diaz (2002) Anal Chem 74:2781

11. Antonisse MMG, Reinhoudt DN (1999) Electroanalysis 11:1035
12. Buck RP, Lindner E (2001) Anal Chem 73:88A
13. Miyahara Y, Yamashita K, Ozawa S, Watanabe Y (1996) Anal Chim Acta 331:85
14. Hogg G, Steiner G, Cammon K (1994) Sens Actuators B 18:376
15. Lindner E, Buck RP (2000) Anal Chem 72:336A
16. Malinowska E, Niedziolka J, Meyerhoff ME (2001) Anal Chim Acta 432:67
17. Shamsipur M, Rouhani S, Mohajeri A, Ganjali MR, Rashidi-Ranjbar P (2000) Anal Chim Acta 418:197
18. Shamsipur M, Soleymanpour A, Akhond M, Shargi H, Naseri MA (2001) Anal Chim Acta 450:37
19. Lizondo-Sabater J, Martinez-Manaz R, Sancenon F, Segui MJ, Soto S (2002) Anal Chim Acta 459:229
20. Malon A, Radu A, Qin W, Ceresa A, Maj-Zurawska M, Bakker E, Pretsch E (2003) Anal Chem 75:3865
21. Lee HJ, Yoon IJ, Yoo CL, Pyun HJ, Cha GS, Nam H (2000) Anal Chem 72:4694
22. Choi YS, Lvova L, Shin JH, Oh SH, Lee CS, Kim BH, Cha GS, Nam H (2002) Anal Chem 74:2435
23. Goff TL, Braven J, Ebdon L, Scholefield D (2002) Anal Chem 74:2596
24. Fibbioli M, Berger M, Schmidtchen FP, Pretsch E (2000) Anal Chem 72:156
25. Wojciechowski K, Wroblewski W, Brzozka Z (2003) Anal Chem 75:3270
26. Zhang XB, Guo CC, Jian LX, Shen GL, Yu RQ (2000) Anal Chim Acta 419:227
27. Hassan SSM, Abou Ghalia MH, Amr AE, Mohamed AHK (2003) 482:9
28. Sanchez-Pedreno C, Ortuno JA, Hernandez J (2000) Anal Chim Acta 415:159
29. Shahrokhian S, Hamzehloei A, Bagherzadeh M (2002) Anal Chem 74:3312
30. Picioreanu S, Poels I, Frank J, van Dam JC, van Dedem GWK, Nagels LJ (2000) 72:2029
31. Prodromidis MI, Veltsistas PG, Karayannis K (2000) Anal Chem 72:3995
32. Ye Q, Meyerhoff ME (2001) Anal Chem 73:332
33. van den Berg A, Bergveld P, Reinhoudt DN, Sudholter EJR (1985) Sens Actuators 8:129
34. Reinhoudt DN, Sudholter EJR (1990) Adv Mater 2:23
35. Bousse L, de Rooij NF, Bergveld P (1983) IEEE Trans Electron Devices ED-30:1263
36. Moss SD, Janata J, Johnson CC (1975) Anal Chem 47:2238
37. McBride PT, Janata J, Comte PA, Moss SD, Johnson CC (1978) Anal Chim Acta 101:239
38. Hofmeister F (1988) Arch Exp Patol Pharmakol 24:247
39. Janata J, Huber RJ (1979) Ion Sel Electrode Rev 1:31
40. Sibbald A (1983) IEE Proc 130:233
41. Sudholter EJR, van der Wal PD, Skowrońska-Ptasińska M, van den Berg A, Bergveld P, Reinhoudt DN (1990) Recl Trav Chim Pays Bas 109:222
42. van der Wal P, Skowrońska-Ptasińska M, van den Berg A, Bergveld P, Sudholter EJR, Reinhoudt DN (1990) Anal Chim Acta 231:41
43. Harrison DJ, Teclemariam A, Cunningham L (1973) Anal Chem 61:246
44. Fiedler U, Ruzicka J (1973) Anal Chim Acta 67:179
45. Fogt EJ, Untereker DF, Norenberg MS, Meyerhoff ME (1985) Anal Chem 57:1995
46. van den Vlekkert HH, Decroux M, de Rooij NF (1987) Proc Transducers '87 Tokyo 730
47. Talma AG, Volders JPGM, Ligtenberg HCG, Dost L, Bouwmeester HJM, Boekema B (1987) Sens Actuators 10:35
48. Sudholter EJR, van der Wal P, Skowrońska-Ptasińska M, van den Berg A, Bergveld P, Reinhoudt DN (1990) Anal Chim Acta 230:59
49. Hara H, Ohkubo H, Sawai K (1993) Analyst 118:549
50. Campanella L, Colapicchioni C, Crestcentini G, Sammartino MP, Su Y, Tommassetti M (1995) Sens Actuators B 26:329

51. Liao Y, Zheng Z, Li C (1994) Chem Abstr 121:98507k
52. Shiramizu B, Janata J, Moss SD (1979) Anal Chim Acta 108:161
53. Bakker E (1996) Sens Actuators B 35:20
54. Hulaniki A, Michalska A, Leuenstraum A (1994) Talanta 41:232
55. Zielinska R, Mulik E, Michalska A, Achmatowicz S, Maj-Zurawska M (2002) Anal Chim Acta 451:243
56. Lawrence NS, Jiang L, Jones TGJ, Compton KG (2003) Anal Chem 75:2054
57. Lvova L, Kim SS, Legin A, Vlasov Y, Yang JS, Cha GS, Nam H (2002) Anal Chim Acta 468:303
58. Janata J (2003) Proc IEEE 91:864
59. Nam H, Cha GS, Strong TD, Ha J, Sim H, Hower W, Martin SM, Brown RB (2003) Proc IEEE 91:870
60. Cané C, Gràcia I, Merlos A (1997) Microelectr J 24:389
61. Statny V, Lhotak P, Michlova V, Stibor I, Sykora J (2002) Tetrahedron 54:7207
62. Pelizzi N, Casnati A, Friggeri A, Ungaro R (1998) J Chem Soc Perkin Trans II 1307
63. Antonisse MMG, Snellink-Reül BHM, Engberson JFJ, Reinhoudt DN (1998) J Org Chem 63:870
64. Gao D, Li D, Yu RQ, Zheng GD (1994) Anal Chem 66:2245
65. Chaniotakis NA, Park SB, Meyerhoff ME (1989) Anal Chem 61:566
66. Xiao KP, Bühlmann P, Nishizawa S, Amemiya S, Umezama Y (1997) Anal Chem 69:1038
67. Meyerhoff ME, Pretsch E, Welitti DH, Simon W (1987) Anal Chem 59:144
68. Smirnova AL, Tarasevitch VN, Akhmaniko EM (1994) Sens Actuators B 18:392
69. Wroblewski W, Chudy M, Dybko A (2000) Anal Chim Acta 416:97
70. Ikarijama Y, Galiatsatos C, Heineman WR (1987) Sens Actuators 12:455
71. Dong S, Sun Z, Lu Z (1988) Analyst 113:1525
72. Dong S, Che G (1988) Talanta 38:111
73. Cadogan A, Lewenstam A, Avaska A (1992) Talanta 39:617
74. Hutchins R, Bachas L (1995) Anal Chem 67:1654
75. Michalska A, Dumanska J, Maksymiuk K (2003) Anal Chem 75:4964
76. Kliza DM, Meyerhoff ME (1992) Electroanalysis 4:841
77. Wang Z, Zhang HS, Mark HB, Rubinson JF (1997) Anal Lett 30:1
78. Nicolas M, Fabre B, Simonet J (1999) Chem Commun 1881
79. Allen JR, Florido SD, Young SD, Daunert S, Bachas LG (1995) Electroanalysis 7:710
80. Winnischofer H, Lima SD, Araki K, Toma HE (2003) Anal Chim Acta 480:97
81. Tse YH, Janda P, Lam H, Lever ABP (1995) Anal Chem 67:981
82. Sjoberg P, Bobacka J, Lewnstam A, Avaska A (1999) Electroanalysis 11:821
83. Ugo P, Moretto LM, Mazzocchin GA, Guerriero P, Martin CR (1998) Electroanalysis 10:1168
84. Bergveld P, van den Berg A, van der Wal PD, Skowrońska-Ptasińska M, Suholter EJR, Reinhoudt DN (1989) Sens Actuators 18:307
85. Fujihira M, Fukui M, Osa TJ (1980) Electroanal Chem 106:413
86. Matsuo T, Nakajima H (1984) Sens Actuators 5:293
87. Kimura J, Kuriyama T, Kawana Y (1980) Sens Actuators 9:373
88. Skowrońska-Ptasińska M, van der Wal P, van den Berg A, Bergveld P, Sudholter EJR, Reinhoudt DN (1990) Anal Chim Acta 230:67
89. Wakida SI, Yamane M, Higashi K, Hiiro K, Ujihiru Y (1990) Sens Actuators B 1:412
90. Ito S, Baba K, Asano Y, Takesako H, Wada H (1996) Talanta 43:1869
91. Kim W, Chung S, Park SB, Lee SC, Kim C, Sung DD (1997) Anal Chem 69:95
92. Linder E, Niegreisz Z, Pungor E, Berube TB, Buck RP (1989) J Electronanal Chem 259:67

93. val der Val PD, Sudholter EJR, Boucamp HJM, Boumester HJM, Reinhoudt DN (1991) J Electroanal Chem 317:153
94. Hogg G, Lutze O, Cammann K (1996) Anal Chim Acta 335:103
95. Antonisse MMG, Lugtenberg RJW, Egberink RJM, Engbersen JFJ, Reinhoudt DN (1996) Anal Chim Acta 332:123
96. Beer PD, Keefe AD (1989) J Organomet Chem 375:C40
97. Beer PD, Drew MGB, Hesek K, Nam KC (1999) Organometallics 18:3933
98. Beer PD, Drew MGB, Hesek K, Jagessar R (1995) J Chem Soc Chem Commun 1187
99. Beer PD, Chen Z, Drew MGB, Kingston J, Ogden MI, Spencer P (1993) J Chem Soc Chem Commun 1187
100. Beer PD, Cadman J, Lloris JM, Martinez-Manez R, Padilla-Tosta ME, Pardo T, Smith DK, Soto J (1999) J Chem Soc Dalton Trans 127
101. Dusemund C, Sandanayake KRAS, Shinkai S (1995) J Chem Soc Chem Commun 333
102. Seki A, Kubo I, Sasabe H, Tomioka H (1994) Appl Biochem Biotechnol 48:205
103. Wu Q, Storrier GD, Pariente F, Wang Y, Shapleigh JP, Abruna HD (1997) Anal Chem 68:4856
104. Conrath N, Grundig B, Huwel S, Cammann K (1995) Anal Chim Acta 309:47
105. Cosnier S, Gondran C, Watelet J-C, de Giovani W, Furreil RPM, Leone FA (1998) Anal Chem 70:3952
106. Gavalas VG, Chaniotakis NA (2001) Anal Chim Acta 427:271
107. Volotovsky V, Kim N (1998) Biosens Bioelectron 13:1029
108. Shan D, Mousty C, Cosnier S (2004) Anal Chem 76:178
109. Johnson DC, Ryan MD, Wilson GS (1988) Anal Chem 60:R147
110. Malinski T, Taha Z (1992) Nature 358:676
111. de Beer D, Schramm A, Santegoeds CM, Kuhl M (1997) Appl Environ Microbiol 63:973
112. Tomcik P, Bustin D (2001) Fresenius J Anal Chem 371:562
113. Liu P, Hu J (2002) Sens Actuators B 84:194
114. Zhao YD, Zhang WD, Luo QM, Li SFY (2003) Microchem J 75:189
115. Giovanelli D, Lawrence NS, Jiang L, Jones TGJ, Compton RG (2003) Sens Actuators B 88:320
116. Seitz WR (1988) Crit Rev Anal Chem 19:15
117. Wise DL (1991) Biosensors with fiber optics. Humana, New York
118. Wolfbeis OS (1991) Fiber optic chemical sensors and biosensors. CRC, Boca Raton
119. Wolfbeis OS (2002) Anal Chem 74:2663
120. Martinez-Manez R, Sancenon F (2003) Chem Rev 103:4419
121. Mohr GJ, Wolfbeis OS (1995) Anal Chim Acta 316:239
122. Mohr GJ, Murkovic I, Lehmann F, Haidler C, Wolfbeis OS (1997) Sens Actuators B 39:239
123. Badr IHA, Johnson RD, Diaz M, Hawthorne MF, Bachas LG (2000) Anal Chem 72:4249
124. Zhang W, Rozniecka E, Malinowska E, Parzuchowski P, Meyerhoff ME (2002) Anal Chem 74:4548
125. Wang E, Meyerhoff ME (1993) Anal Lett 26:1519
126. Ambrose TM, Meyerhoff ME (1999) Anal Chim Acta 378:119
127. Huber C, Klimant I, Krause C, Werner T, Mayr T, Wolfbeis OS (2000) Fresenius J Anal Chem 368:196
128. Huber C, Klimant I, Krause C, Wolfbeis OS (2001) Anal Chem 73:2097
129. Huber C, Klimant I, Krause C, Werner T, Mayr T, Wolfbeis OS (2001) Anal Chim Acta 449:81
130. Peper S, Tsagkatakis I, Bakker E (2001) Anal Chim Acta 442:25
131. Ma A, Rosenweig Z (2004) Anal Chem 76:569
132. Aylott JW, Richardson DJ, Russell DA (1997) Analyst 122:77

133. Salins LLE, Deo SK, Daunert S (2004) Sens Actuators B 97:81
134. Black CB, Andrioletti B, Try AC, Ruiperez C, Sessler JL (1999) J Am Chem Soc 121:10438
135. Niikura K, Bisson AP, Anslyn EV (1999) J Chem Soc Perkin Trans II 1111
136. Gale PA, Twyman LJ, Handlin CI, Sessler JL (1999) J Chem Soc Chem Commun 1851
137. Sancenon F, Descalzo AB, Martinez-Manez R, Miranda MA, Soto J (2001) Angew Chem Int Ed 40:2640
138. Shao S, Gao Y, He L, Jiang S, Yu X (2003) Tetrahedron Lett 44:2175
139. Piatek P, Jurczak J (2002) J Chem Soc Chem Commun 2450
140. Guo Y, Shao S-J, Xu J, Shi Y-P, Jiang S-X (2004) Inorg Chem Commun 7:333
141. Palomares E, Vilar R, Green A, Durrant JR (2004) Adv Funct Mater 14:111
142. Lee JK, Na JH, Kim TH, Kim YS, Park WH, Lee TS (2004) Mater Sci Eng C 24:261
143. Kubo Y, Yamamoto M, Ikdea M, Takeuchi M, Shinkai S, Yamaguchi Y, Tamao K (2003) Angew Chem Int Ed 42:2036
144. Lee DH, Lee HJ, Hong J-I (2002) Tetrahedron Lett 43:7273
145. Wooster TJ, Bond AM, Honeychurch MJ (2003) Anal Chem 75:586
146. Yang Y-J, Huang H-J (2001) Anal Chem 73:1377
147. Zhang Y, Ji H-F, Brown GM, Thundat T (2003) Anal Chem 75:4773
148. Galloway M, Stryjewski W, Henry A, Ford SM, Llopis S, McCarley RL, Soper SA (2002) Anal Chem 74:2407
149. Hebert GN, Odom MA, Bowman SC, Strauss SH (2004) Anal Chem 76:781
150. www.hydrolab.com
151. www.nexsens.com
152. www.analyticaltechnology.com

Top Curr Chem (2005) 255: 125–162
DOI 10.1007/b101165


# Anion Sensing by Metal-Based Receptors

Paul D. Beer (✉) · Simon R. Bayly

Department of Chemistry, Inorganic Chemistry Laboratory, University of Oxford, South Parks Road, Oxford, OX1 3QR, UK
*paul.beer@chem.ox.ac.uk*



**Abstract** Metal-based receptors are an important and expanding class of molecules in anion sensing. This review describes the subject with specific focus on the metal-containing subunit's function in determining their anion sensing properties. In general the metal ion is used as a photo- and/or electrochemical reporter group, an anion recognition site and/or as a simple structural component. Metal-based receptors, particularly where the metal ion performs a combination of these roles, exhibit a range of functionality not available to purely organic structures. The most recent advances in the field are also detailed including the use of metallo-receptors in dendrimer, functionalised nanoparticle and surface-bound anion sensors.

**Keywords** Anion recognition · Organometallic receptors · Transition metal receptors · Electrochemical sensors · Optical sensors

# 1 Introduction

Metal ions, in the form of organometallic and coordination complexes, possess many physical and chemical properties desirable for use in anion sensors. Amongst the many examples of metal-based anion sensors that have been developed over the last 15 years the metal is most commonly incorporated as a reporter group, whose photochemical or redox response is changed upon proximal binding of an anion. Alternatively the metal can act directly as a selective Lewis acid binding site for anions or simply serve as a structural component, allowing the self-assembly of binding sites of different topologies. Metal-based receptors, particularly where the metal ion combines these roles, show a range of functionality not available to purely organic structures.

Many aspects of anion recognition by metal-based receptors have been covered in previous reviews [1–5]. This review is not intended to be comprehensive; rather, it provides an overview of the subject with particular reference to the specific properties that metal-containing sub-units impart to this class of anion sensor. The most recent advances in the field are also detailed including the use of metallo-receptors in dendrimer, functionalised nanoparticle and surface-bound anion sensors.

# 2 Organometallic Anion Receptors

## 2.1 Cobaltocenium-Based Receptors

Anion receptors incorporating cobaltocenium have been studied extensively due to the combination of an accessible redox couple and potential favourable electrostatic interactions of the cationic organometallic metallocene complex with anions. The first anion receptor based on this species was reported by Beer and co-workers in 1989 [6]. The macrocyclic bis-cobaltocenium receptor **1** was shown to bind (via electrostatic interaction) and to electrochemically sense bromide in acetonitrile solvent media.

Augmentation of cobaltocenium receptors with hydrogen-bond donor groups, such as amides in receptors **2** and **3**, generates both stronger and more selective anion binding [7]. Proton NMR anion titration studies in $CD_3CN$ reveal **2** and **3** to have selectivity for dihydrogenphosphate over chloride by approximately an order of magnitude [8]. This is attributed to the greater basicity of the dihydrogenphosphate anion. In these receptors the electrochemical sensing of anions is also enhanced. For **2** and **3** chloride causes cathodic shifts of 30 and 85 mV, respectively, in the cobaltocene/cobaltocenium redox couple. The more strongly bound anion dihydrogenphosphate correspondingly generates larger cathodic shift magnitudes of 200 and 240 mV, respectively. It

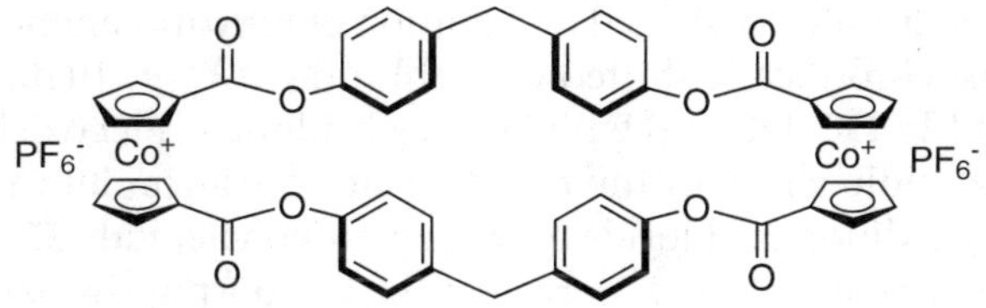

1

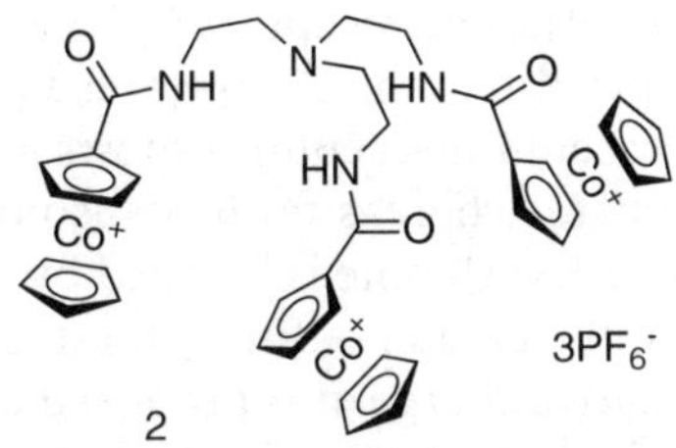

2

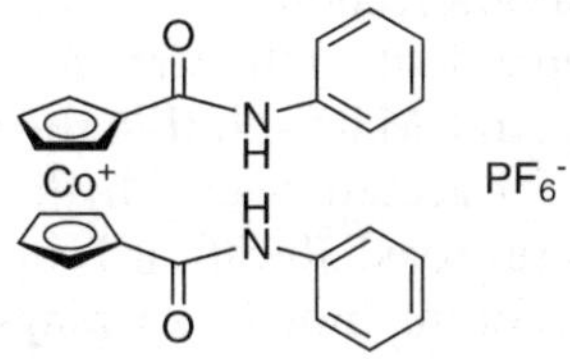

3

is thought that these cathodic shifts are due to the stabilisation of the cationic cobaltocenium species relative to the neutral cobaltocene species by the closely bound anion guest.

A tetra-cobaltocenium receptor **4** has been synthesised that is supported by a porphyrin skeleton [9]. This exhibits parallel photo- and electrochemical sensing of anions with a trend in selectivity of $Cl^- > Br^- \gg NO_3^-$. Proton NMR titrations in $CD_3CN$ showed chloride and bromide to be bound in 1:1 stoichiometry with stability constants of 860 and 820 $M^{-1}$, respectively, whereas nitrate exhibited weaker binding with $K=190\ M^{-1}$. In the UV–vis spectrum of **4** in acetonitrile solution the Soret band ($\lambda_{max}=425$ nm) of the porphyrin was signif-

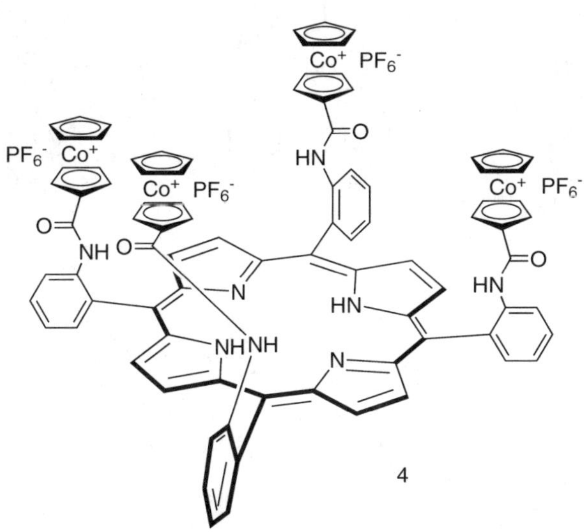

4

icantly bathochromically shifted on addition of dihydrogenphosphate ($\Delta\lambda_{max}$= 15 nm), hypsochromically shifted with chloride ($\Delta\lambda_{max}$= 10 nm) and split into two maxima ($\lambda$=430, 440 nm) with hydrogensulphate. Electrochemical studies displayed cathodic shifts in the cobaltocene/cobaltocenium redox couple of 35–75 mV on addition of chloride or hydrogensulphate, and 225 mV for dihydrogenphosphate. Smaller shifts were seen in the porphyrin oxidation wave. It should be noted that these anion sensing effects are specific to the *cis*-$\alpha,\alpha,\alpha,\alpha$-atropisomer.

The anion recognition properties of calix[4]arene receptors **5–7** were found to be dependent on the structure of the upper rim of the calix[4]arene [10]. For instance, in DMSO solution the bis-cobaltocenium receptor **5** shows a greater affinity for acetate over dihydrogenphosphate, whereas for **6** (its isomer) the trend is reversed. Receptor **7**, in which the calix[4]arene is bridged by a single cobaltocenium moiety, displays significantly greater affinity for the above anions despite possessing only a single positive charge. It is proposed that this selectivity preference is due to the upper-rim bidentate amide hydrogen-bond donor cavity of **7** being of complementary topology to bidentate anions such as carboxylates. The cobaltocene/cobaltocenium redox couple of **7** was found to undergo an anodic shift of 155 mV in the presence of acetate.

The incorporation of crown ether units into a cobaltocenium receptor has been shown to allow the switchable sensing of anions. Proton NMR titrations of receptor **8** in $CD_3CN$ solution gave log *K* values of 3.1 for chloride and 3.0 for

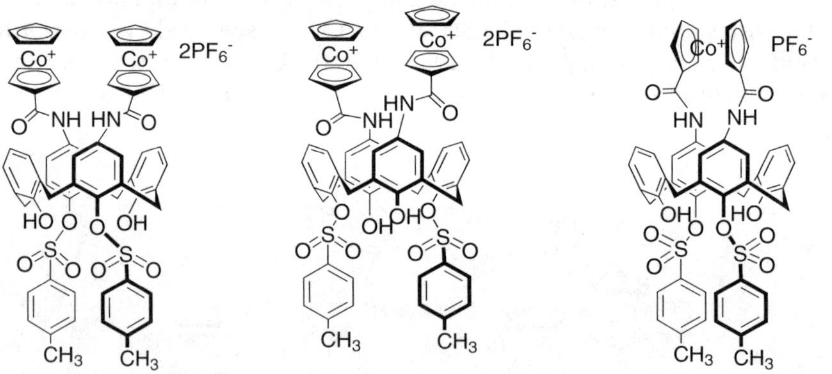


5 6 7

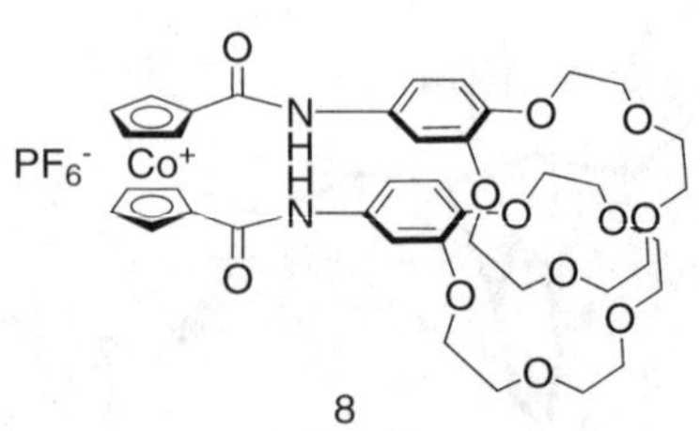


8

bromide [11]. Electrochemical titrations showed cathodic shifts of the cobaltocene/cobaltocenium redox couple of 60 and 30 mV for the two anions, respectively. However, when either the NMR or electrochemical titrations were carried out in the presence of $K^+$ no significant shifts were observed. It is postulated that the $K^+$ ions form a 1:1 intramolecular sandwich complex with the two benzocrown ether units of the receptor. The concomitant change in conformation of the amide groups reduces their availability for anion binding.

## 2.2 Ferrocene-Based Receptors

The accessible ferrocene/ferrocenium redox couple of ferrocene has led to its frequent use in electrochemical anion sensors. The chemical and structural similarity between ferrocene and cobaltocenium has meant that receptors based on these complexes often share the same design. The most relevant difference is that the ferrocene derivatives are neutral (until oxidised to ferrocenium), have no inherent electrostatic interaction with anions and therefore their complexes with anions exhibit lower stability constants.

Molecules **9–13** are a selection of ferrocene-based receptors which include secondary amide groups for the hydrogen bonding of anions [12, 13]. Measured

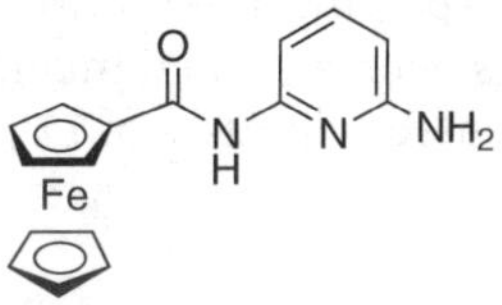

9

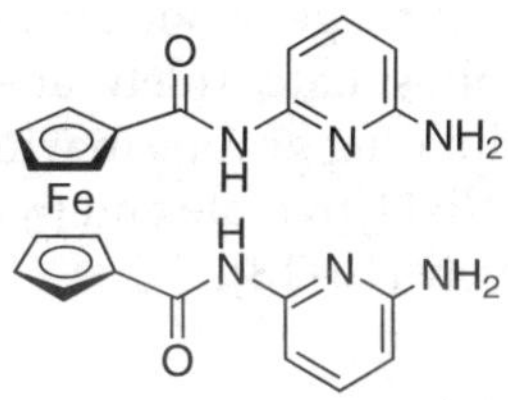

10

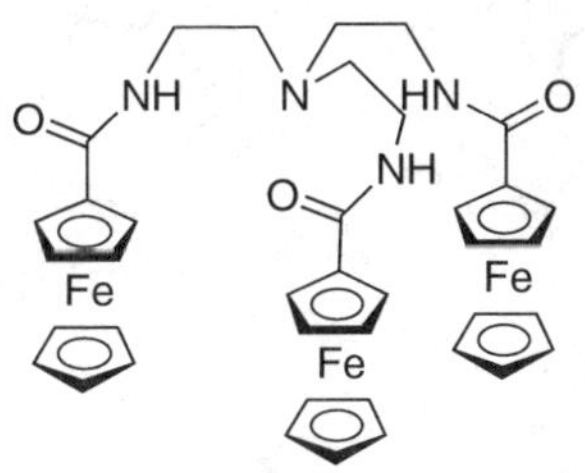

11

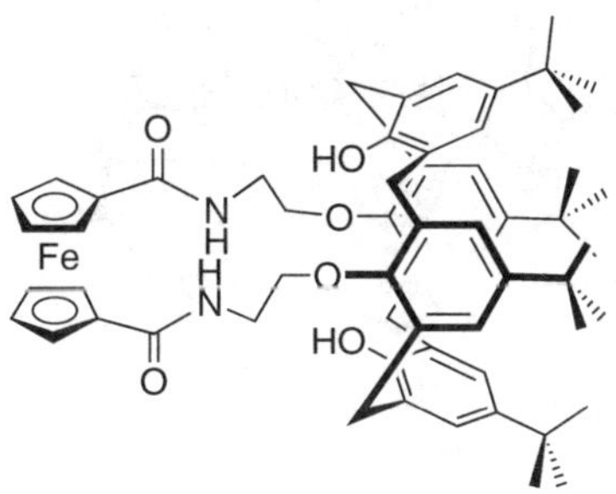

12

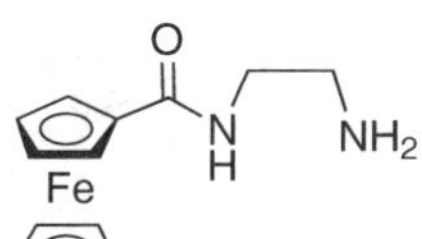

13

in acetonitrile **9–12** showed cathodic shifts in the ferrocene/ferrocenium oxidation wave of up to 240 mV induced by dihydrogenphosphate. Competition experiments demonstrated the same shift even in the presence of a tenfold excess of chloride or hydrogensulphate. A cathodic shift is expected in this class of compounds since the coordination of an anion guest close to the ferrocene stabilises the cationic ferrocenium redox state, facilitating the oxidation process. It is therefore largely the stability of the electrostatically enhanced anion-ferrocenium complex which determines the magnitude of the redox shift. Receptor **13** has the opposite selectivity, displaying a hydrogensulphate-induced shift of 220 mV which does not change in the presence of excess dihydrogenphosphate. In this case the $HSO_4^-$ anion donates a proton to the basic amine function of the receptor and the resulting cationic complex then binds the residual $SO_4^{2-}$ anion.

The cyclotriveratrylene amides **14** and **15** described by Moutet and co-workers include closely spaced multiple ferrocene reporter units [14]. In $CH_2Cl_2$ or acetonitrile two-wave electrochemical behaviour and cathodic shifts of up to 260 mV are observed in the presence dihydrogenphosphate or $ATP^{2-}$. The increased electrostatic affinity of the multiply charged oxidised species for these anions eliminates the irreversible electrode adsorption processes which occur with related monoferrocenyl receptors. A very simple cationic receptor **16** was investigated by the same group [15]. It is able to sense dihydrogenphosphate and $ATP^{2-}$ in a range of solvents, displaying a shift of 470 mV in $CH_2Cl_2$ with dihydrogenphosphate, solely due to a strong ion-pairing interaction. Other anion coordinating groups to which ferrocene has been attached include imidazolium, calix[4]pyrrole and pyridinium-substituted tripods, e.g. **17** , **18** and **19**, respectively [16–18].

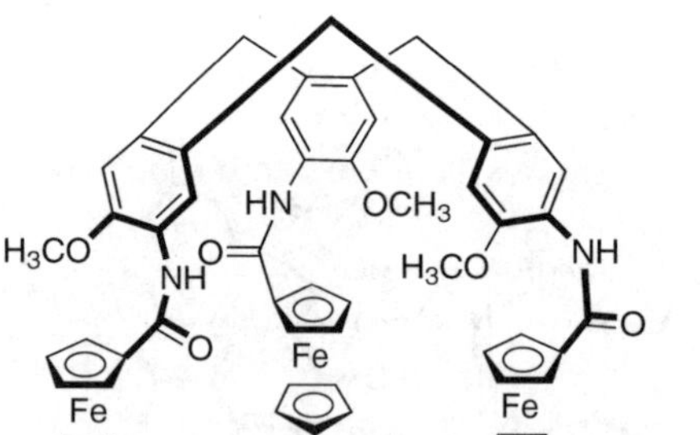


14

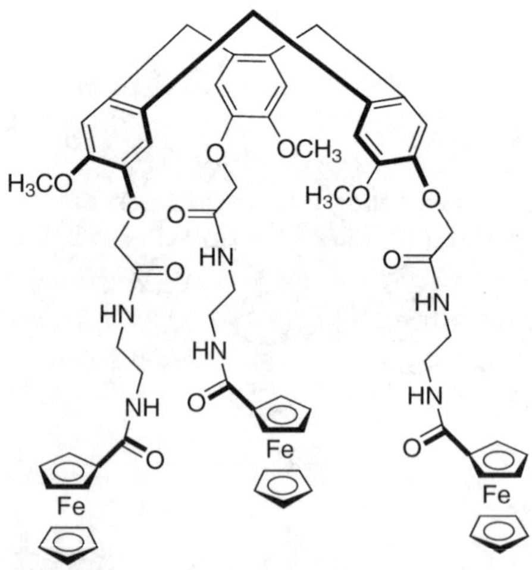


15

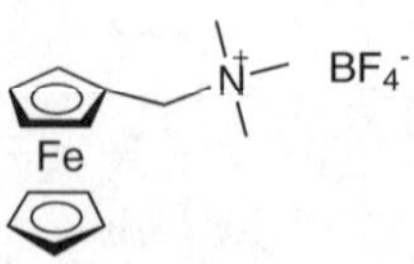


16

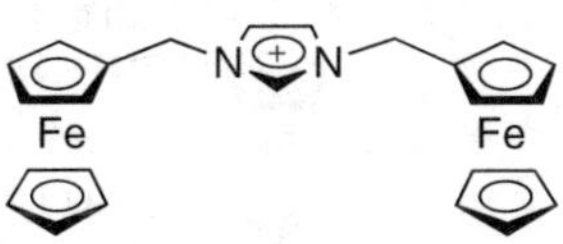

17

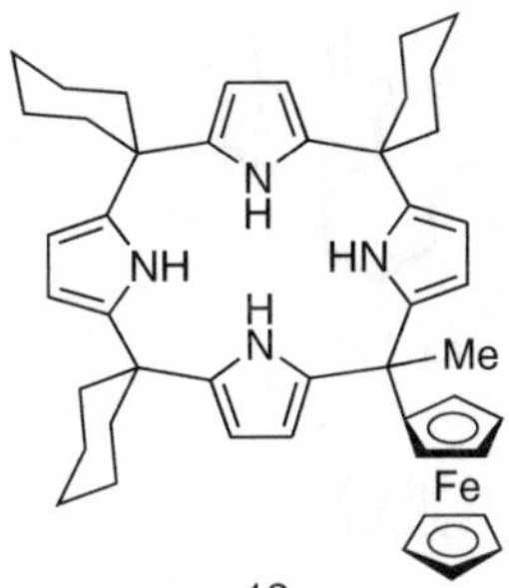

18

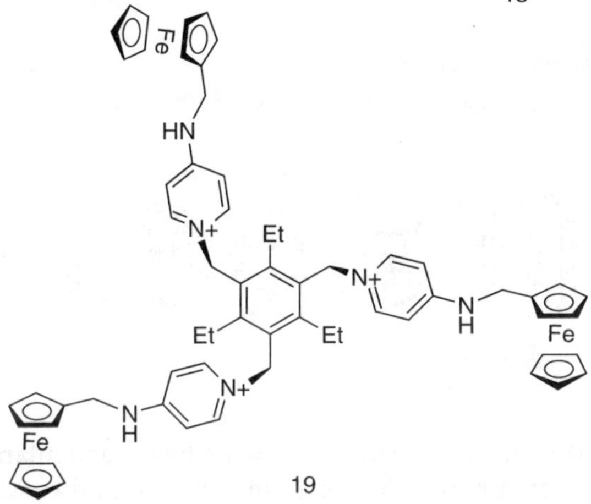

19

Molecules **20–22** in which a ferrocene-1,1′-bisamide is bridged across the upper rim of a calix[4]arene show interesting anion sensing selectivity [19]. The results of $^{1}H$ NMR studies in $CD_3CN$ show that the receptors preferentially bind carboxylate anions (acetate and benzoate) over dihydrogenphosphate and chloride.

Organometallic receptors are also capable of operating in an aqueous environment, an important consideration in the sensing of biologically relevant anions. Strong electrostatic interactions are once again the key. Beer et al. have generated a series of ferrocene-based receptors appended with various open chain and cyclic amine functional groups, e.g. molecules **23** and **24** bind $ATP^{2-}$ and dihydrogenphosphate in water [20–22]. The selectivity of this class of receptors is pH dependent. In the 1:1 complexes formed at pH 6.5 at least two of the amines are protonated. Anion-induced cathodic shifts of 60–80 mV were observed in the ferrocene/ferrocenium couple. Metallacyclic receptors **25** and **26** were shown to be able to quantitatively determine phosphate and sulphate in the presence of other anions. The electrochemistry of these receptors was studied in 70:30 THF/$H_2O$ over a range of pH values. A maximum selective redox shift of 54 mV for phosphate over sulphate was observed at pH 4 for **25**, whereas **26** gave a maximum selective redox shift of 50 mV for phosphate over sulphate at pH 7.

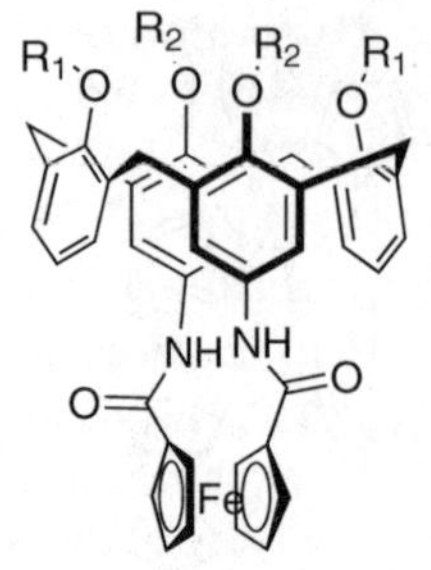

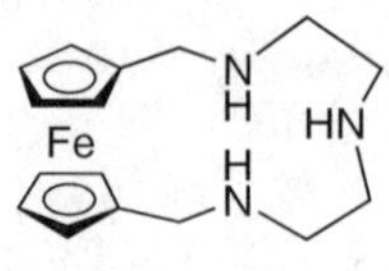

23

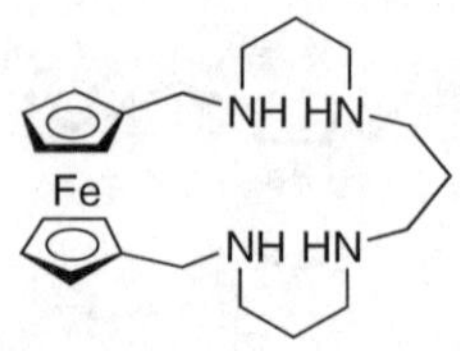

24

20 $R_1 = R_2 = CH_3$

21 $R_1 = CH_3$, $R_2 = CH_2COOEt$

22 $R_1 = R_2 = CH_2COOEt$

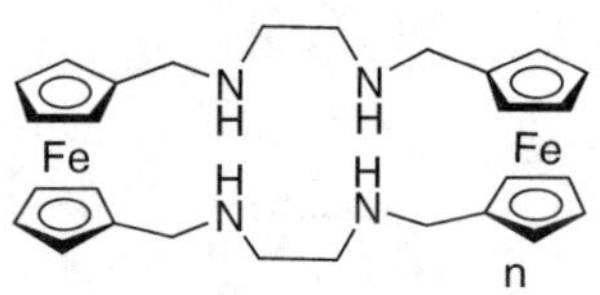

25

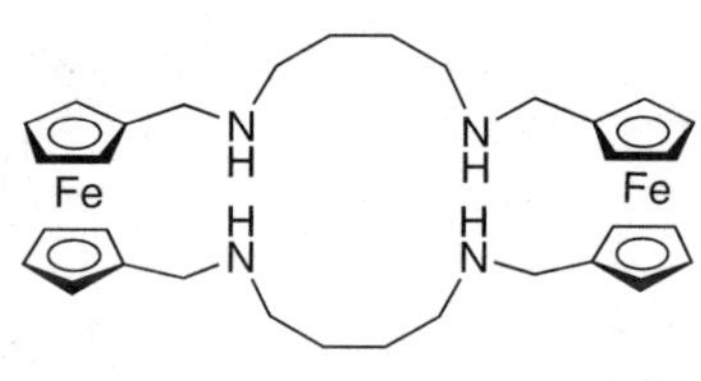

26

Recognition of fluoride in aqueous media is particularly difficult due to the strongly hydrated nature of the anion. Shinkai and co-workers have demonstrated that ferrocene-boronic acid **27** acts as a selective redox sensor for fluoride which operates in $H_2O$ [23]. The favourable interaction between boron and fluoride (a hard acid and hard base, respectively) generates a stability constant of 700 $M^{-1}$ for the fluoride-ferrocenium complex. Stability constants for both the bromide and chloride complexes are <2 $M^{-1}$.

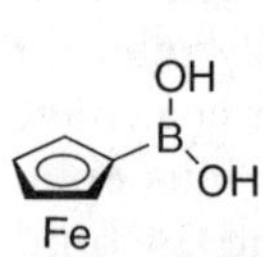

27

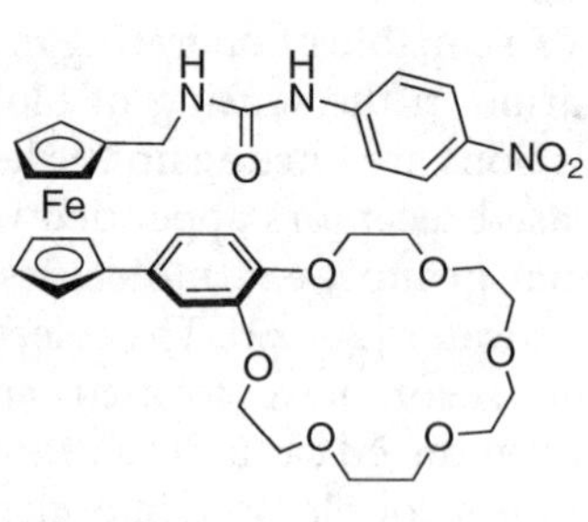

28

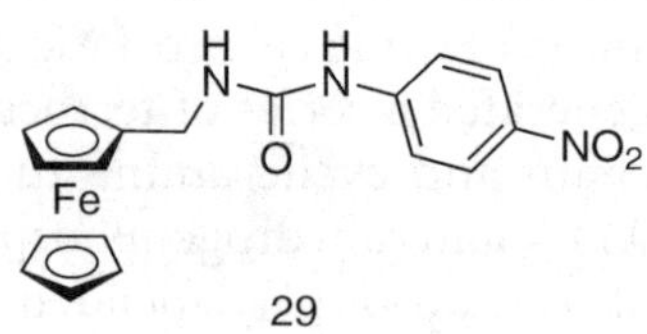

29

Tucker and co-workers have recently reported a ferrocene ion-pair receptor which acts as a chromogenic molecular switch [24]. Appended to one cyclopentadienyl ring of the ferrocene of molecule **28** is a phenyl urea unit for anion binding and to the other a crown ether for cation binding. On addition of fluoride to a solution of **28** in acetonitrile, a significant perturbation of the UV–vis spectrum was observed including the appearance of a new absorption at 472 nm. The stability constant for the 1:1 anion/receptor complex was determined as 9,340 $M^{-1}$. The addition of 10 equivalents of $KPF_6$ to the solution of **28** containing 10 equivalents of fluoride caused the complete disappearance of the 475 nm absorption. The $K_a$ value of the receptor with $K^+$ in the presence of fluoride was calculated as 1,460 $M^{-1}$. It should be noted that the model receptor **29**, lacking the cation binding site, exhibited similar switching properties. The $K_a$ value of this receptor for fluoride is equivalent (9,660 $M^{-1}$) but the $K_a$ for $K^+$ in the presence of fluoride was understandably far lower (230 $M^{-1}$). Inhibition by $K^+$ of the response of these receptors to fluoride is therefore thought to be due to the ion-pairing interaction between fluoride and $K^+$.

## 2.3
## Other Organometallic Receptors

If one of the cyclopentadienyl rings of ferrocene is replaced with an arene moiety, the Fe(II) complex becomes cationic and hence gains the potential for electrostatic interaction with anions. This concept was first explored by Beer and co-workers using the simple amido[CpFe(arene)]$^+$, **30** [25]. Simi-

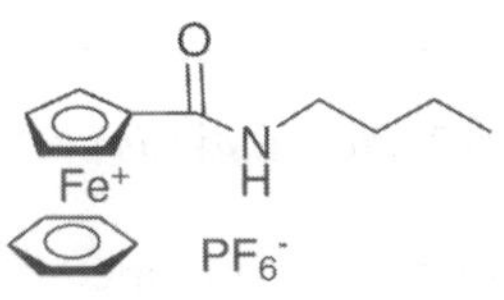


30

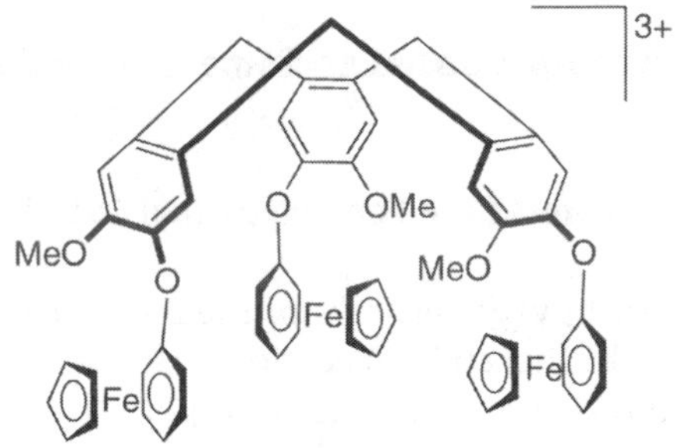


31

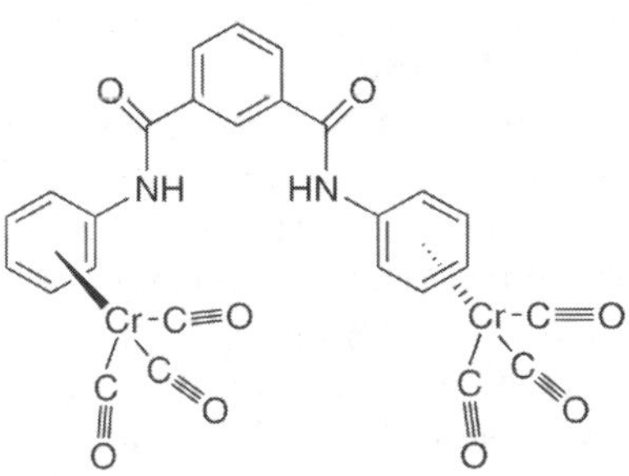


32

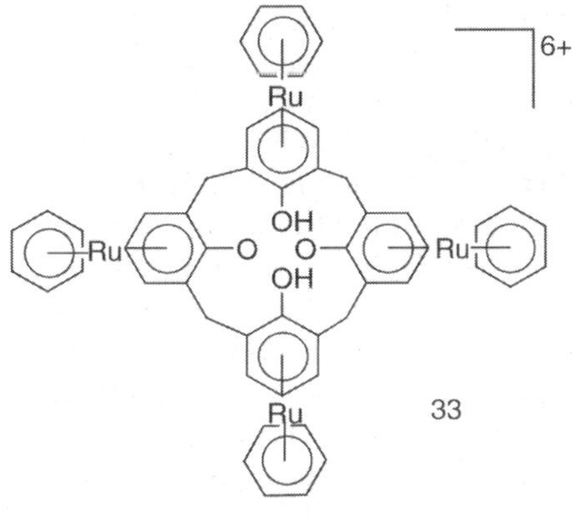


33

larly, Atwood and co-workers have derivatised a cyclotriveratrylene with the $[CpFe(arene)]^+$ unit to generate receptor **31** [26]. Qualitative NMR studies demonstrate that **30** binds chloride and bromide in acetonitrile, and **31** binds halides in acetone. It is proposed that the strength of the interaction in **31**, which does not possess amides or other H-bonding groups, is due to the arrangement of positive charges around the upper rim of the hydrophobic cavity. No electrochemical data are reported for either receptor.

In receptor **32** Gale and co-workers have exploited the electron-withdrawing properties of $Cr(CO)_3$ [27]. The coordination of this metal unit to the arene substituents of the isophthalamide increases the acidity of the NH protons and hence their affinity for anions. $^1$H NMR titrations with **32** in $CD_3CN$ provided stability constants of the same order of magnitude as those for the non-metallated receptor in the far less competitive solvent $CD_2Cl_2$.

The anion binding properties of a series of highly cationic metallated calix[4]arenes have also been investigated. Host **33**, in which the calix[4]arene is coordinated to four Ru($\eta^6$-*p*-cymene) units, was found (using $^1$H NMR titrations) to bind chloride ($K_1$=551 $M^{-1}$), bromide ($K_1$=133), iodide ($K_1$=51) and nitrate ($K_1$=49) in aqueous solution [28]. No electrochemical data were reported.

## 3 Receptors Based on Complexes of Transition Metals

### 3.1 Receptors Based on Polypyridyl Complexes of Group 8 and 9 Cations

$[Ru(bpy)_3]^{2+}$ is perhaps the most extensively studied transition metal complex in the literature. Its cationic character, chemical stability, rich electrochemistry and luminescent emission [29, 30] have led to its widespread use as a reporter group in anion sensing. This combination of properties has allowed the development of anion receptors capable of both optical and electrochemical sensing of anions, as demonstrated by a series of $[Ru(bpy)_3]^{2+}$ amide receptors **34–37** prepared by Beer and co-workers [31–33]. The single crystal X-ray structure of the chloride complex of **34** clearly indicates that the anion is bound tightly within the bpy amide ligand by six hydrogen bonds. It forms hydrogen bonds not only to the two NH groups but also to four aromatic CH groups with H–Cl distances ranging from 2.51 to 2.71 Å. $^1$H NMR titrations in DMSO-$d_6$ revealed strong binding of chloride and dihydrogenphosphate (stability constants are shown in Table 1). Four reversible redox couples (one metal-centred oxidation and three ligand-centred reductions are expected for $[Ru(bpy)_3]^{2+}$ species) were observed in electrochemical studies. Of these only the least cathodic ligand reduction was significantly shifted in the presence of anionic guests. The assignment of this redox process to the relatively electron-poor amide-substituted bipyridyl reaffirms the X-ray structure evidence that anion recognition

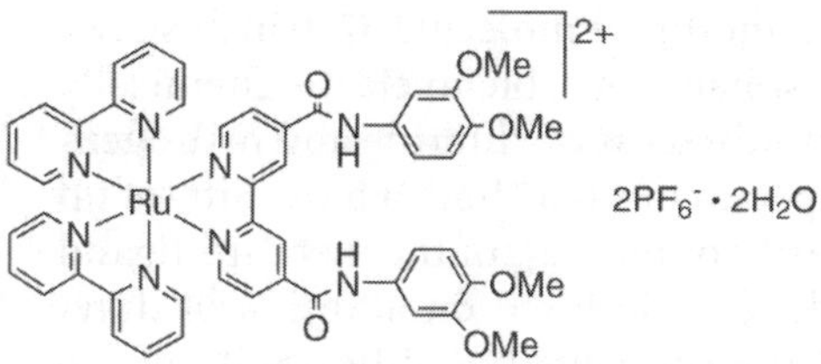

34

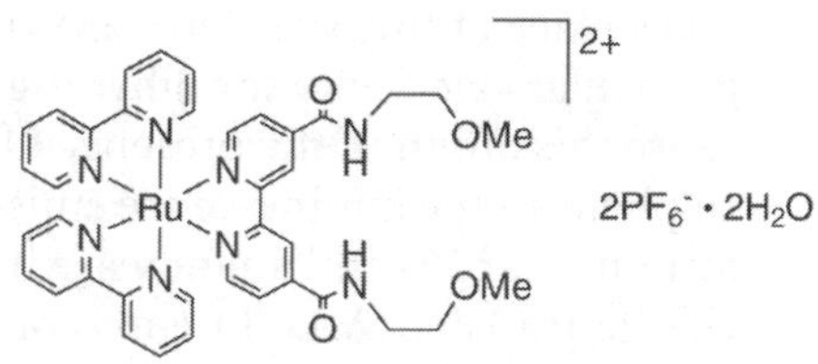

35

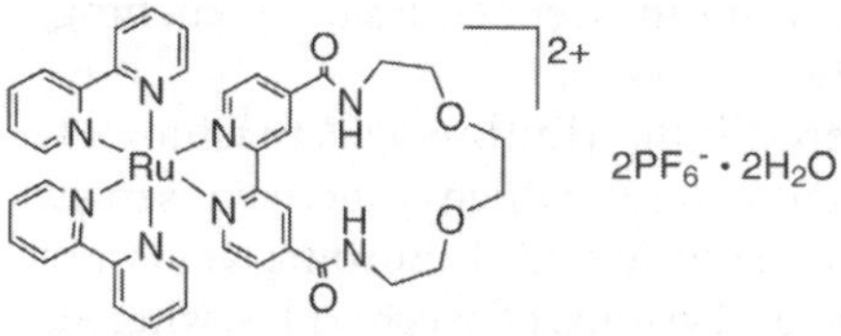

36

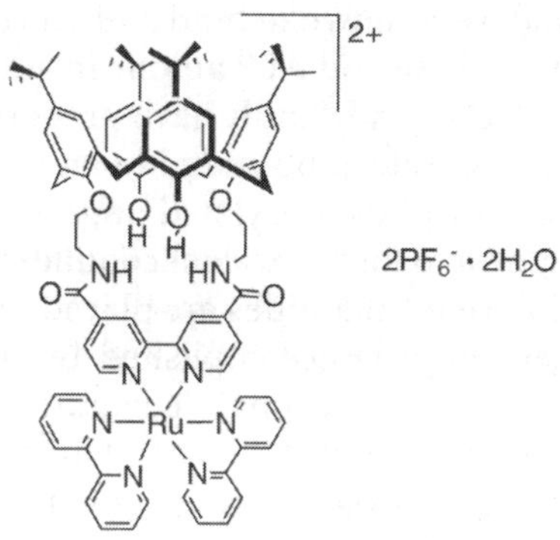

37

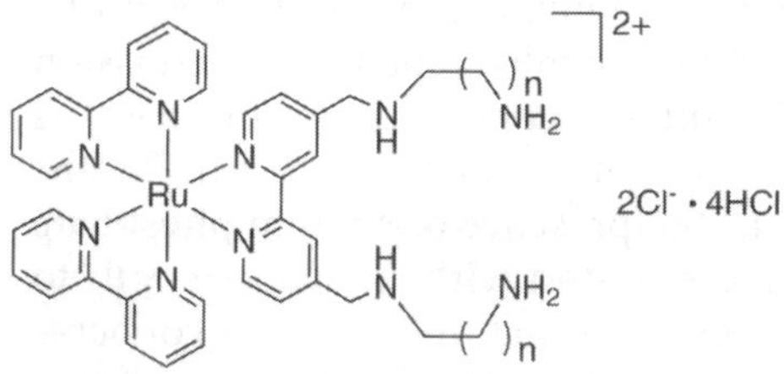

38 n = 1

39 n = 2

40 n = 3

**Table 1** Stability constant data for **34–37** with chloride and dihydrogenphosphate anions in DMSO-$d_6$ solution, errors ≤5%

| Receptor | $K$ ($M^{-1}$) | |
|---|---|---|
| | $Cl^-$ | $H_2PO_4^-$ |
| **34** | $5.0\times10^2$ | $8.0\times10^3$ |
| **35** | $1.8\times10^2$ | $1.6\times10^3$ |
| **36** | $4.2\times10^2$ | $5.6\times10^3$ |
| **37** | $1.6\times10^3$ | $2.8\times10^4$ |

takes place at this site. The calix[4]arene-modified molecule **37**, which shows particular selectivity for dihydrogenphosphate, was able to electrochemically sense this anion in the presence of a tenfold excess of chloride and hydrogensulphate. In the luminescence emission spectra of **34–37** both a blue shift (of up to 16 nm for **37**) and an increase in intensity of the $\lambda_{max}$ of the metal-to-ligand charge-transfer (MLCT) emission band were observed on addition of dihydrogenphosphate. It was proposed that the conformational flexibility of the receptors is decreased by complexation of the anion guest, thus reducing the rate of non-radiative decay through vibrational and rotational relaxation. Water-soluble receptors **38–40**, based on polyaza bipyridines, have been prepared and were shown to bind and sense, via MLCT luminescent emission quenching, phosphate and ATP anions in aqueous solution [34].

Deetz and Smith have prepared a heteroditopic $[Ru(bpy)_3]^{2+}$ receptor **41**, incorporating both amide and boronic acid groups, which selectively senses certain phosphorylated sugars in aqueous solution [35]. Boronic acids are known to form covalent complexes with the diol groups of saccharides, whereas the adjacent amides are placed to complex the anionic phosphate component. Sensing was accomplished by measuring luminescence enhancement, with fructose-6-phosphate generating the highest stability constant (log $K_a$=3.1). Non-phosphorylated saccharides gave much smaller changes in emission intensity (log $K_a$<1.2), showing that the anionic component of the guest is essential for strong binding. However, a covalent attachment between the anion and the saccharide is not required. In the presence of sodium phosphate buffer non-phosphorylated saccharides are bound with similar strength to their phosphorylated counterparts. It is reported that this apparent cooperativity is a result of favourable hydrogen bonding between the phosphate anion and the saccharide. Watanabe and co-workers have also shown that anionic and neutral phosphodiesters can be sensed in acetone by the imidazole-functionalised receptor **42** [36].

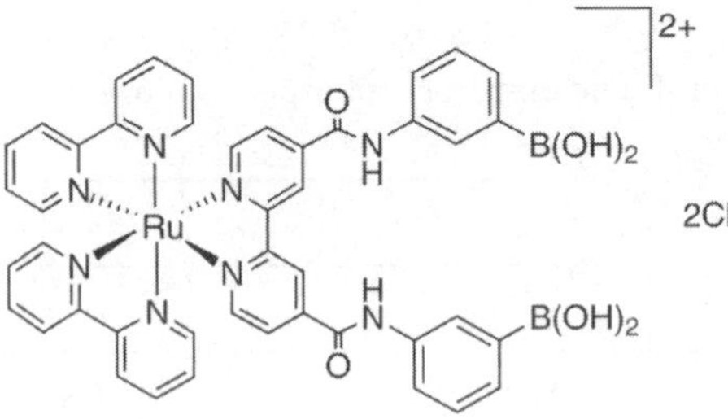


41

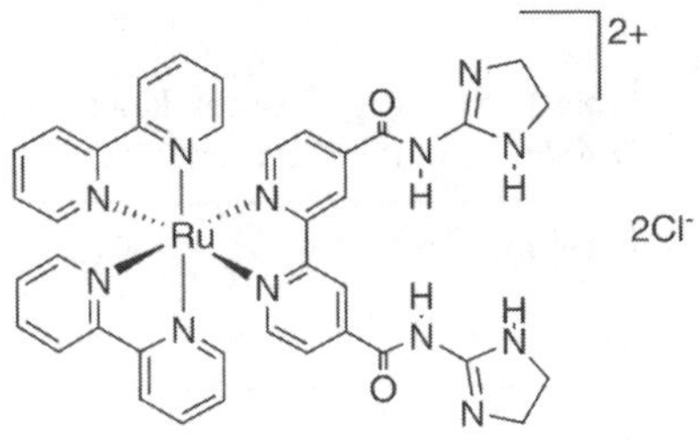


42

The Ru(II)bipyridylcalix[4]diquinone receptor **43** selectively binds and senses acetate anions (from $^1$H NMR titrations in DMSO-$d_6$ solution $K$= 9,990 $M^{-1}$) [37]. This receptor exhibited only weak luminescence because calix[4]diquinone is an electron acceptor, quenching the $[Ru(bpy)_3]^{2+}$ emission

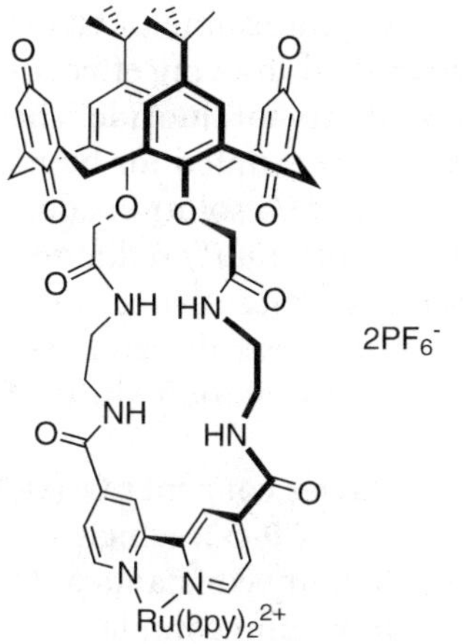


43

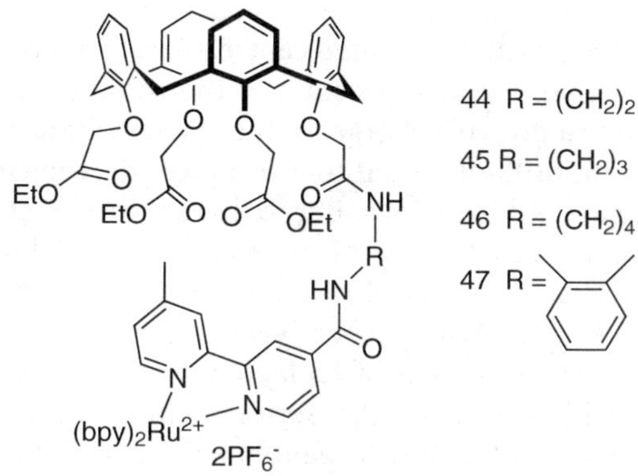


by oxidative electron transfer. Addition of acetate to acetonitrile solutions of **43** resulted in a 500% increase in luminescence intensity (60% for chloride), concomitant with a slight blue shift of the emission maximum. It is clear that the presence of the acetate anion in the binding pocket between the $[Ru(bpy)_3]^{2+}$ moiety and the quencher interrupts the oxidative electron transfer process.

The sensing of ion-pair species can be achieved by the attachment of a calix[4]arene tetraester to $[Ru(bpy)_3]^{2+}$ [38]. The oxygen-rich lower rim of the macrocycle in **44–47** was designed to act as a cation binding site. $^1H$ NMR titrations on **44–47** in $CD_3CN$ were carried out with bromide and iodide in the presence and absence of one equivalent of $Li^+$ or $Na^+$. The receptors showed a preference for bromide over iodide and significantly enhanced anion binding in the presence of the alkali metal. With iodide the greatest enhancement is with the co-bound sodium cation, which correlates with the known selectivity preference of lower-rim tetrasubstituted ethyl ester for this alkali metal.

Sessler and co-workers have recently prepared complexes **48** and **49** which are capable of selectively sensing fluoride, giving stability constants for fluoride of 12,000 and 54,000 $M^{-1}$, respectively, by UV-vis spectrophotometry in DMSO

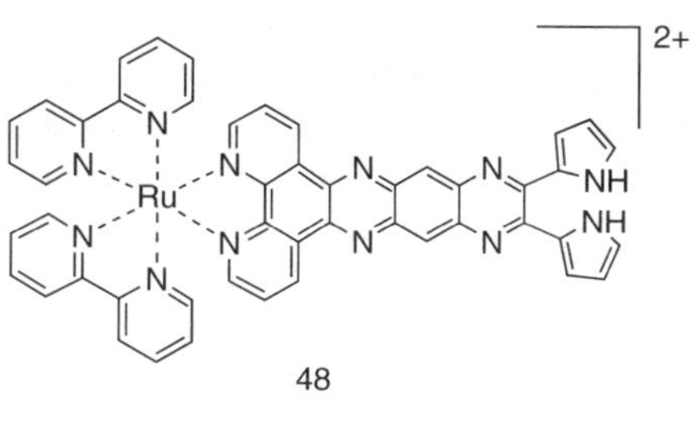


48

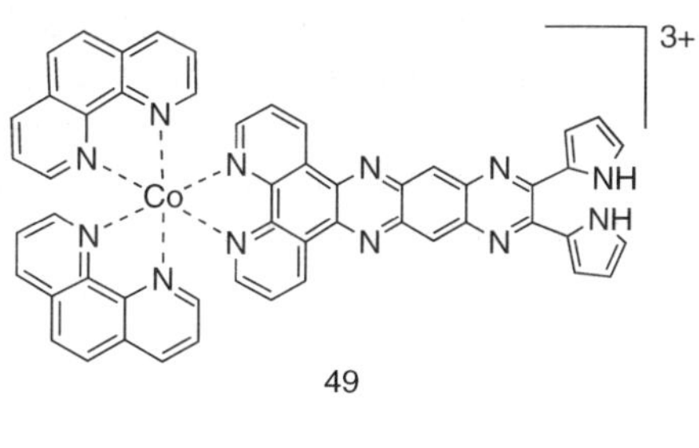


49

[39]. It is proposed that the pyrrole NH protons of the quinoxaline phenanthroline ligand are rendered more acidic by the electron-withdrawing effect of the coordinated metal centre, thereby increasing their affinity for fluoride. The higher stability constant of the cobalt(III) complex **49** is accounted for by its extra positive charge. In the cyclic voltammetry of **49** in DMSO solution addition of fluoride led to a complete disappearance of the Co(II)/(III) reduction wave. Addition of water to this solution restored this redox process, suggesting that the presence of a strongly bound fluoride anion renders the complex redox inactive. Chloride and dihydrogenphosphate produced cathodic shifts of 160 and 70 mV, respectively.

The influence of replacing the Ru(II) centre for its heavier congener Os(II) is illustrated by the series of bimetallic 'cleft' complexes **50–52**, which are selective for dihydrogenphosphate over chloride [40]. Stability constants were measured by $^1$H NMR in DMSO-$d_6$, revealing that the diosmium complex (**52**, $K$>30,000 $M^{-1}$) has the highest affinity for dihydrogenphosphate, followed by the mixed metal (**51**, $K$=22,150 $M^{-1}$) and the diruthenium (**50**, $K$=15,480 $M^{-1}$) receptors. This suggests that the Lewis acidic {$(bpy)_2Os^{2+}$} moiety is a particularly efficient centre for dihydrogenphosphate binding.

$[Ru(bpy)_3]^{2+}$ macrocycles **53–56** bind anions with similar strength to the cleft complexes **50–52** [41]. The results of $^1$H NMR studies in DMSO-$d_6$ demonstrated that **53**, with the smallest macrocyclic cavity, binds acetate preferentially to chloride, whereas the larger macrocycles **54–56** show the opposite trend. Remarkably, **53–56** were also found to bind chloride preferentially to dihydrogenphosphate. The presence of a second positively charged metal moiety leads to increased overall anion stability constants for **55** and **56**. Receptor **55** displays outstanding selectivity, with a stability constant for chloride of 40,000 $M^{-1}$ and no measurable affinity for dihydrogenphosphate. The substitution of one Ru(II) centre for Os(II) in **56** results in both an increased stability constant for the chloride complex and a diminished affinity for acetate – in effect an increase in selectivity between the two anions. The results of electrochemical measurements on the complexes in acetonitrile solution confirmed the pattern of anion selectivity, generating a maximum anion-induced cathodic shift in the first bpy redox wave of 125 mV for **56** in the presence of chloride. Complexes **53–56** were also found to sense chloride by luminescence enhancement. The rate of non-radiative decay of the $[Ru(bpy)_3]^{2+}$ excited state is thought to be lessened by the rigidifying effect of the chloride anion binding to the macrocycle.

Bis-terpyridine iridium(III) has been used as a reporter group in anion sensing. In aqueous solution isomers **57** and **58** both exhibit halide-induced luminescence quenching with selectivity for chloride [42]. Receptor **58**, although less sensitive than **57**, is reported to possess good characteristics for sensing chloride at physiologically relevant concentrations.

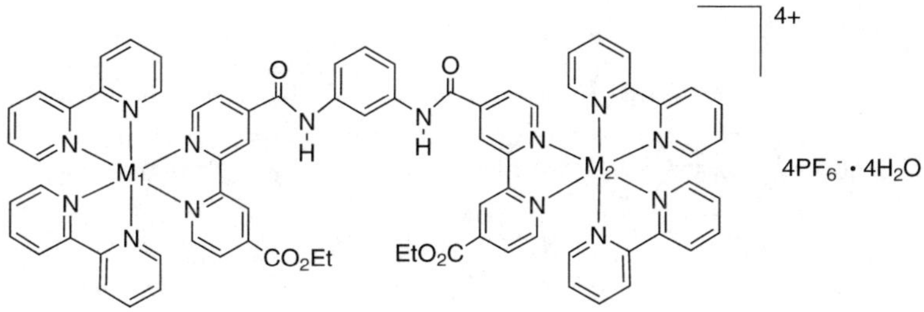

50 $M_1 = M_2 = Ru$

51 $M_1 = Ru$, $M_2 = Os$

52 $M_1 = M_2 = Os$

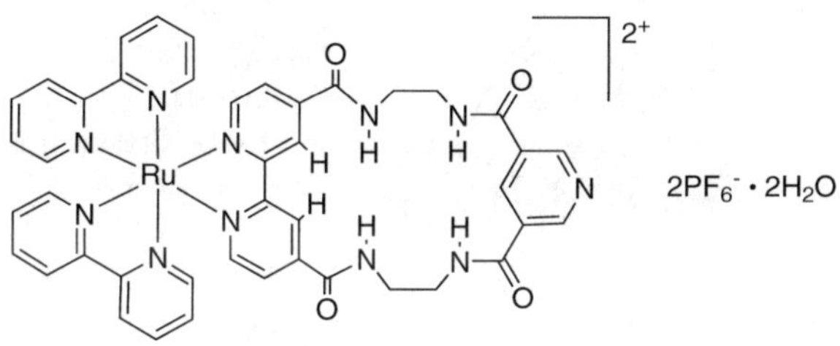

53

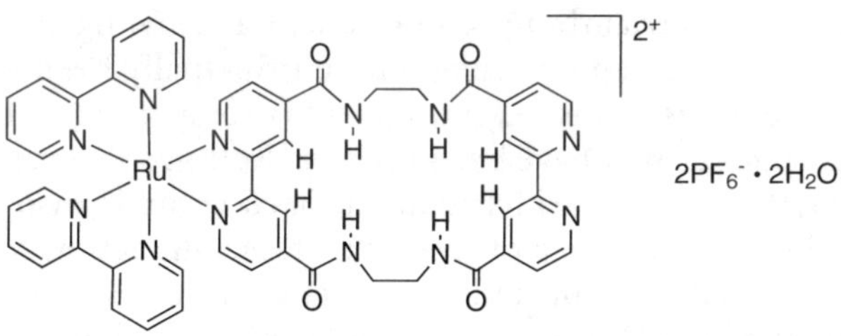

54

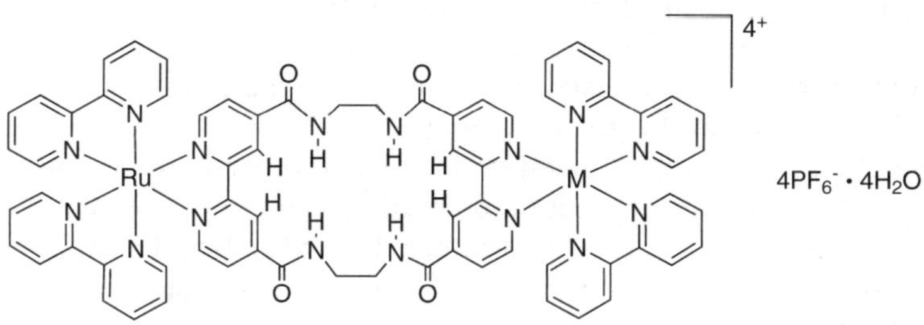

55 M = Ru
56 M = Os

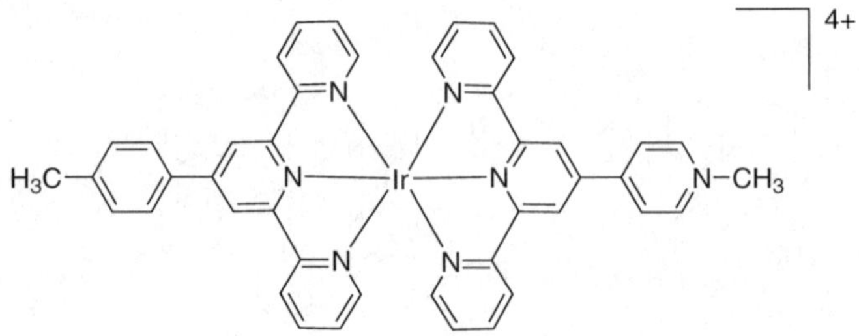

57

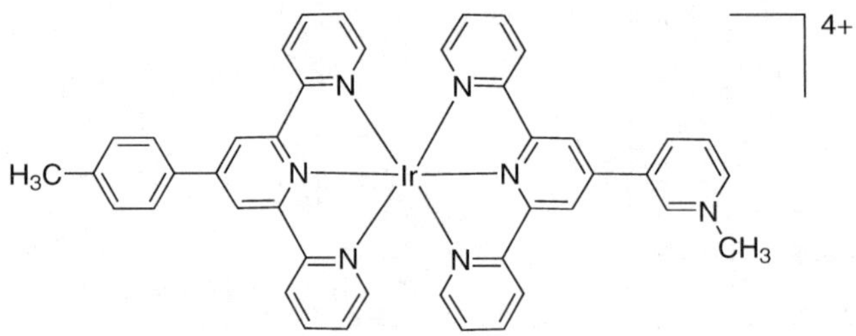

58

## 3.2
## Rhenium(I) Tricarbonyl Chloride-Based Receptors

Complexes of rhenium(I) tricarbonyl chloride with pyridyl ligands are often luminophores and hence in anion sensing have principally been used as reporter groups. In many cases this Re(I) fragment is compatible with the same bipyridyl receptor ligands that have been used in $[Ru(bpy)_3]^{2+}$ examples.

For instance **59**, the Re(I) bipyridyl analogue of receptor **43**, also selectively senses acetate anions [37]. The lack of an electrostatic interaction accounts for a significantly lower stability constant for acetate (from $^1$H NMR titrations $K$=1,790 $M^{-1}$ in deuterated DMSO solution) and hence a smaller luminescence response than its $[Ru(bpy)_3]^{2+}$ counterpart.

Receptor **60** possesses similar ion-pair sensing properties to its $[Ru(bpy)_3]^{2+}$ analogue **47** [38]. The neutral receptor naturally showed lower overall stability constants for halides, but the anion binding enhancement observed on the addition of $Li^+$ or $Na^+$ was increased, with a 60-fold enhancement in the stability constant for bromide in the presence of $Li^+$.

The Re(I) bipyridyl unit has been exploited in another series of ion-pair sensors. Molecules **61–64** incorporate crown ether components to act as cation receptors [43, 44]. $^1$H NMR titrations revealed **61–64** to be selective for acetate over chloride. In all the receptors (except **62** and **64** with acetate) enhanced binding was observed in the presence of $K^+$ cations. The degree of enhancement is lower in the xylyl-spaced molecules **62** and **64** (40–50%) than in **61** and **63** (80–110%). It is also small compared to that seen in the calix[4]arene receptor **60**.

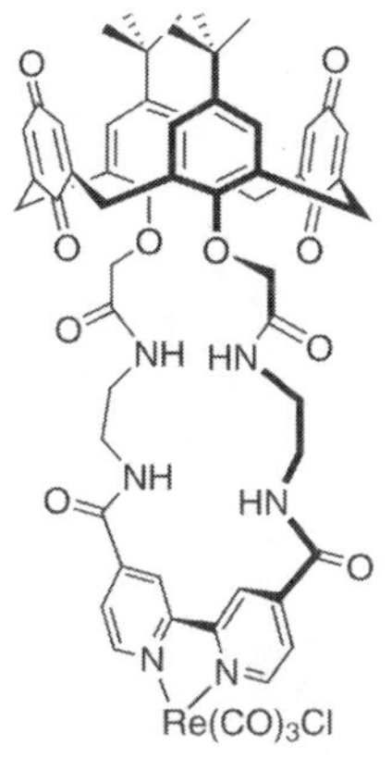

59

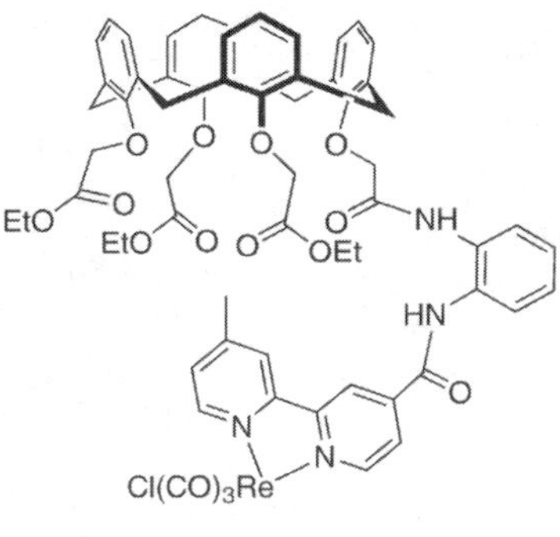

60

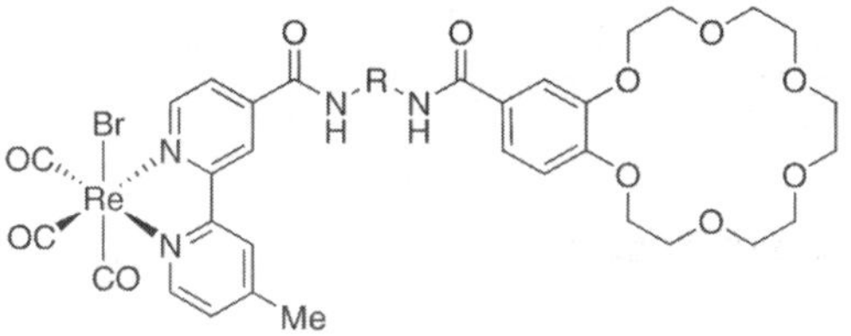

61 R = $CH_2CH_2$

62 R = 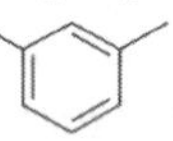

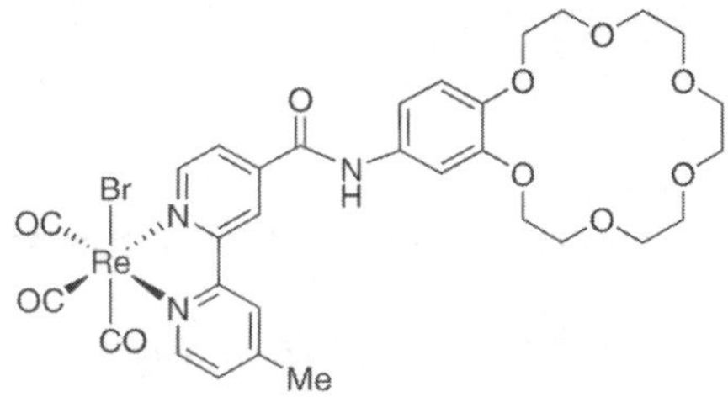

63

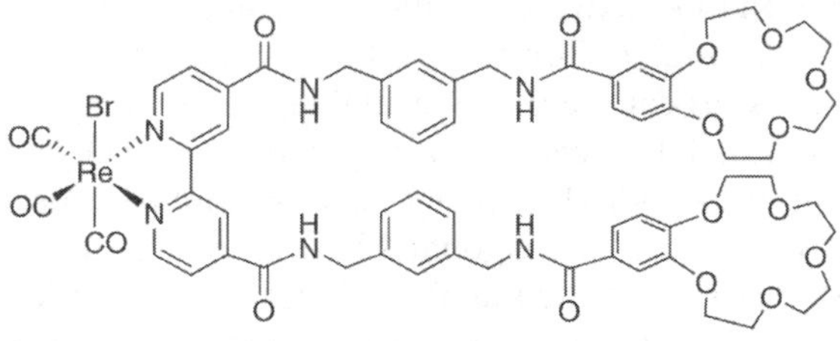

64

Lees and co-workers have investigated Re(I) bipyridyl anion hosts based on aryl bisamide skeletons **65–67** [45]. Measurement of anion-induced luminescence quenching in $CH_2Cl_2$ showed **65** to have strong binding affinities for halides, acetate and cyanide, weaker affinity for dihydrogenphosphate, and even less affinity for nitrate and perchlorate. The iso- and terephthalamide receptors **66** and **67** possess smaller stability constants for all the anions tested. It is proposed that the anion sensing efficiency of **65** is due to intramolecular hydrogen bonding holding the receptor in a 'cleft' conformation.

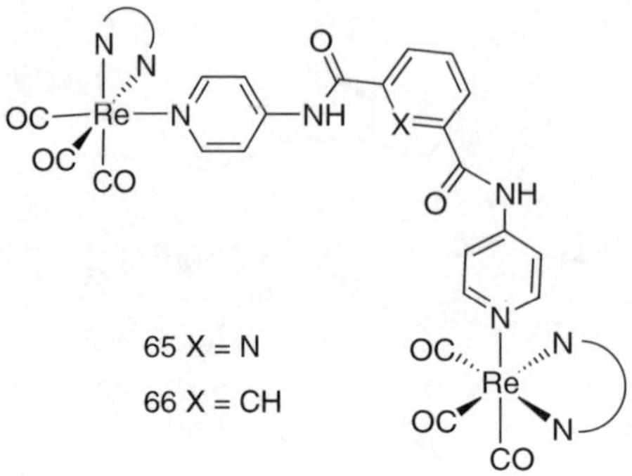

65 X = N

66 X = CH

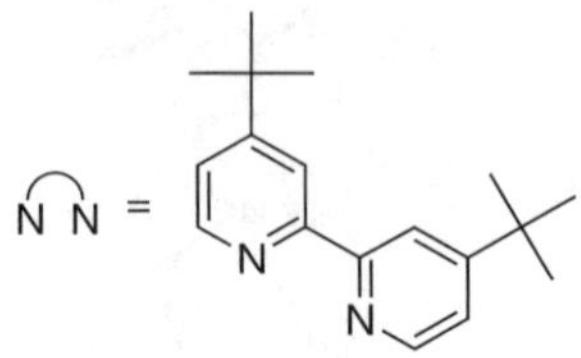

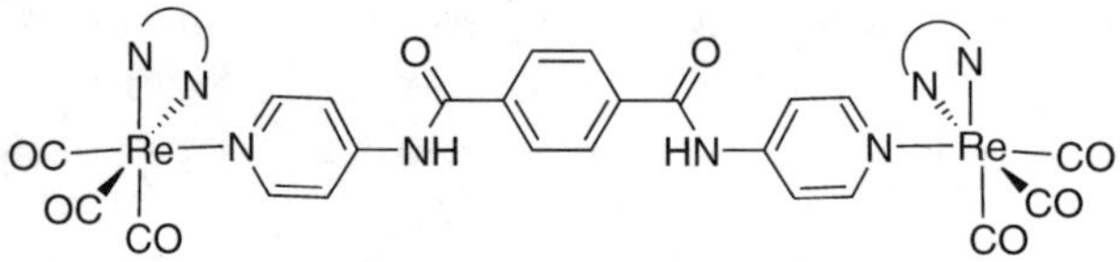

67

## 3.3 Receptors Based on Other Transition Metals (Including Group 12)

Complexes of the late transition metals are well studied in anion sensing; much of the work has been pioneered by Fabbrizzi and co-workers. As well as providing an electrochemical or optical signalling function the Lewis acid metal ion can act as a binding site for the anion. In addition, the metal ion often forms an organisational unit designed to create a receptor of a specific shape. In order to harness metal-anion interactions for anion sensing, the binding properties of the metal must be modulated by an ancillary ligand. In this way one or two vacant coordination sites can be made available for the incoming anion, and other elements appended to allow signalling or modified selectivity.

In complexes with simple tripodal amines such as **68** and **69**, zinc(II) forms five-coordinate metal complexes of trigonal bipyramidal geometry, which leaves one of the axial positions of the metal available for anion binding. The Zn(II) complex of **68** was found to undergo quenching (by photoinduced electron transfer) of the anthracene fluorescent emission in the presence of aromatic carboxylate anions such as 4-*N*,*N*-dimethylaminebenzoate in ethanol solution [46]. Complete quenching was observed at the 1:1 anion-to-receptor ratio and log *K* of the association equilibrium was measured as 5.45. Likewise the Zn(II) complex of **69** was found to form 1:1 adducts with carboxylate anions in methanol solution, with log *K* values ranging from 4 to 5 [47]. Only aromatic carboxylates induced quenching of the ligand luminescence.

The bis(boradiazaindacene)-substituted bipyridine ligand **70** is highly fluorescent in organic solvents whereas its Zn(II) complex is not [48]. It was found that the complex progressively regained its fluorescent emission when it

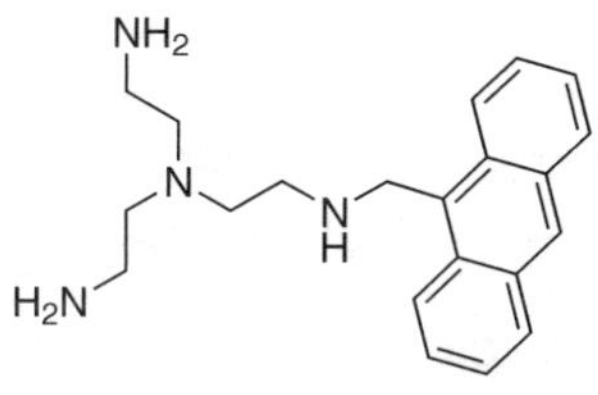

68

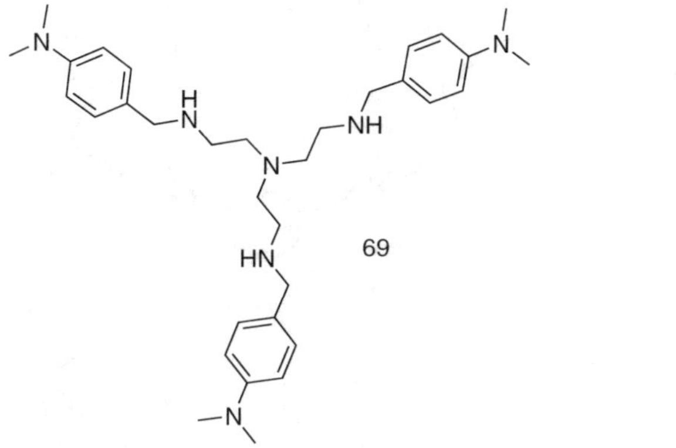

69

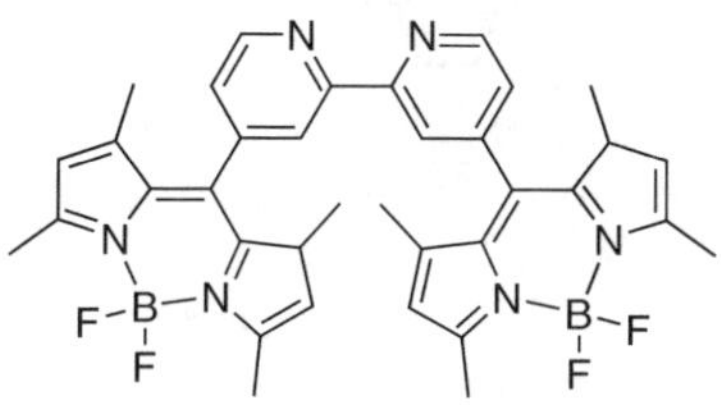

70

was titrated with various anions in acetonitrile. Stability constants for the anion–receptor complexes were calculated for fluoride (4,160 $M^{-1}$), chloride (3,230 $M^{-1}$), bromide (2,500 $M^{-1}$), acetate (4,760 $M^{-1}$) and phosphate (4,000 $M^{-1}$). Decomplexation of the chelated Zn(II) ion from **70** by the weakly coordinating anions was ruled out as the sensing mechanism. It is proposed that the quenching of the ligand luminescence by electron transfer to the Zn(II) centre is inhibited by anion coordination.

Fabbrizzi and co-workers have demonstrated the use of bis-copper(II) cryptates to sense ambidentate anions [49]. On titrating molecule **71** with $NaN_3$ in aqueous solution, the colour changed from pale blue to bright green and an anion–metal LMCT absorption appeared at 400 nm. X-ray diffraction studies have shown that the azide anion is held colinear with the two Cu(II) centres, coordinated through the two terminal $sp^2$-hybridised nitrogen atoms. Stability constants for **71** with a variety of anions in aqueous solution were calculated and the selectivity of this anion sensor was found to be controlled by the bite distance between the two copper atoms (Fig. 1).

Cryptate **72**, in which the aryl spacer of **71** is replaced with a furanyl unit, acts a colorimetric sensor for anions. UV-vis titrations in aqueous solution gave log $K$ values for the 1:1 halide/receptor adducts of 3.98 for chloride, 3.01 for bromide and 2.39 for iodide. X-ray diffraction studies confirm that bromide is held between the two copper atoms. Under the same conditions **72** also interacts strongly with azide (log $K$=4.7) and thiocyanate (log $K$=4.28) anions. This receptor is interesting because of its lack of selectivity compared to **71**. The complex appears to be able to expand and contract its bite length in order to accommodate anions of various sizes.

The use of this class of receptors in practical applications is limited by the small changes in UV-vis absorption which indicate anion binding. To over-

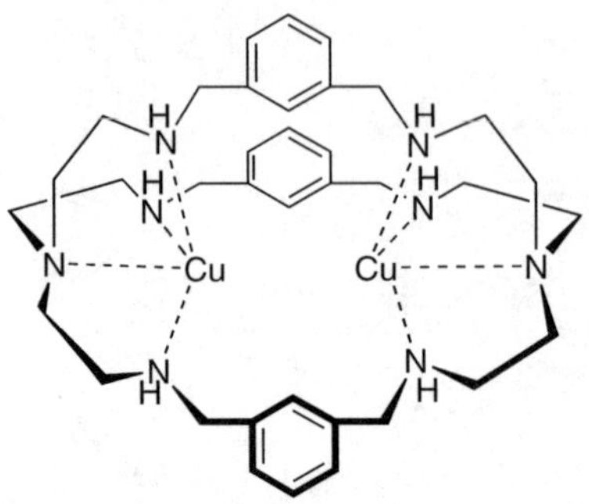

71

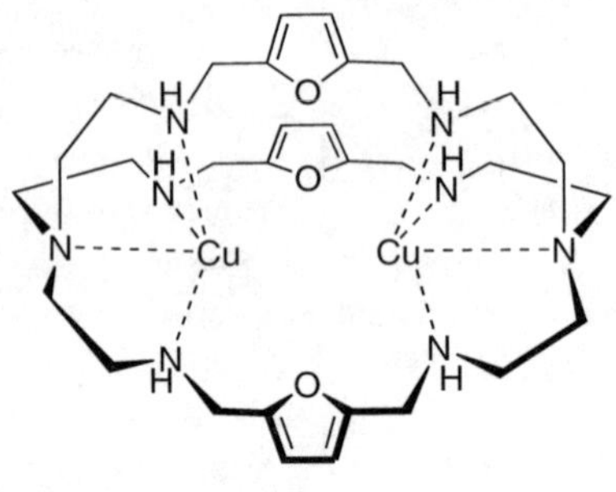

72

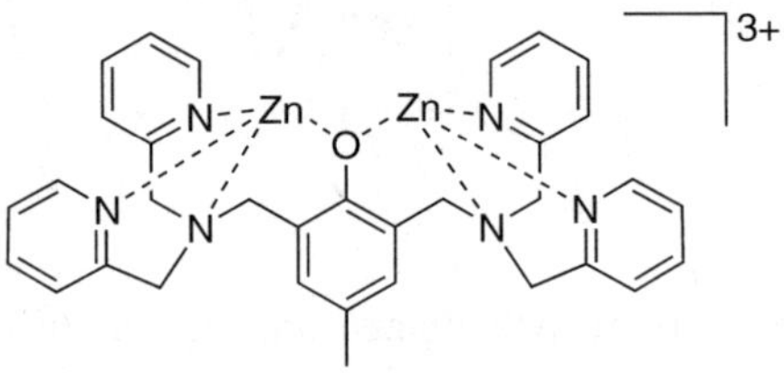

73

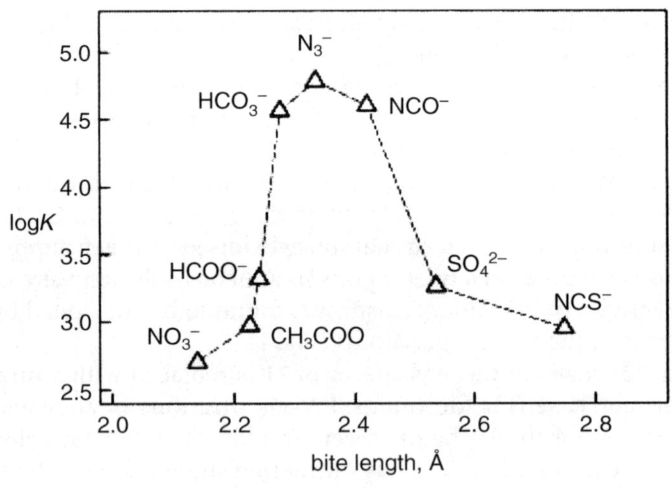


**Fig. 1** Geometrical selectivity in the recognition of polyatomic anions by **71**. Log *K* values of stability constants determined in an aqueous solution buffered at pH 8. Bite length refers to the distance of the two coordinating donor atoms of the ambidentate anion. Reproduced with permission from [49]

come this problem of sensitivity a chemosensing ensemble approach has been applied. The fluorescent indicator coumarin 343 carries a carboxylate group, which allows it to be bound by **71** in a 1:1 complex (log $K_{ass}$=4.8) with complete quenching of the luminescent emission [50]. Titration of a solution containing 0.2 mM **71** and 0.1 µM coumarin 343 with hydrogencarbonate, azide or cyanate anions resulted in complete recovery of the indicator luminescence. Anions with a lower affinity for the receptor were unable to displace the coumarin 343 and produced only a slight luminescence enhancement. The usefulness of this chemosensing ensemble was demonstrated by the quantitative determination of carbonate in mineral water. Using the same principle Han and Kim have recently reported a chemosensing ensemble made up of the dizinc complex **73** and pyrocatechol violet which is selective for phosphate [51].

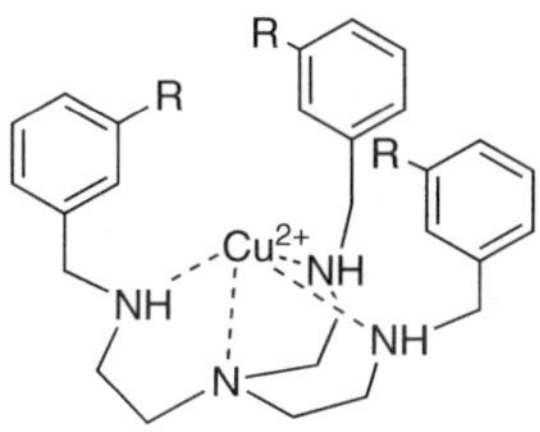


74 R = $CH_2NH_3^+$

76 R = H

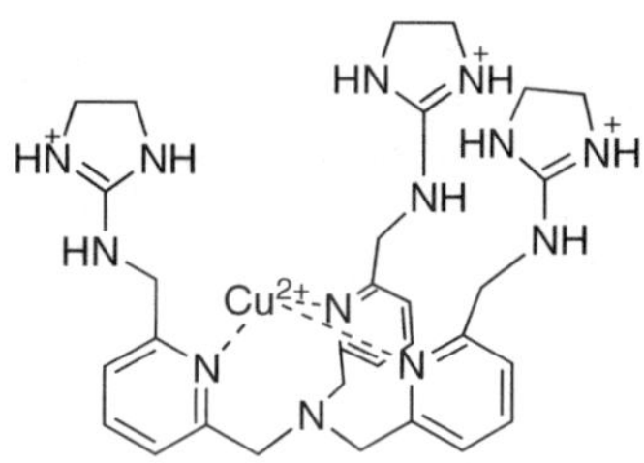


75

Anslyn and co-workers have developed a series of tripodal Cu(II) complexes **74** and **75** in which the metal ion and three cationic organic groups form a tetrahedral cavity designed to host phosphate [52]. Receptor **74** is built from the tris(2-ethylamino)amine skeleton with appended benzylamine groups. UV-vis anion titrations were carried out in aqueous solution at pH 7.4 where it can be assumed the terminal amines are all protonated. Hydrogenphosphate and its congener hydrogenarsenate were found to bind strongly in a 1:1 anion/host ratio, both with a log *K* value of 4.40. Perrhenate was bound an order of magnitude less strongly and the affinity for chloride was too small to measure. Model complex **76**, which has no ammonium groups, gave a log *K* value of 2.95 with hydrogenphosphate, indicating that the Cu(II)-anion interaction contributes significantly to anion binding. Receptor **75** follows the same design principle, this time incorporating guanidinium binding units with a tris-[(2-pyridyl)methyl]amine skeleton. Log *K* values for hydrogenphosphate and hydrogenarsenate were found to be 4.18 and 4.23, respectively. Other anions, including perrhenate, had no significant affinity for **75**. It is apparent that the guanidinium groups are responsible for the improved selectivity of this receptor for phosphate. In a separate study the driving force for hydrogenphosphate binding was found to be entropic for receptor **74**, but enthalpic for receptor **75** [53]. Partnered with the colorimetric indicator 5-(and 6)-carboxyfluorescein, receptor **75** provides an effective chemosensing ensemble for the determination of inorganic phosphate in serum and saliva [54].

Cadmium(II), the heavier congener of zinc(II), can also act as a coordination site for anion binding. Mizukami et al. employed a novel approach with receptor **77**, by covalently attaching the coumarin indicator group of a chemosensing ensemble to a macrocyclic Cd(II) anion receptor [55]. It is designed such that anions will displace the nitrogen of the aromatic fluorophore from the coordination sphere of the Cd(II) centre. In aqueous solution as the receptor was titrated with pyrophosphate, the luminescent emission of the molecule was observed to shift gradually from 342 to 383 nm. Receptor **77** shows a high degree of selectivity for pyrophosphate, citrate, ATP and ADP with log $K$ values in the range 4–5. It worthy of note that the Zn(II) analogue of **77** was found to be ineffective for anion sensing.

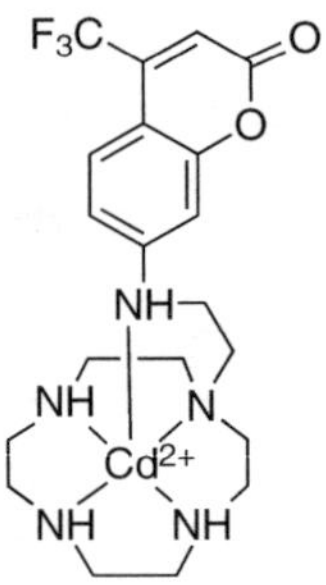


77

An interesting development has been the self-assembly of metallo-dithiocarbamate macrocyclic receptors for electrochemical anion sensing. The naphthyl-based Cu(II) macrocycle **78** displays a cathodic shift of 85 mV in the Cu(II)/(III) redox couple in the presence of hydrogenphosphate or perrhenate, but gives no response to halides [56]. It is thought that, like the Cu(II) cryptate systems, selectivity is controlled by cavity size. Dithiocarbamate receptors incorporating thiourea and hydrogen-bond donor groups **79–82** have also been investigated [57]. UV-vis titrations in DMSO indicated that the Ni(II) macrocycle **79** forms strong complexes with acetate with 1:1 stoichiometry, giving a log $\beta_1$ value of 5.36. Hydrogenphosphate, however, was found to bind in a 2:1 anion/host ratio with a log $\beta_1$ of 4.10 and a log $\beta_2$ of 9.07. Electrochemical measurements on the Cu(II) macrocycle **80** revealed cathodic shifts in the Cu(II)/(III) couple of up to 160 mV with hydrogenphosphate. The open-chain thiourea complex **82** preferentially binds carboxylates with typical log $\beta_1$ values of 6. The electrochemical anion sensing properties of dithiocarbamate cryptands **83** and **84** in dichloromethane solution have also been investigated [58]. The irreversible Co(IV)/Co(III) redox couple of the complexes was found to undergo significant anion-induced cathodic perturbation: up to a maximum of 125 mV for **84** with dihydrogenphosphate.

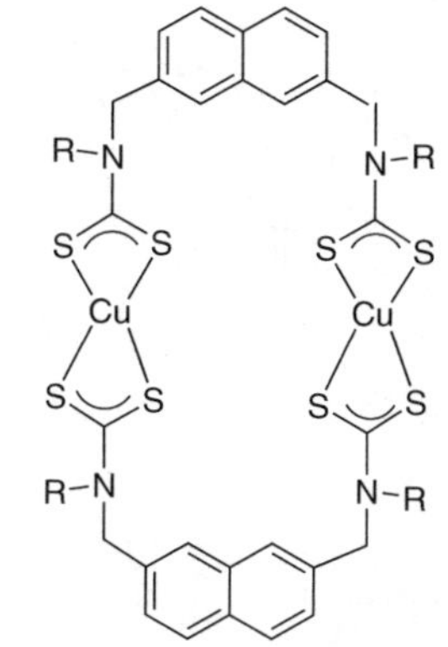

R = $^{n}$butyl

78

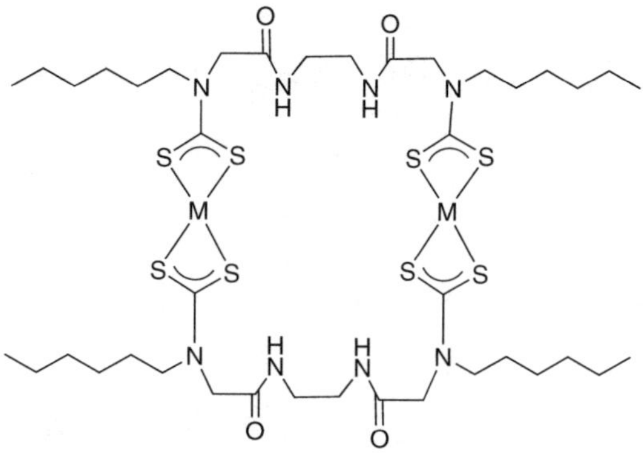

79 M = Ni

80 M = Cu

81 M = Zn

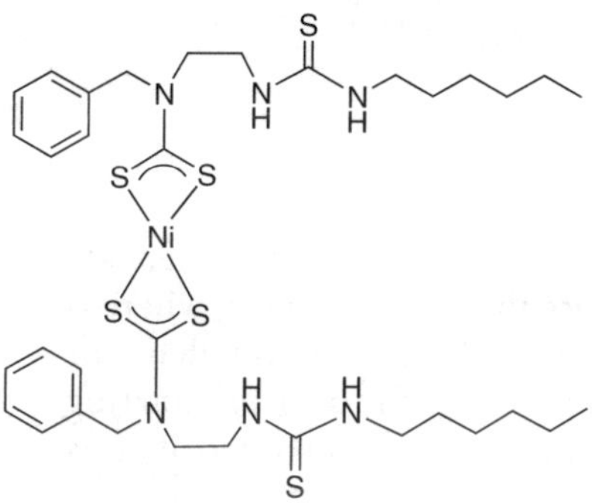

82

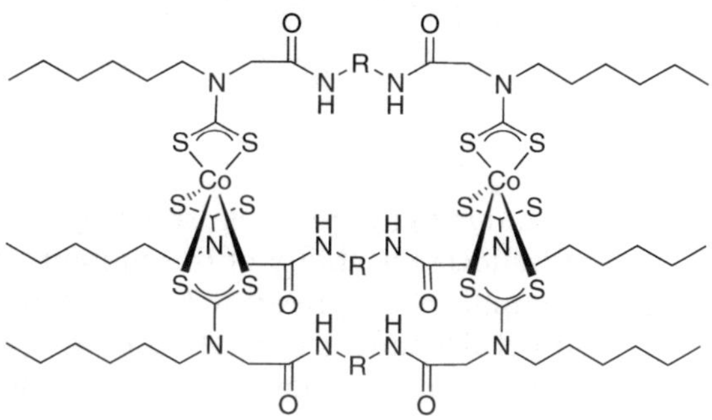

83 R = *o*-phenyl

84 R = *m*-phenyl

## 3.4 Receptors Based on Lanthanide(III) Complexes

Given their favourable luminescence properties and propensity to coordinate oxy-anions, it is perhaps surprising that complexes of lanthanide(III) ions have received comparatively little attention as anion sensors.

Parker and co-workers have reported a novel method for the selective detection of carboxy anions by time-delayed luminescence using the Eu(III) and Tb(III) complexes of **85** [59]. Luminescence measurements on the coordinatively unsaturated complexes in aqueous solution showed significant increases in lifetime and emission intensity in the presence of anions. This behaviour is consistent with the anions displacing water from the non-ligated coordination sites at the metal centre. Studies allowed the number of water molecules ($q$) remaining coordinated in the presence of added anions to be estimated. For the triflate salt of the Eu(III) receptor $q$=2.14, whereas with hydrogencarbonate

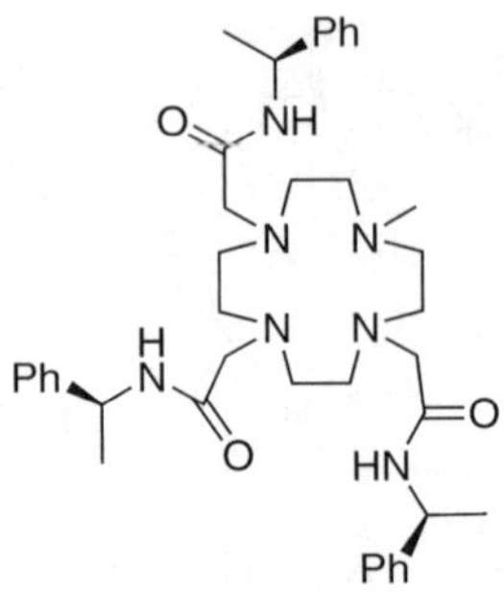

85

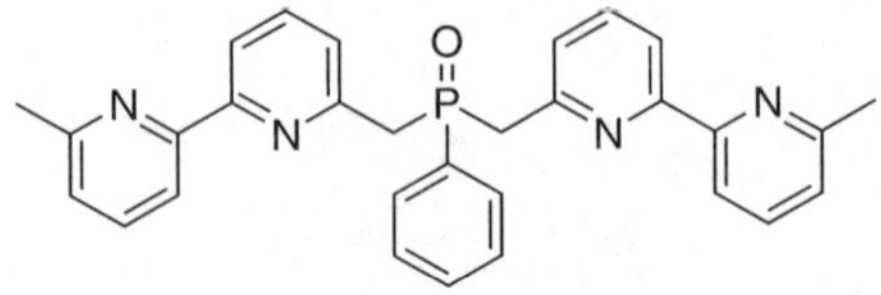

86

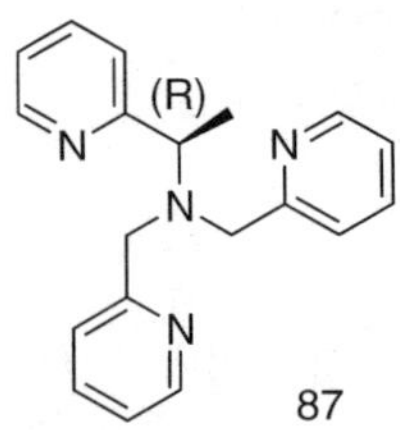

87

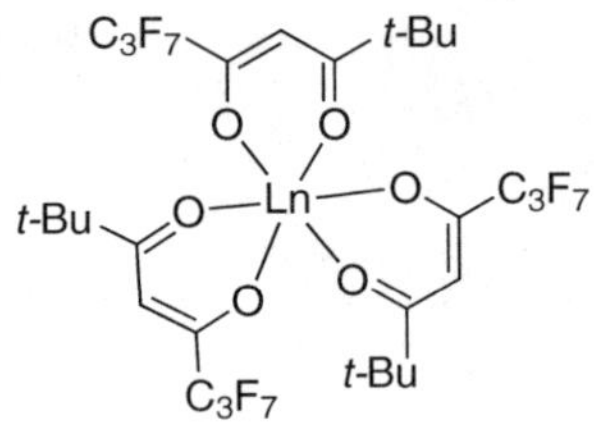

88

$q$=0.34 and with hydrogenphosphate $q$=0.74. The selectivity for hydrogencarbonate reflects the ability of this anion to chelate the Eu(III) centre, whereas hydrogenphosphate prefers to bind in a monodentate fashion. This behaviour is also pH-dependent (p$K_a$ $HCO_3^-/H_2CO_3^-$=6.38) so that for a given starting bicarbonate concentration, a pH-dependent lifetime and emission intensity was observed [60].

Other groups have subsequently reported anion receptors that work on the same principle. For instance, an Eu(III) complex of the bis-bipyridinephenylphosphine oxide ligand **86** made by Ziessel and co-workers is able to sense anions by luminescence enhancement in acetonitrile, with stability constants which follow the trend fluoride>acetate>chloride>nitrate [61]. Tsukube and co-workers have investigated the properties of the Eu(III) and Tb(III) complexes of the chiral ligand **87** [62]. Anion binding was assessed by profiling luminescence enhancement in acetonitrile, and it was found that the different metal centres provided different selectivities. The emission at 548 nm of the Tb(III) complex was increased by 5.5 times in the presence of 3 equivalents of chloride compared to 2.2 for nitrate and 1.1 for acetate. Conversely the emission at 618 nm of the Eu(III) complex was increased 8.3 times by 3 equivalents of nitrate, 2.5 times for chloride and 1.0 times for acetate. Stability constants were not reported.

The same group most recently reported the use of neutral lanthanide(III) tris-diketonates of type **88** for the determination of chloride [63]. The response in luminescence of the Eu(III) complex for chloride in acetonitrile solution was large enough to be seen by the naked eye. Incorporation of the complexes in

PVC membrane electrodes allowed measurement of potentiometric selectivity coefficients. These showed the Eu(III) complex to be more selective for chloride than the Pr(III), Dy(III) or Yb(III) analogues.

## 4 Mixed-Metal Receptors

Combining the different properties of the types of metallo-receptor described above can lead to added anion-sensing functionality. For instance, the incorporation of phosphine groups into ferrocene-amide receptor **89** allowed the coordination of various transition metals to generate mixed-metal complexes **90–92** [64]. $^1$H NMR titrations were carried out and the stability constants of these receptors with various anions in $CD_2Cl_2$ were calculated. The neutral molecules **90** and **91** were found to bind chloride approximately an order of magnitude more strongly than the parent phosphine (**89**). The complexes of the metallated molecules with chloride, bromide, iodide and hydrogenphosphate were also found to be more stable, but to a lesser extent. These results demonstrate that the presence of a Lewis acid metal centre enhances anion binding. The stability constants of cationic complex **92** for the same anions was found to be an order of magnitude higher again, illustrating once more the influence of electrostatic attraction. In the electrochemistry of the complexes with chloride, bromide and hydrogensulphate in acetonitrile/dichloromethane solution, significant cathodic shifts of both the ferrocene/ferrocenium couple and oxidation wave of the metal were observed.

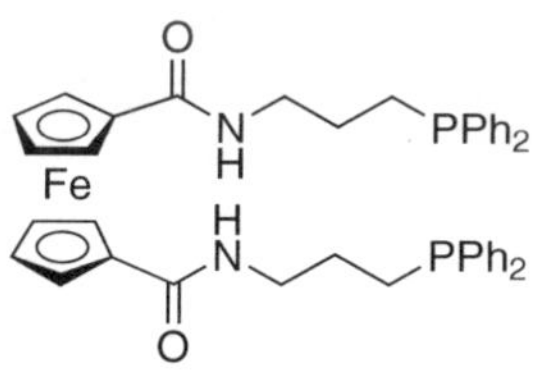


89

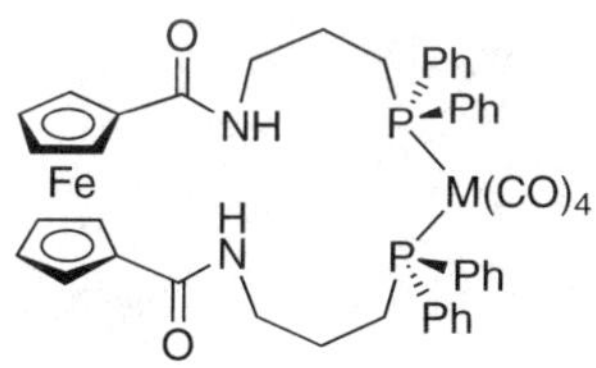


M = Mo, Cr

90

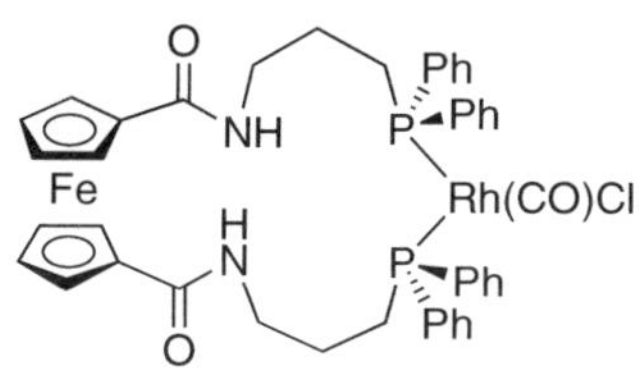


91

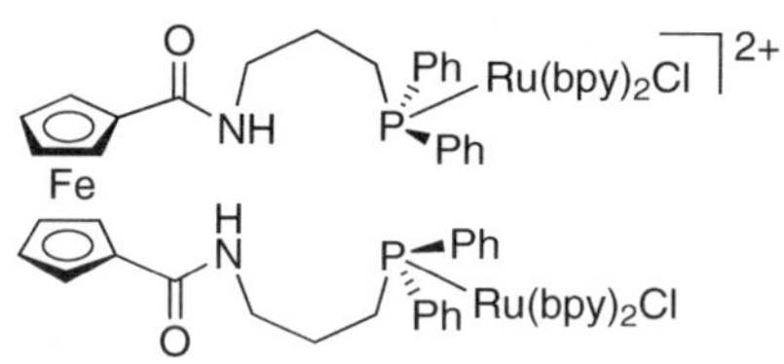


92

Receptor **93** incorporates a zinc porphyrin backbone with four ferrocene amides [65]. This shares the design of the cobaltocenium receptor **4**, except that now a zinc atom occupies the centre of the porphyrin. The Lewis acid metal centre provides an additional binding site for anion recognition. In dichloromethane solution no significant anion-induced shifts in the $^{1}$H NMR signals of the amide protons were seen in the free-base precursor of **93**, whereas the

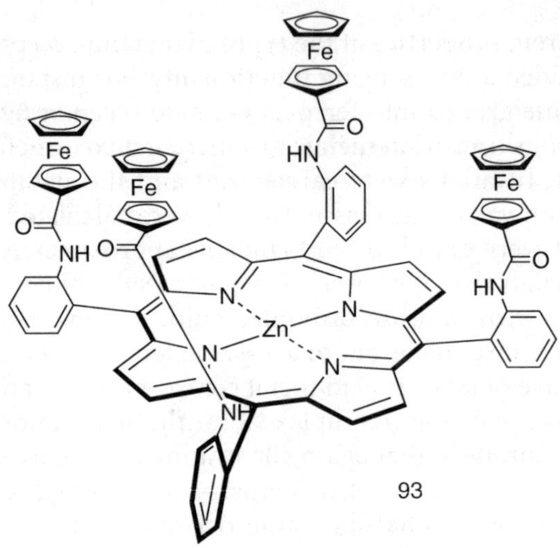

93

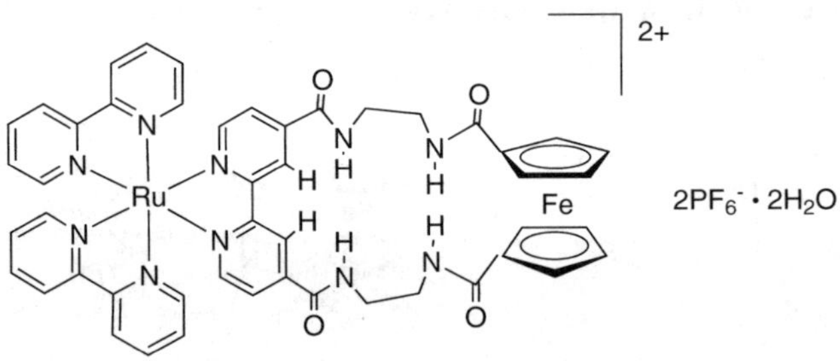

94

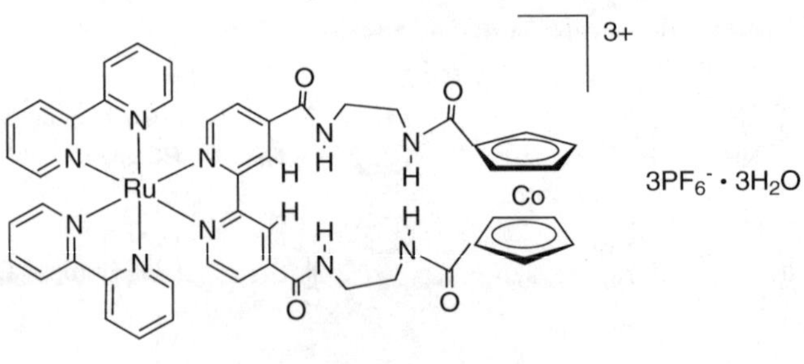

95

metalloporphyrin binds bromide ($K$=6,200 $M^{-1}$), nitrate ($K$=2,300 $M^{-1}$) and hydrogenphosphate ($K$=2,100 $M^{-1}$). Electrochemical studies in 3:2 dichloromethane/acetonitrile revealed anion-induced cathodic shifts in both the porphyrin ($\Delta E$=85–115 mV) and tetraferrocene ($\Delta E$=20–60 mV) oxidation waves. The trend in magnitude of $\Delta E$ for the porphyrin oxidation wave is hydrogensulphate>chloride>bromide>nitrate, reflecting the polarising power, i.e. charge-to-radius ratio of the anionic guest species. Atropisomers of **93** (other than the $\alpha,\alpha,\alpha,\alpha$-atropisomer) were also studied. These displayed different selectivity and overall lower stability constants.

Macrocycles **94** and **95** incorporate the $[Ru(bpy)_3]^+$ moiety with a bridging ferrocene or cobaltocenium unit, respectively [41]. In DMSO-$d_6$ solution these receptors were found to exhibit similar anion sensing properties to their bis-Ru(II) and Ru(II)/Os(II) analogues **55** and **56**, with stability constants of comparable magnitude and selectivity for chloride over acetate and dihydrogenphosphate. The $[Ru(bpy)_3]^+$ luminescence emission quantum yield, however, is about five times lower for **94** and **95** in acetonitrile solution, indicating intramolecular quenching due to the appended organometallic unit. For **95** this results in a 100% increase in emission intensity in the presence of chloride. The ferrocene/ferrocenium and cobaltocene/cobaltocenium redox couples of **94** and **95** were determined to be more sensitive to anion binding than the bpy redox couple in the same receptor. For instance, chloride induces a cathodic shift of 60 mV in the ferrocene/ferrocenium couple of **94** in acetonitrile solution, whereas the bpy couple changes by only 40 mV.

The mixed Re(I)/Pd(II) molecular square **96** has been found to sense perchlorate in acetone (giving $K$=900 $M^{-1}$) by enhancement of the Re(I) luminescent emission [66]. In this case luminescence quenching by oxidative transfer to the Pd(II) ion is inhibited by the bound anion. The Pd(II) ion also plays a role as a structural element and charge carrier. Squares **97–101** are very similar, but incorporate a bis-phosphinylferrocene-supporting ligand [67]. Again the normally strong luminescence of the Re(I) component is partially quenched by the bimetallic corners. Binding studies of the squares and different inorganic anions were carried out by luminescence titrations in acetone solution. Of the anions investigated, viz., perchlorate, acetate, trifluoroacetate, hexafluorophosphate and tetrafluoroborate, only the latter two induced significant changes in luminescence. As these anions were added an initial decrease in emission intensity was followed by an increase to a plateau. This is taken to indicate the presence of two competing quenching pathways which are inhibited to different extents by anion binding.

A very unusual mixed-metal receptor has been prepared by Jurkschat and co-workers [68]. Macrocycle **102** comprises two ferrocene reporter units linked together by two Lewis acidic organotin spacers. Electrochemical measurements in dichloromethane solution show anion-induced cathodic shifts of the ferrocene/ferrocenium redox couple of 130 mV for chloride, 210 mV for fluoride and 480 mV for dihydrogenphosphate.

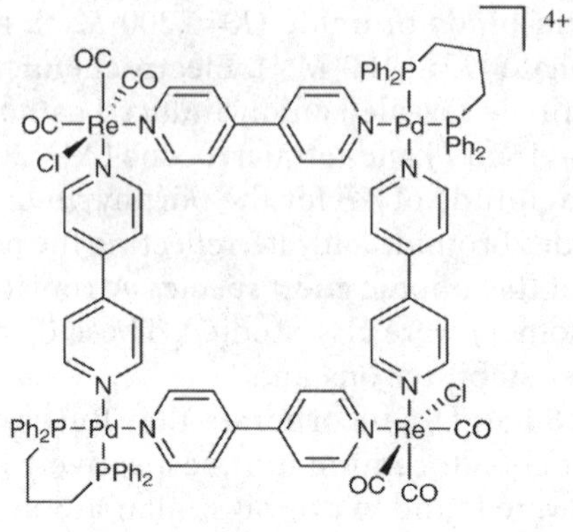

96

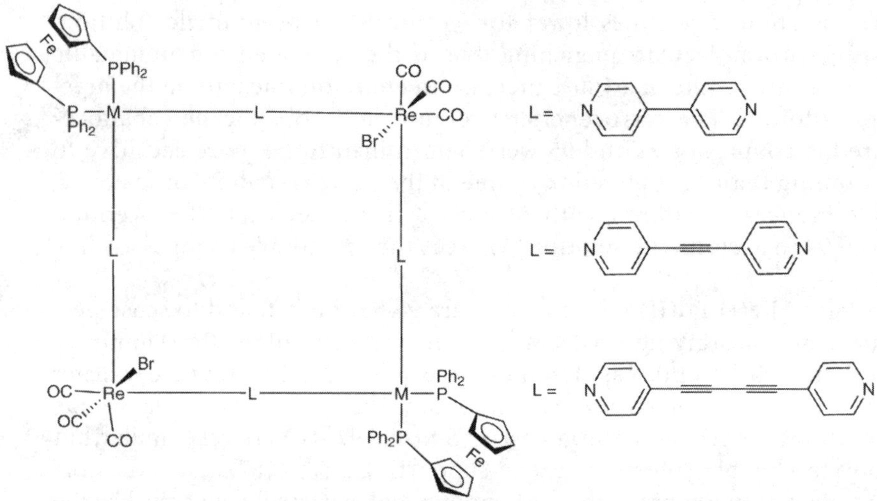

97 M = Pd
98 M = Pt
99 M = Pd
100 M = Pd
101 M = Pt

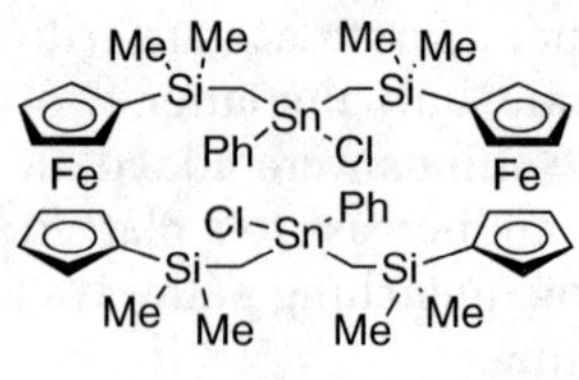

102

# 5
# Metallo-Dendrimers, Functionalised Nanoparticles and Surface Effects

A recent significant advance in the development of metal-based anion sensors has been their incorporation into nanostructured surfaces. Preorganised groups of closely spaced metallo-receptors, e.g. at the periphery of a dendrimer molecule or gold nanoparticle, have been found to exhibit significantly enhanced anion binding. The attachment of redox-active metallo-receptors to electrode surfaces allows similarly augmented anion binding and represents a potential route for the fabrication of anion-sensing devices.

## 5.1
## Metallo-Dendrimers

Astruc and co-workers are pioneers in this area and have synthesised dendrimers **103–105**, incorporating up to 18 amido-ferrocene units, which are able to electrochemically sense anions in dichloromethane solution [69]. In the higher dendrimer the selectivity trend is dihydrogenphosphate>hydrogensulphate>chloride>nitrate. Evidence of a dendritic effect was observed in the redox response of the consecutive dendrimer generations in the presence of dihydrogenphosphate or hydrogensulphate. As the number of amido-ferrocene units increases, so does the magnitude of the cathodic shift in the ferrocene/ferrocenium couple. Calculated stability constants for the hydrogensulphate complexes of generations **104** and **105** in dichloromethane solution are reported to be 9,390 and 216,900 $M^{-1}$, respectively.

Kaifer and co-workers have produced a similar dendrimer series based on a commercial DSM polyamine core with 4, 8, 16 and 32 peripheral ferrocenyl

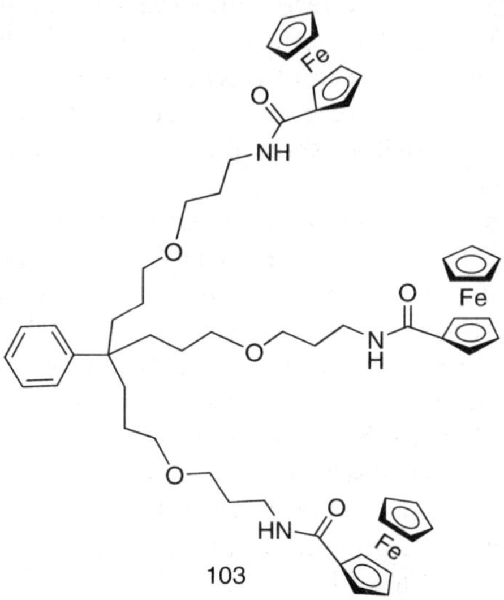


103

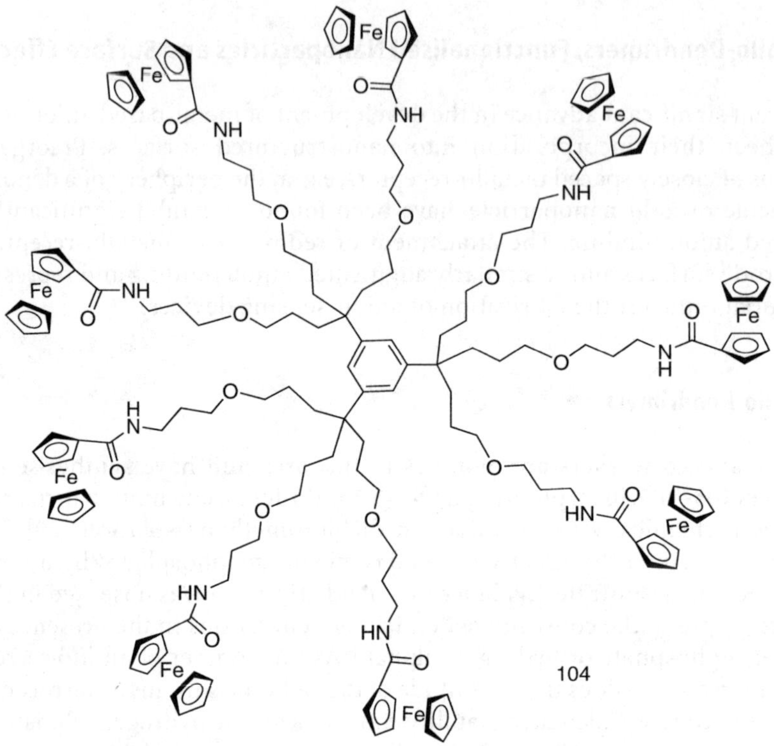

104

urea groups as the anion-sensing component [70]. No stability constants are reported, but cathodic shifts in the redox response of the ferrocene/ferrocenium couple with various anions in DMSO show a similar selectivity trend to **103–105**. In this case the data suggest that two ferrocene urea arms bind a single dihydrogenphosphate anion. The dendritic effect was observed in the change from the first (four ferrocene units) to second (eight ferrocene units) generation dendrimers. No further increase in response was seen in the third generation and the fourth generation dendrimer showed a decreased response, presumably due to steric crowding.

Astruc and co-workers have recently investigated five generations of pentamethylamidoferrocene dendrimers using the DSM polyamine core [71]. The pentamethyl-substituted ferrocene was chosen to overcome the irreversible electrochemistry and electrode adsorption occurring with **103–105**. In this new series the dendritic effect seen in the electrochemistry in DMF solution varied according to the anion studied. In changing from lower to higher dendrimer generations, modest increases in the anion-induced cathodic shift of the ferrocene/ferrocenium couple were observed with dihydrogenphosphate, whereas with $ATP^{2-}$ anion binding progressed from weak to relatively strong. This is perhaps due to the binding stoichiometry of the $ATP^{2-}$ anion being 1:2 anion/ferrocene unit compared to 1:1 for dihydrogenphosphate.

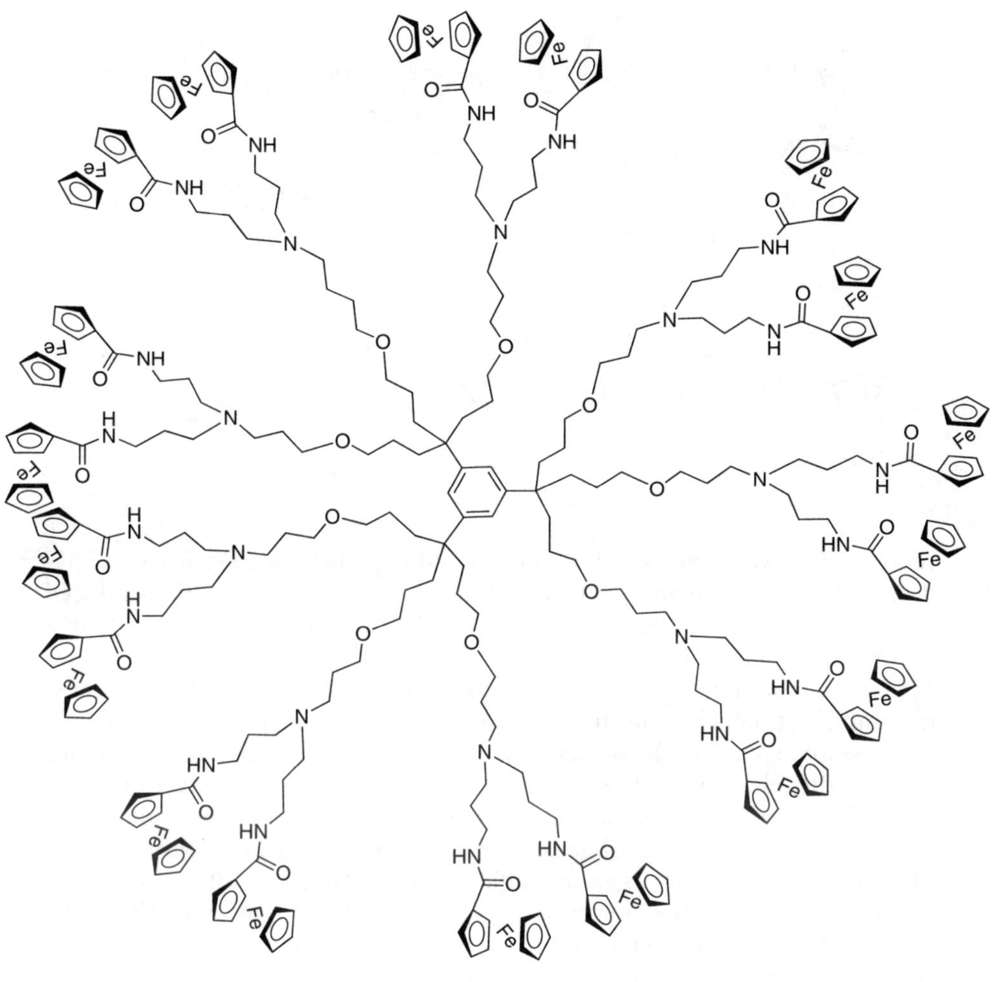

105

## 5.2 Metal-Based Receptors at Electrode Surfaces

Preorganisation of redox anion sensors on electrode surfaces is a promising new technique for electrochemical anion sensing. Self-assembled monolayers or thin polymer films of metal-based receptors have the potential to generate an amplified response to anion binding akin to the dendritic effect.

Beer and co-workers have investigated this concept using self-assembled monolayers of the 1,1′-bis(alkyl-*N*-amido)ferrocene **106** on gold electrodes [72]. The pendant disulphide groups serve to covalently anchor the receptor to the gold surface. In electrochemical experiments on **106** in acetonitrile/dichloromethane solution, anion-induced cathodic shifts of the ferrocene/ferrocenium

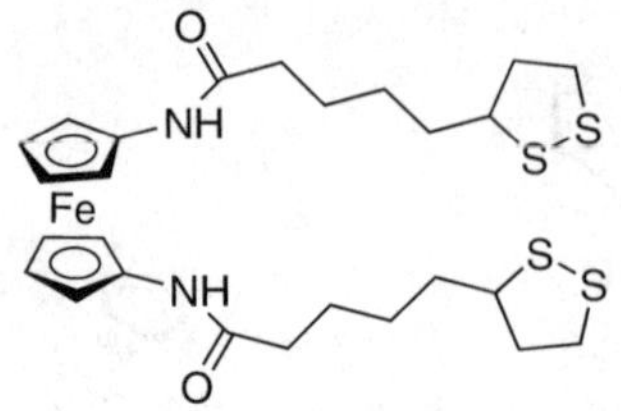

106

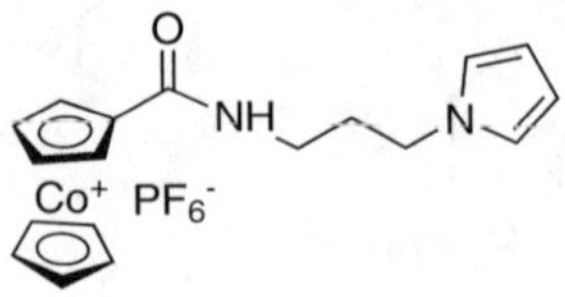

107

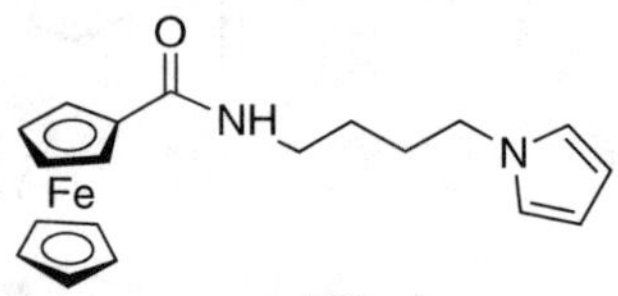

108

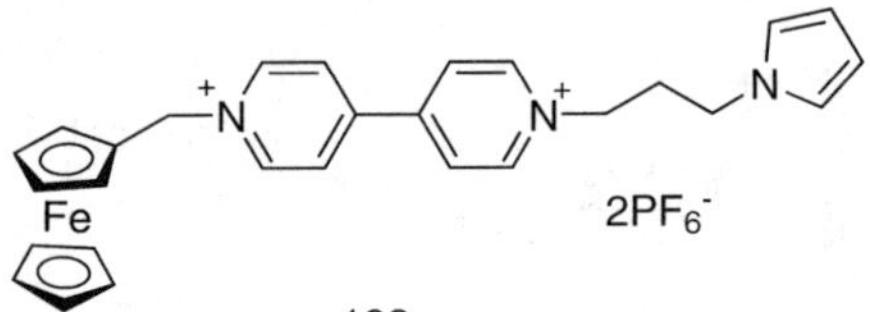

109

redox couple were observed for chloride (40 mV), bromide (20 mV) and hydrogenphosphate (210 mV). When confined to a monolayer the anion-induced shifts measured were 100 mV for chloride, 30 mV for bromide and 300 mV for hydrogenphosphate in the same solvent system – consistently greater than for the solution-phase receptor. This represents a significant 'surface sensing amplification'. The modified electrodes were also able to selectively detect hydrogenphosphate in the presence of a 100-fold excess of halide. In aqueous solution the selectivity of the system was altered, enabling the detection of the poorly hydrated anion perrhenate in the presence of hydrogenphosphate.

A number of groups have been exploring the anion sensing of thin polymer films which incorporate organometallic receptors. Monomer **107** incorporates a cobaltocenium amide reporter with a polymerisable pyrrole unit. Thin films of the receptor were prepared by electropolymerisation on a platinum or carbon electrode [73]. In electrochemical experiments on **107** in acetonitrile solution significant anion-induced cathodic shifts of the cobaltocene/cobaltocenium redox couple were observed only for dihydrogenphosphate (45 mV) and hydrogensulphate (20 mV). When confined to a polymer film the anion-induced shifts measured were 210 mV for dihydrogenphosphate and 250 mV for hydrogensulphate. Chloride and bromide could also be detected, both giving shifts of 20 mV. Polymerisation of **107**, as well as giving rise to surface sensing amplification, also results in a change in selectivity from dihydrogenphosphate to hydrogensulphate. It was also found that film thickness influences the sensitivity of the sensor. Thin films ($\Gamma$=1.8×10$^{-9}$ mol cm$^{-2}$) exhibit higher sensitivity to hydrogenphosphate at low concentrations (<50 μM), whereas thick films ($\Gamma$=2.7×10$^{-8}$ mol cm$^{-2}$) extend the measurable concentration range to high concentrations (up to 2 mM).

Films of the analogous ferrocene monomer **108** have been examined by Moutet and co-workers [74]. Anion-induced shifts of the ferrocene/ferrocenium couple were measured in acetonitrile for hydrogensulphate (30 mV), $ATP^{2-}$ (180 mV) and hydrogenphosphate (220 mV). Again this represents a surface sensing amplification. The same group has recently explored the anion sensing properties of viologen **109** in thin polymer films [75]. In aqueous solution poly-**109** registered small anion-induced cathodic shifts of the ferrocene/ferrocenium redox couple, with hydrogensulphate (20 mV), $S_2O_4^{2-}$ (10 mV) and $ATP^{2-}$ (35 mV).

## 5.3 Nanoparticle-Based Sensors

Metal-based receptors which are able to form preorganised monolayers on planar electrodes can be expected to do the same on the surface of nanoparticles, leading to a surface sensing amplification effect. The very large surface area of nanoparticles could also allow greater overall sensitivity to anions.

Astruc and co-workers have prepared the amidoferrocenylalkylthiol (AFAT)-gold nanoparticle system depicted as **110** [76]. The proportion of AFAT to dodecanethiol obtained by ligand substitution on different batches of dode-

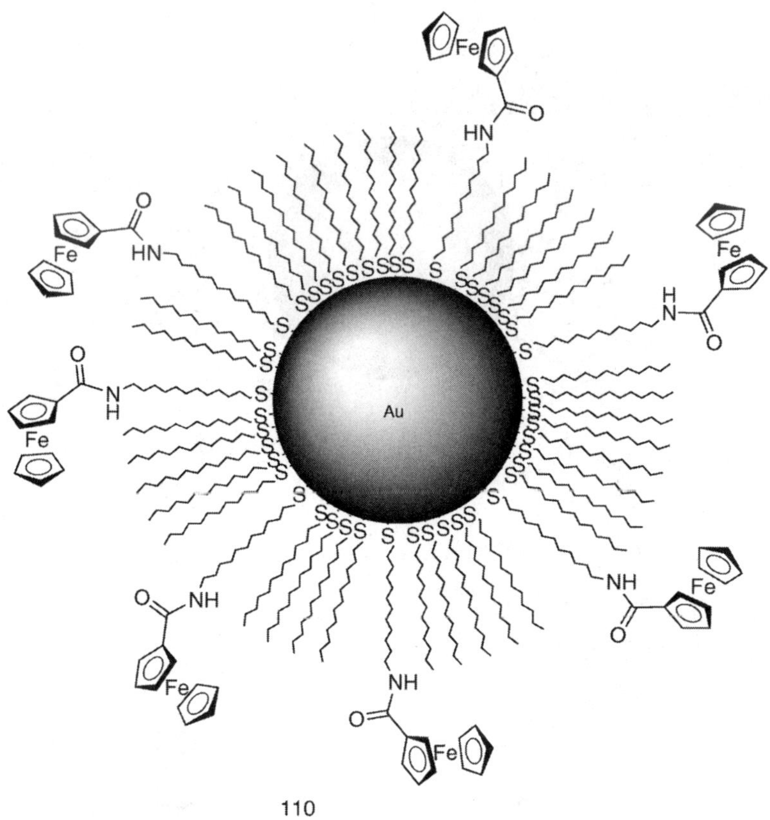

**110**

canethiol-stabilised nanoparticles ranged from 7 to 38%, corresponding to an average of 8–39 AFAT units per nanoparticle. Electrochemical measurements in dichloromethane solution show a single reversible redox wave for the ferrocene/ferrocenium couple at identical potential in each case. Addition of dihydrogenphosphate led to the appearance of a new redox wave (220 mV cathodically shifted) at the expense of the initial wave, which was completely replaced at 1 equivalent of anion per AFAT branch, indicative of 1:1 anion/branch binding. The cathodic shift is the same irrespective of the AFAT loading and is considerably larger than that observed for the comparable amidoferrocene monomer $FcCONHCH_2CH_2OPh$ (45 mV) or even a representative ferrocene tripod $PhC(CH_2CH_2CH_2NHCOFc)_3$ (110 mV).

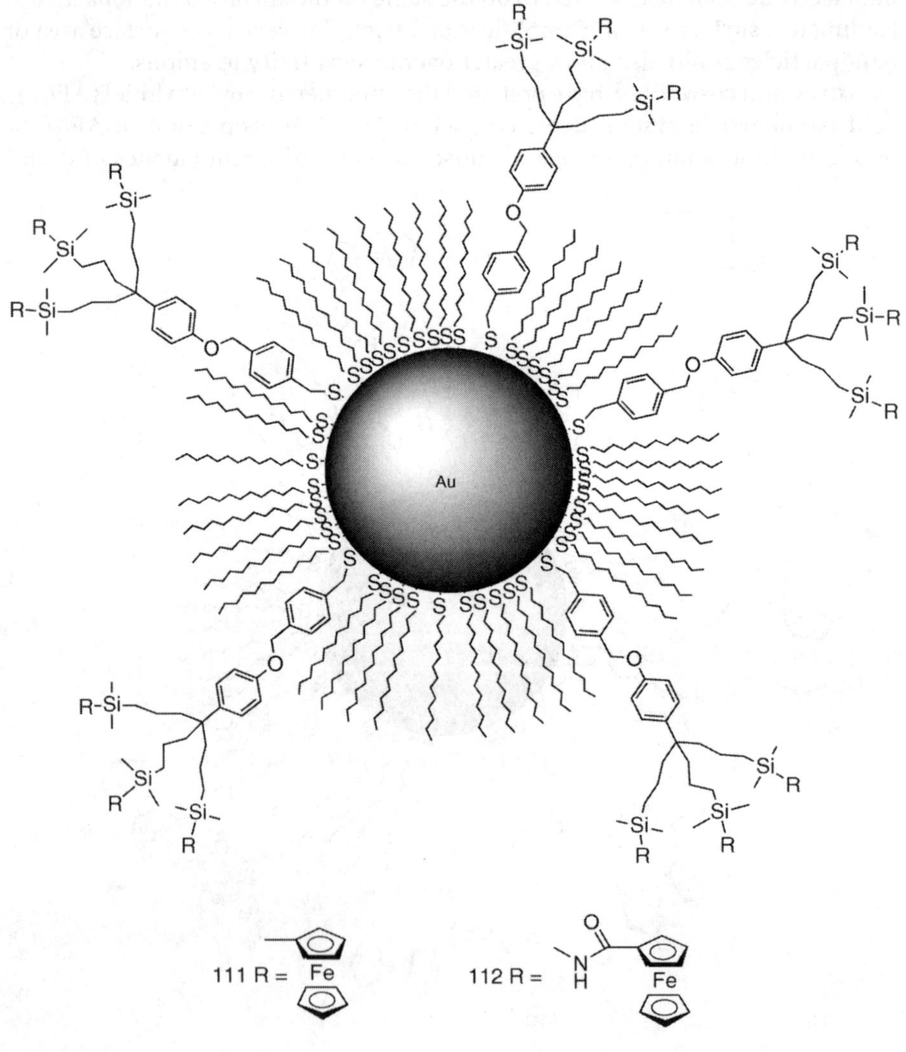

111 R =

112 R =

The same group has also investigated the anion sensing properties of gold nanoparticles **111** and **112** substituted with dendrons comprising three amido-ferrocene or silyl ferrocene branches [77]. The surface loadings of **111** and **112** were 3 and 4.8%, respectively, corresponding to an average of three and five dendrons per nanoparticle. Nanoparticles of type **111** show very similar properties to the AFAT-modified nanoparticles **110**, with a dihydrogenphosphate-induced cathodic shift of 210 mV in dichloromethane solution. Despite lacking any hydrogen bonding groups, the nanoparticles dendronised with silyl ferrocenes **112** gave a dihydrogenphosphate-induced cathodic shift of 110 mV in the same solvent.

Very recently Beer and co-workers have developed a surface-enhanced optical anion sensor based on gold nanoparticles [78]. Dodecanethiol-stablised gold nanoparticles were modified by ligand substitution with a disulphide-substituted zinc porphyrin **113** to provide 30 and 80 receptors per nanoparticle. Titration of both the free receptor and the modified nanoparticles with various

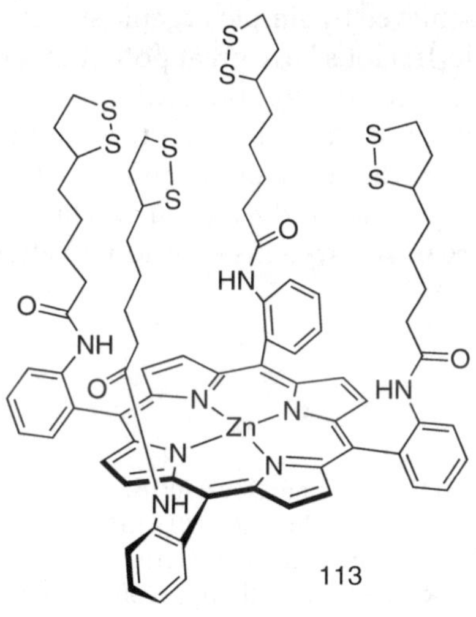

113

**Table 2** Stability constant (log $K$) data for free and nanoparticle- bound **113** in dichloromethane solution, errors ±0.1

| Anion | **113** | **113** nanoparticles[a] |
|---|---|---|
| $Cl^-$ | >6 | >6 |
| $Br^-$ | 4.1 | 5.0 |
| $I^-$ | 3.2 | 4.0 |
| $NO_3^-$ | 2.4 | 3.2 |
| $H_2PO_4^-$ | >6 | >6 |
| $ClO_4^-$ | 0 | 0 |

[a] Stability constant values for the 1:1 porphyrin:anion complex on the nanoparticle surface.

anions in dichloromethane or DMSO solution revealed significant changes in the intense porphyrin absorption bands. Calculated stability constants are given in Table 2 and reveal highly enhanced anion binding affinities (up to two orders of magnitude with chloride and dihydrogenphosphate in DMSO solution) for the surface-bound porphyrin receptor with respect to the free metalloporphyrin.

## 6 Conclusion

Metal-based receptors continue to play a fundamental part in anion sensing chemistry. The useful physicochemical properties of metal ions in the form of organometallic or coordination complexes have allowed them to be used as multifunctional components in anion receptors. Metal moieties are able to combine roles such as reporter groups, anion binding sites and structural components in ways which cannot be achieved by simple organic structures. Although relatively little studied, lanthanide(III) ions have great potential for use in anion receptors and are due to receive far greater attention. Another significant recent advance has been the incorporation of metallo-receptors into nanostructures such as dendrimers, nanoparticles, thin polymer films and self-assembled monolayers. The origin and development of the anion 'surface sensing amplification' observed in these preorganised systems is certain to attract great interest in the future.

## References

1. Beer PD, Hayes EJ (2003) Coord Chem Rev 240:167
2. Beer PD, Cadman J (2000) Coord Chem Rev 205:131
3. Suksai C, Tuntulani T (2003) Chem Soc Rev 32:192
4. Fabbrizzi L, Licchelli M, Rabaioli G, Taglietti A (2000) Coord Chem Rev 205:85
5. Sun S-S, Lees AJ (2002) Coord Chem Rev 230:171
6. Beer PD, Keefe AD (1989) J Organomet Chem 375:C40
7. Beer PD (1989) Chem Commun 689
8. Beer PD, Hesek D, Kingston JE, Smith DK, Stokes SE, Drew MGB (1995) Organometallics 14:3288
9. Beer PD, Drew MGB, Hesek R, Jagessar R (1995) Chem Commun 1187
10. Beer PD, Drew MGB, Hesek D, Nam KC (1999) Organometallics 18:3933
11. Beer PD, Stokes SE (1995) Polyhedron 14:2631
12. Beer PD, Graydon AR, Johnson AOM, Smith DK (1997) Inorg Chem 36:2112
13. Beer PD, Chen GZ, Goulden AJ, Graydon AR, Stokes SE, Wear T (1993) Chem Commum 107
14. Reynes O, Maillard F, Moutet J-C, Royal G, Saint-Aman E, Stanciu G, Dutasta J-P, Gosse I, Mulatier J-C (2001) J Organomet Chem 637–639:356
15. Reynes O, Maillard F, Moutet J-C, Pecaut J, Royal G, Saint-Aman E (2002) New J Chem 26:9

16. Thomas J-L, Howarth J, Hanlon K, McGuirk D (2000) Tetrahedron Lett 41:413
17. Gale PA, Hursthouse MB, Light ME, Sessler JL, Warriner CN, Zimmerman RS (2001) Tetrahedron Lett 42:6759
18. Abouderbala LO, Belcher WJ, Boutelle MG, Cragg PJ, Dhaliwal J, Fabre M, Steed JW, Turner DR, Wallace KJ (2002) Chem Commun 358
19. Tomapatanaget B, Tuntulani T, Chailapakul O (2003) Org Lett 5:1539
20. Beer PD, Chen Z, Drew MGB, Kingston JE, Ogden M, Spencer P (1993) Chem Commun 1046
21. Beer PD, Chen Z, Drew MGB, Johnson AOM, Smith DK, Spencer P (1996) Inorg Chim Acta 246:143
22. Beer PD, Cadman J, Lloris JM, Martinez-Manez R, Padilla-Tosta ME, Pardo T, Smith DK, Soto J (1999) J Chem Soc Dalton Trans 1805
23. Dusemund C, Sandanayake KRAS, Shinkai S (1995) Chem Commun 333
24. Miyaji H, Collinson SR, Prokeŝ, Tucker JHR (2003) Chem Commun 64
25. Beer PD, Dickson CAP, Fletcher N, Goulden AJ, Grieve A, Hodacova J, Wear T (1993) Chem Commun 828
26. Holman KT, Orr W, Steed JW, Atwood JL (1998) Chem Commun 2109
27. Camiolo S, Coles SJ, Gale PA, Hursthouse MB, Mayer TA, Paver MA (2000) Chem Commun 275
28. Staffilani M, Hancock KSB, Steed JW, Holman KT, Atwood JL, Juneja RK, Burkhalter RS (1997) J Am Chem Soc 119:6324
29. Balzani V, Barigeletti F, Cola LD (1990) Top Curr Chem 158:31
30. Juris A, Balzani V, Barigeletti F, Campagna S, Belser P, Zelewsky AV (1988) Coord Chem Rev 84:85
31. Szemes F, Hesek D, Chen Z, Dent SW, Drew MGB, Goulden AJ, Graydon AR, Grieve A, Mortimer RJ, Wear T, Weightman JS, Beer PD (1996) Inorg Chem 35:5868
32. Beer PD (1998) Acc Chem Res 31:71
33. Beer PD, Dent SW, Wear T (1996) J Chem Soc Dalton Trans 2341
34. Beer PD, Cadman J (1999) New J Chem 23:347
35. Deetz MJ, Smith BD (1998) Tetrahedron Lett 39:6841
36. Watanabe S, Onogawa O, Komatsu Y, Yoshida K (1998) J Am Chem Soc 120:229
37. Beer PD, Timoshenko V, Maestri M, Passaniti P, Balzani V (1999) Chem Commun 1755
38. Cooper JB, Drew MGB, Beer PD (2001) J Chem Soc Dalton Trans 392
39. Mizuno T, Wei W-H, Eller LR, Sessler JR (2002) J Am Chem Soc 124:1134
40. Beer PD, Dent SW, Hobbs GW, Wear T (1997) Chem Commun 99
41. Beer PD, Szemes F, Balzani V, Salà CM, Drew MGB, Dent SW, Maestri M (1997) J Am Chem Soc 119:11864
42. Goodall W, Williams JAG (2000) J Chem Soc Dalton Trans 2893
43. Redman JE, Beer PD, Dent SW, Drew MGB (1998) Chem Commun 231
44. Uppadine LH, Redman JE, Dent SW, Drew MGB, Beer PD (2001) Inorg Chem 40:2860
45. Sun S-S, Lees AJ, Zavalij PY (2003) Inorg Chem 42:3445
46. De Santis G, Fabbrizzi L, Licchelli M, Poggi A, Taglietti A (1996) Angew Chem Int Ed Engl 35:202
47. Fabbrizzi L, Licchelli M, Parodi L, Poggi A, Taglietti A (1999) Eur J Inorg Chem 35
48. Coksun A, Baytekin BT, Akkaya EU (2003) Tetrahedron Lett 44:5649
49. Fabbrizzi L, Licchelli M, Rabaioli G, Taglietti A (2000) Coord Chem Rev 205:85
50. Fabbrizzi L, Leone A, Tagliettie A (2001) Angew Chem Int Ed 40:3066
51. Han MS, Kim DS (2002) Angew Chem Int Ed 41:3809
52. Tobey SL, Jones BD, Anslyn EV (2003) J Am Chem Soc 125:4026
53. Tobey SL, Anslyn EV (2003) J Am Chem Soc 125:14807

54. Tobey SL, Anslyn EV (2003) Org Lett 5:2029
55. Mizukami S, Nagano T, Urano Y, Odani A, Kikuchi K (2002) J Am Chem Soc 124:3920
56. Beer PD, Berry N, Drew MGB, Fox OD, Padilla-Tosta ME, Patell S (2001) Chem Commun 199
57. Berry NG, Pratt MD, Fox OD, Beer PD (2001) Supramol Chem 13:677
58. Beer PD, Berry NG, Cowley AR, Hayes EJ, Oates EC, Wong WWH (2003) Chem Commun 2408
59. Dickins RS, Gunnlaugsson T, Parker D, Peacock RD (1998) Chem Commun 1634
60. Bruce JI, Dickins RS, Govenlock LJ, Gunnlaugsson T, Lopinski S, Lowe MP, Parker D, Peacock RD, Perry JJB, Aime S, Botta M (2000) J Am Chem Soc 122:9674
61. Montalti M, Prodi L, Zaccheroni N, Charbonnière L, Douce L, Ziessel R (2001) J Am Chem Soc 123:12694
62. Yamada T, Shinoda S, Tsukube H (2002) Chem Commun 1218
63. Mahajan RK, Kaur I, Kaur R, Uchida S, Onimaru A, Shinoda S, Tsukube H (2003) Chem Commun 2238
64. Kingston JE, Ashford L, Beer PD, Drew MGB (1999) J Chem Soc Dalton Trans 251
65. Beer PD, Drew MGB, Jagessar R (1997) J Chem Soc Dalton Trans 881
66. Slone RV, Yoon DI, Calhoun RM, Hupp JT (1995) J Am Chem Soc 117:11813
67. Sun S, Anspach JA, Lees AJ, Zavalij PY (2002) Organometallics 21:685
68. Altmann R, Gausset O, Horn D, Jurkschat K, Schurmann M (2000) Organometallics 19:430
69. Valero C, Fillaut J-L, Ruiz J, Guittard J, Blais J-C, Astruc D (1997) J Am Chem Soc 119:2588
70. Alonso B, Casado C, Cuadrado I, Morán M, Kaifer AE (2002) Chem Commun 1778
71. Daniel M-C, Ruiz J, Blais J-C, Daro N, Astruc D (2003) Chem Eur J 9:4371
72. Beer PD, Davis JJ, Drillsma-Milgrom DA, Szemes F (2002) Chem Commun 1716
73. del Peso I, Alonso B, Lobete F, Casado CM, Cuadrado I, del Barrio JL (2002) Inorg Chem Commun 5:288
74. Reynes O, Gulon T, Moutet J-C, Royal G, Saint-Aman E (2002) J Organomet Chem 656:116
75. Reynes O, Bucher C, Moutet J-C, Royal G, Saint-Aman E (2004) Chem Commun 428
76. Labande A, Astruc D (2000) Chem Commun 1007
77. Daniel M-C, Ruiz J, Nlate S, Blais J-C, Astruc D (2003) J Am Chem Soc 125:2617
78. Beer PD, Cormode DP, Davis JJ (2004) Chem Commun 414

Top Curr Chem (2005) 255: 163–198
DOI 10.1007/b101166


# Chromogenic Anion Sensors

Chomchai Suksai · Thawatchai Tuntulani (✉)

Department of Chemistry, Supramolecular Chemistry Research Unit, Faculty of Science, Chulalongkorn University, Bangkok 10330, Thailand
*tthawatc@chula.ac.th*



**Abstract** Chromogenic sensors for anions generally consist of two parts: anion receptors and chromophores. In this review, six types of chromogenic anion sensors are described, namely, NH-based hydrogen bonding, Lewis acid, metal-ion template, transition metal complexes, chromogenic guest displacement, and chromoreactands. The first four types possess anion receptors attached directly to the chromophores, while the guest displacement techniques employ indicators that are replaced by specific anions. The last type has emerged recently and uses specific reactions between chromogenic hosts or indicators and particular anions to cause dramatic color changes.

**Keywords** Chromogenic anion sensor · Colorimetric anion sensor

**Abbreviations**

| | |
|---|---|
| $AcO^-$ | Acetate anion |
| ATP | Adenosine triphosphate |
| ADP | Adenosine diphosphate |
| $A^-$ | Anion |
| DMSO-$d_6$ | Deuterated dimethyl sulfoxide |
| DMF | Dimethylformamide |
| DMSO | Dimethyl sulfoxide |
| EDOT | Ethylenedioxythiophene |
| $EDTA^{4-}$ | Ethylenediaminetetraacetate |
| GMP | Guanosine monophosphate |
| h | Hour(s) |
| $^1$H NMR | Proton nuclear magnetic resonance spectroscopy |
| H-bpmp | 2,6-Bis{[bis(2-pyridylmethyl)amino]methyl}-4-methylphenol |
| HEPES | 2-[4-(2-Hydroxyethyl)-1-piperazinyl]ethanesulfonic acid |
| $K_a$ | Association constant |
| $M^{-1}$ | Per mole |
| $M^+$ | Metal ion |
| Me | Methyl |
| MeOH | Methanol |
| MLCT | Metal-to-ligand charge transfer |
| mV | Millivolt |
| nm | Nanometer |
| Ph | Phenyl |
| SEM | Monotrimethylsilylethoxymethyl |
| SHE | Standard hydrogen electrode |
| TCNQ | Tetracyanoquinodimethane |
| THF | Tetrahydrofuran |
| UV–vis | UV–visible spectroscopy |
| v/v | Volume by volume |
| $\lambda_{max}$ | Maximum absorption wavelength |
| $\lambda_{em}$ | Emission wavelength |
| $\lambda_{ex}$ | Excitation wavelength |

# 1 Introduction

In the field of supramolecular chemistry, the progress in synthetic receptors for anions [1–5] has attracted considerable attention in recent decades due to the fact that a large number of biological processes involve molecular recognition of anionic species. It is thus important to develop techniques for quantifying or sensing such anions. Nowadays, the development of colorimetric anion sensing [6–11] is particularly challenging, since visual detection can give immediate qualitative information and is becoming increasingly appreciated in terms of quantitative analysis.

One approach to reach an effective anion sensor for biological anions involves the construction of optical anion sensors. Such a system generally

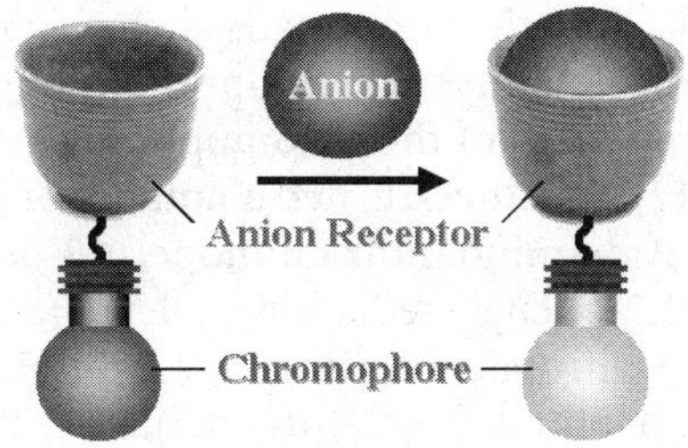


**Fig. 1** Operating principle of chromogenic anion sensors

consists of two parts. One part is an anion binding site employing various combinations of anion receptor units. Anion receptors can be mainly divided into two categories: neutral anion receptors and positive-charge anion receptors. Neutral anion receptors employed hydrogen bonding NH-based donors such as pyrroles, amides, and urea/thioureas or Lewis acids. Positive-charge anion receptors use ammonium derivatives or guanidinium centers for binding negative-charge anions. The other is the chromophore part, which converts the binding events or recognition phenomena to optical signals. These two parts can be either covalently attached or intermolecularly linked to each other.

The concept of chromogenic anion sensors is illustrated in Fig. 1. The crystal in the figure can be compared to analytes such as anions, and the cup acts like a receptor that is connected to a light bulb. When the crystal goes into the cup that fits its shape, the electronic part of the light bulb is perturbed, and the light bulb will give a different color.

Chromogenic anion sensors can be divided into two main categories: metal and nonmetal chromogenic hosts. For metal-involved chromogenic hosts, the changing color comes from the color of metals or metal complexes, especially transition metal ions, as their electronic properties are perturbed upon coordination to anions. The other way to generate a chromogenic sensor is to displace the coordinated chromophore by specific anions. Besides coordination aspects, the color changes can stem from reactions between chromogenic hosts or indicators and anions. This type of chromogenic anion sensor can be called a "chromoreactand". When the reaction occurs, the conformed host will definitely change its electronic properties. This, therefore, results in an observable color change.

## 2 NH-Based Hydrogen Bonding Chromogenic Hosts

These hosts include simple anion sensor systems containing urea, thiourea, amine, amide, alcohol, and pyrrole groups linked to chromophores. Chromophores used in this type of chromogenic anion sensors are mainly organic dyes such as azobenzene, nitrobenzene, indoaniline, and anthraquinone or extensively conjugated aromatic compounds such as quinoxaline, oxadiazole, and porphyrin. The color change occurs upon binding of anion guests that affect the electronic properties of the chromophores.

Sessler and Miyaji [12] have investigated a number of commercially available compounds such as 1,2-diaminoanthraquinone, 1,8-diaminoanthraquinone, 4-nitroaniline, 4-nitro-1,2-phenylenediamine, L-leucine-4-nitroanilide, 1-(4-nitrophenyl)-2-thiourea, 4-nitrophenol, alizarin, 2,2′-bis(3-hydroxy-1,4-naphthoquinone), acid blue 45, naphthol AS, 9(10*H*)-acridone, and Direct Yellow 50 as anion sensors in organic solvents such as dichloromethane. These compounds contain hydrogen bond donors such as amine, amide, alcohol, and urea as anion binding sites. The compounds also contained chromophore subunits (acting as an electron acceptor) whose electronic properties were modified as a result of interactions with a bound anionic substrate (acting as an electron donor). The NH-based hydrogen bonding chromogenic hosts can be classified as follows.

## 2.1 Chromogenic Hosts Containing Nitrophenyl Units

Thiourea is an especially good hydrogen bond donor and is an excellent anion receptor for carboxylate anions. Teramae and coworkers [13] prepared thiourea-based chromophores with *p*-nitrophenyl units, **1** and **2**. As a result, compounds **1** and **2** were found to be more highly selective for $AcO^-$ than other anions in 1% water:MeCN. The binding properties of **1** and **2** with $AcO^-$ showed that increasing the concentration of $AcO^-$ produced a significant bathochromic shift in the UV–vis spectra. The stability constant of the $AcO^-$ complex of **1** was $5.6\times10^3$ $M^{-1}$, much weaker than that of **2** ($K_a$=$3.5\times10^5$ $M^{-1}$). It was evident that introducing a *p*-nitrophenyl group into the (*p*-nitrophenyl)thiourea moiety enhanced the hydrogen bonding ability. Likewise, chromophore **2** was then applied to the colorimetric determination of acetic acid in commercially available brands of vinegar. The results agreed well with the specification provided by the suppliers.

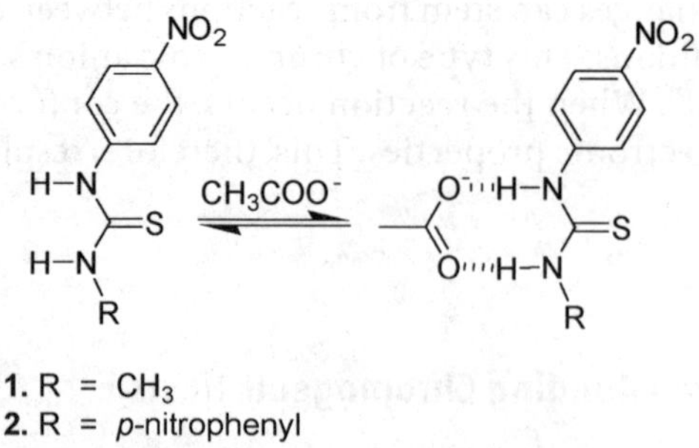


**1.** R = $CH_3$
**2.** R = *p*-nitrophenyl

Besides urea and thiourea, amide groups can also form effective chromogenic anion sensors. Jurczak and coworkers [14] have demonstrated the use of amide groups containing nitrophenyl-macrocycle **3** as a selective colorimetric sensor for $F^-$. A DMSO solution of compound **3** showed dramatic color changes upon addition of $F^-$, $AcO^-$, and $H_2PO_4^-$ ions. It was found that a colorless solution of **3** turned dark blue ($\lambda$=593, 708 nm), yellow ($\lambda_{max}$=375 nm), and yellow ($\lambda_{max}$=384 nm) when exposed to $F^-$, $H_2PO_4^-$, and $AcO^-$, respectively. However, compound **3** formed a substantially stable 2:1 anion-to-ligand ratio with $F^-$ in DMSO-$d_6$ and $CD_3CN$ with $K_a$ values of $7.8\times10^6$ and $7.5\times10^6$ $M^{-1}$, respectively.

The 3,5-dinitrophenyl derivatives of pyrrole 2,5-diamide **4** [15] acted as a selective naked-eye sensor for $F^-$ in MeCN solution, which gives rise to a deep blue color ($\lambda_{max}$=598 nm). This color change was due to a deprotonation process caused by $F^-$ acting as a base and subsequent charge-transfer interactions between the deprotonated pyrrole and the nitroaromatic groups.

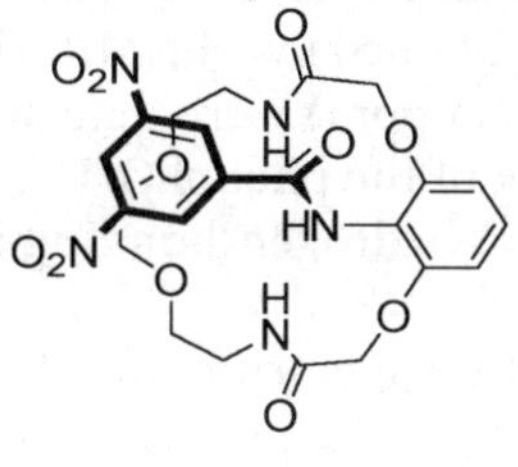

3

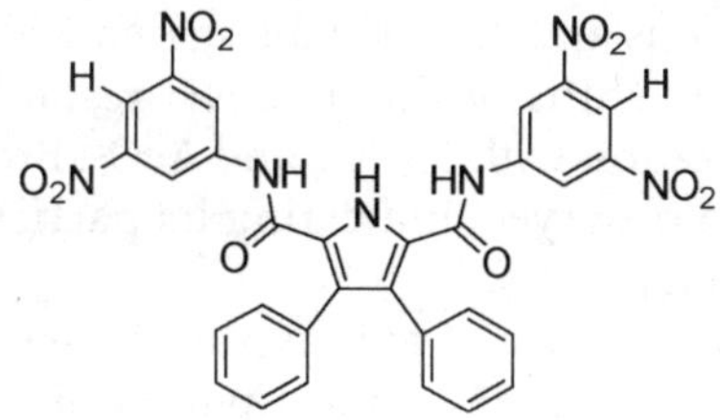

4

The calix[4]pyrroles functionalized with nitrobenzenes **5** and **6** [16] can act as $F^-$ sensors. These systems bear an appended chromophore directly linked to the calix[4]pyrrole skeleton through a conjugated C≡C triple bond. Upon addition of tetrabutylammonium $F^-$, solutions of **5** and **6** in $CH_2Cl_2$ turned from pale yellow ($\lambda_{max}$=391 nm) to intense yellow ($\lambda_{max}$=433 nm) for **5** and from yellow ($\lambda_{max}$=441 nm) to red ($\lambda_{max}$=498 nm) for **6**. In addition, the color changed from yellow ($\lambda_{max}$=441 nm) to orange when $Cl^-$ ($\lambda_{max}$=483 nm) and $H_2PO_4^-$ ($\lambda_{max}$=478 nm) were added to a solution of **6**.

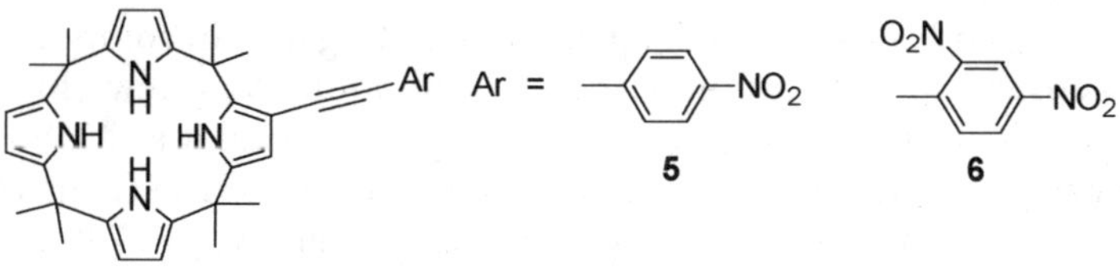

## 2.2 Chromogenic Hosts Containing Azo Dye Units

Hong and coworkers [17] have recently reported anion coordination with a nitro-azophenol thiourea-based sensor, compound **7**. Association constants for anion binding were determined by $^1H$ NMR and UV–vis titrations in $CDCl_3$. $H_2PO_4^-$ ($K_a$=$2.6\times10^4$ $M^{-1}$) and $AcO^-$ ($K_a$=$1.9\times10^4$ $M^{-1}$) gave stronger complexes

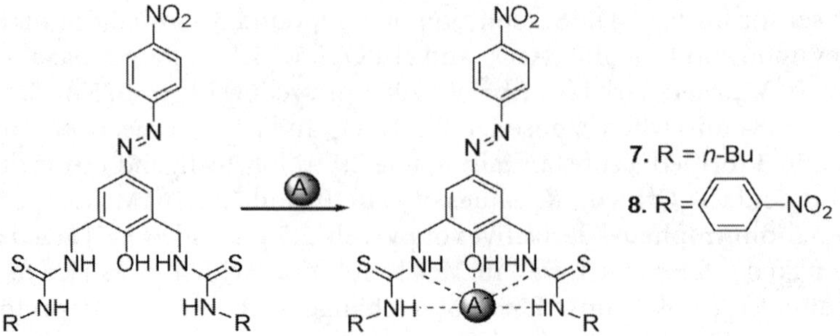


with 7 than other anions due to their high basicity. Moreover, $H_2PO_4^-$ with four oxygens makes the strongest complexes via multitopic hydrogen-bonding interactions and these results agreed with those obtained by $^1$H NMR. The large downfield shifts of thiourea NH resonances (>2.5 ppm) were detected upon complexation with $H_2PO_4^-$ and $AcO^-$. Broadening of the phenol OH resonance was also observed, indicating its participation in hydrogen-bonding interactions with anions.

In the absence of anions, the UV–vis absorption spectrum of 7 showed an absorption $\lambda_{max}$=376 nm. Upon addition of $H_2PO_4^-$, the peak at 376 nm decreased while a new peak appeared at 529 nm, concomitant with a solution color change from light yellow to deep red. This may be due to electronic excitation through charge transfer from the oxygen donor of the phenol to an acceptor substituent ($-NO_2$) of the chromophore. The excited state would be more stabilized by anion binding, resulting in a bathrochromic shift in the absorption maxima as well as a color change. Changing the substituent at the thiourea moiety of 7 from butyl groups to nitrobenzene groups in compound **8** [18] allowed the easy colorimetric differentiation of $F^-$, $H_2PO_4^-$, and $AcO^-$ which have similar basicity. The degree of red shift for **8** was determined to be $H_2PO_4^- \gg AcO^- \approx F^- > Br^- \approx Cl^- > HSO_4^- \approx I^-$ in $CHCl_3$. The maximum red-shift value ($\lambda_{max}$=538 nm) for $H_2PO_4^-$ can be understood on the basis of the guest basicity and the structure of the complex.

Later, Hong and coworkers changed the signaling unit in compound 7 from nitroazobenzene to indoaniline [19]. The compound **9**, a new chromogenic indoaniline–thiourea-based sensor, showed significant color and UV–vis spectral changes upon binding anions. Upon the addition of $H_2PO_4^-$ or $HSO_4^-$, the color of the $CHCl_3$ solution changed from blue-green ($\lambda_{max}$=678 nm) to deep blue ($\lambda_{max}$=632 nm). The association constants obtained from UV–vis titrations for complexes of **9** with $H_2PO_4^-$ and $HSO_4^-$ in $CHCl_3$ are $1.1\times10^4$ and $2.5\times10^4$ $M^{-1}$, respectively. However, addition of $AcO^-$ or $F^-$, which are more basic anions, caused a less intense color change. In addition, in the case of $Cl^-$, $Br^-$, and $I^-$, no detectable color changes were observed. In the same manner as in compound 7, compound **9** possesses four NH urea moieties and allows the selective colorimetric detection of tetrahedral oxoanions such as $HSO_4^-$ and $H_2PO_4^-$.

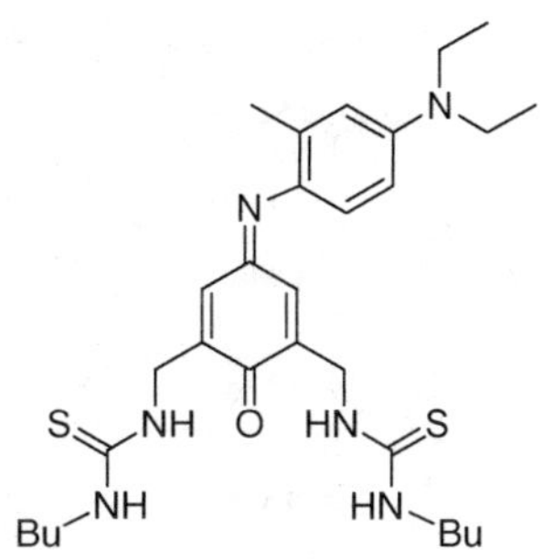

**9**

## 2.3
## Chromogenic Hosts Containing Naphthalimide Units

The receptors **10–13**, which contain naphthalimide groups, can act as chromophores and show recognition of anion especially $F^-$ in DMSO. Upon addition of $F^-$ to the solution of **10** [20], a 1:2 complex was found concomitant with the color change from light yellow to deep purple, while receptors **11–13** [21] showed color changes from green to purple. These results can be attributed to strong hydrogen-bonding interactions between the 4-amino moiety of the naphthalimide group and $F^-$, or more likely a complete deprotonation. Surprisingly, color changes of receptors **11–13** reversed gradually with time due to the fixation of $CO_2$ (as $HCO_3^-$) by the receptors as 1:1 adducts. This was confirmed by X-ray crystallography.

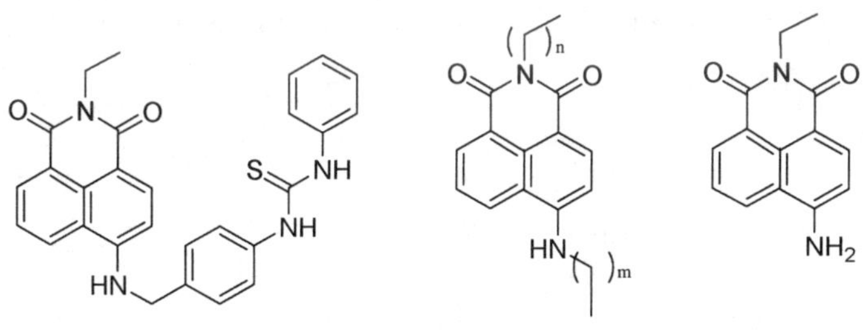

**10** **11.** n = 1, m = 3 **12.** n = m =3 **13**

## 2.4
## Chromogenic Hosts Containing Naphthalene Units

The fluoride-selective chromogenic and fluorescent sensor **14** containing naphthalene ureas was synthesized by Lee and coworkers [22]. From UV-visible studies, the sensor **14** showed a characteristic band at 325 nm. Upon addition of $F^-$, a new peak at 379 nm occurred in MeCN:DMSO (9:1 v/v). Furthermore, $\lambda_{em}$ of **14** at 379 nm was shifted to 445 nm upon addition of $F^-$ ($K_a$=14,200 $M^{-1}$).

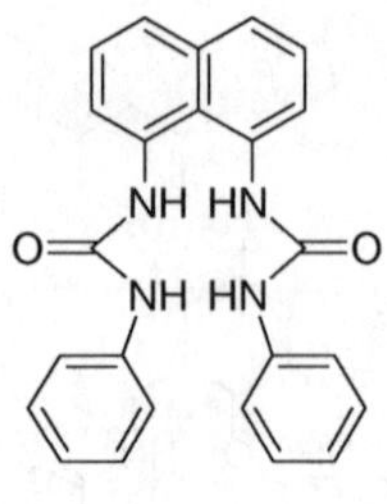

**14**

Addition of other anions showed the same $\lambda_{em}$ as free **14**. $^1$H NMR results suggested that $F^-$ bound four urea NH protons via hydrogen-bonding interactions, which resulted in the downfield shift of the NH proton signals.

The other dual-sensing receptor based on binaphthalene skeleton **15** [23] bearing two thiourea groups showed sensing abilities toward $F^-$, $AcO^-$, and $H_2PO_4^-$ in MeCN. A bathochromic shift with unique isosbestic points was observed upon addition of these anions. From fluorescence titrations, the intensity of $\lambda_{em}$ of **15** at 459 nm decreased and a weak broad emission band around 650 nm occurred upon addition of $F^-$ ($K_a$=2.1×10$^6$ M$^{-1}$), $AcO^-$ ($K_a$=1.1×10$^5$ M$^{-1}$), and $H_2PO_4^-$ ($K_a$=5.5×10$^4$ M$^{-1}$). X-ray crystallography showed that two naphthyl rings are placed in the same plane and two thiourea groups are located in the *anti* position. The thiourea groups are in the *syn–anti* conformation due to formation of intermolecular hydrogen bonds in the solid state. In the solution state, two naphthyl rings of **15** can freely rotate. The complexation with anions then induced the rotation of naphthyl rings to form a planar conformer as shown in Fig. 2, which caused the emission spectrum to move to a longer wavelength.

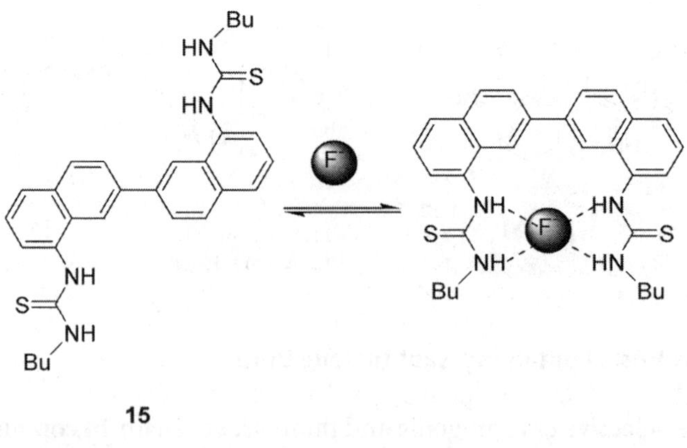

**15**

**Fig. 2** Proposed hydrogen-bonding interactions between fluoride and receptor 15

## 2.5
## Chromogenic Hosts Containing Anthraquinone Units

Anthraquinone derivatives **16** and **17** showed potential as off-the-shelf anion sensors [12]. It was found that **16** had a more dramatic color change than **17** upon addition of anions. The solution of **16** in $CH_2Cl_2$ changed from yellow ($\lambda_{max}$=478 nm) to dark purple ($\lambda_{max}$=555 nm), red ($\lambda_{max}$=519 nm), reddish orange ($\lambda_{max}$=513 nm), orange ($\lambda_{max}$=499 nm), purple ($\lambda_{max}$=548 nm), and orange ($\lambda_{max}$=493 nm) when exposed to $F^-$, $Cl^-$, $Br^-$, $I^-$, $H_2PO_4^-$, and $HSO_4^-$, respectively. Proposed binding modes of sensors **16** and **17** with anions are presented in Fig. 3.

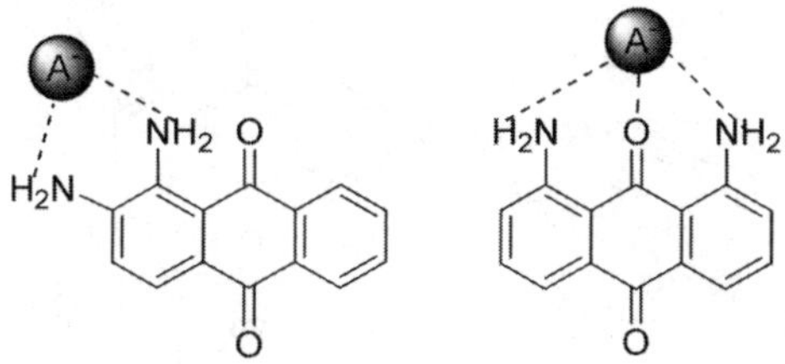


**Fig. 3** Proposed binding modes of an anion for sensors **16** and **17**

Jiménez et al. [24] have synthesized receptors 1,5-bis-*N*-(9,10-dioxo-9,10-dihydroanthracen-1-yl)-*N'*-butylthiourea **18** and its urea analogue, 1,5-bis-*N*-(9,10-dioxo-9,10-dihydroanthracen-1-yl)-*N'*-butylurea **19**, by condensation of 1,5-diaminoanthraquinone with butyl isothiocyanate and butyl isocyanate, respectively.

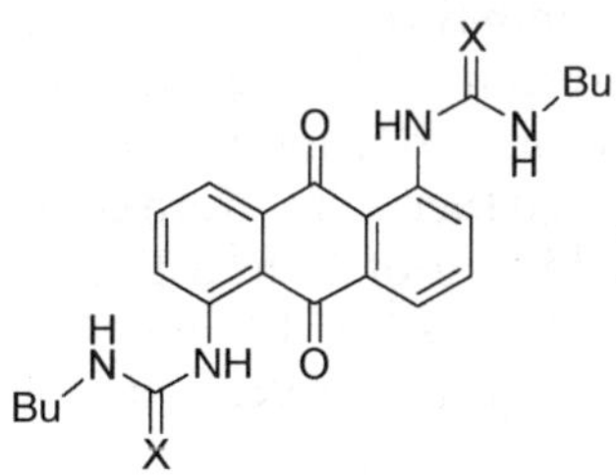

**18**. X = S
**19**. X = O

Compound **18** showed a remarkable color change from orange to brown ($\lambda_{max}$=670 nm) in DMSO upon adding $F^-$. Color changes are most probably due to a charge-transfer process and electron-rich formation of hydrogen bonds between thiourea-bound $F^-$ and the electron-deficient anthraquinone moiety. The anion was believed to form a 2:1 anion-to-ligand ratio as shown in Fig. 4.

Similar to compounds **5** and **6**, anthraquinone-functionalized calix[4]pyrroles **20** and **21** can act as $F^-$ sensors [25]. Upon adding $F^-$ to solutions of **20** and **21** in $CH_2Cl_2$, the color of **20** changed from yellow ($\lambda_{max}$=467 nm) to red ($\lambda_{max}$=518 nm),

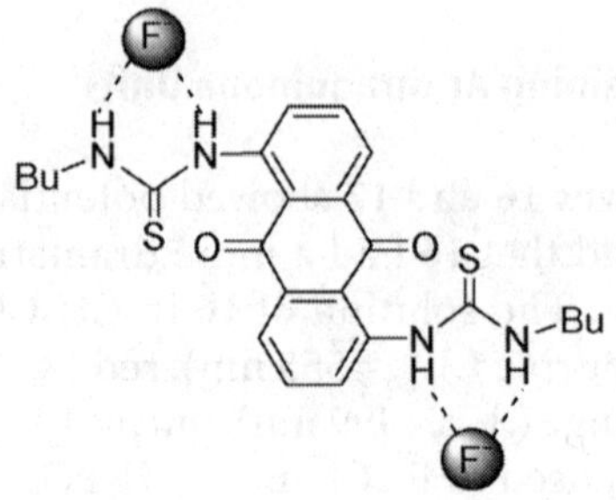

**Fig. 4** Proposed hydrogen bond formation between fluoride and sensor **18**

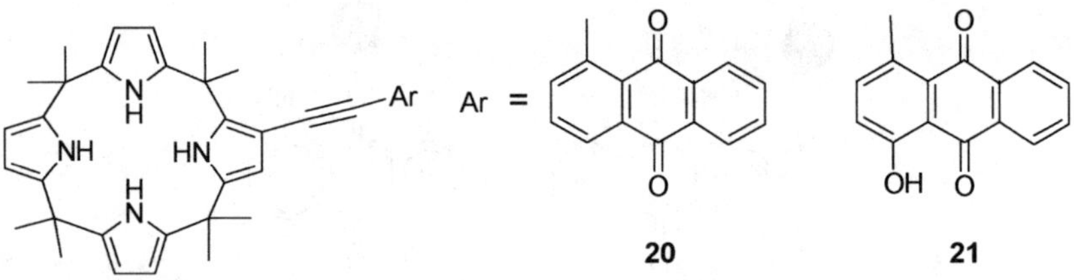

while **21** turned from red ($\lambda_{max}$=526 nm) to blue ($\lambda_{max}$=613 nm). The dramatic color changes were tentatively ascribed to charge-transfer interactions between electron-rich calix[4]pyrrole-bound anions and electron-deficient anthraquinone moieties.

## 2.6 Chromogenic Hosts Containing Oxadiazole and Benzoxazole Units

Wang and colleagues [26] have synthesized two derivatives of oxadiazole **22** and **23** that were used as anion-fluorescent and -colorimetric sensors. The two compounds showed high selectivity for $H_2PO_4^-$ and $F^-$ over $Cl^-$ in DMF. It was found that compounds **22** and **23** changed their color from colorless ($\lambda_{max}$=376 nm) to yellow ($\lambda_{max}$=400 nm) upon addition of $F^-$ and $H_2PO_4^-$ in DMF. Association constants of **22** with $F^-$ and $H_2PO_4^-$ are $8.6\times10^4$ and $7.9\times10^5$ $M^{-1}$, respectively. Furthermore, compound **23** showed high selectivity toward $H_2PO_4^-$ with $K_a$= $1.8\times10^6$ $M^{-1}$.

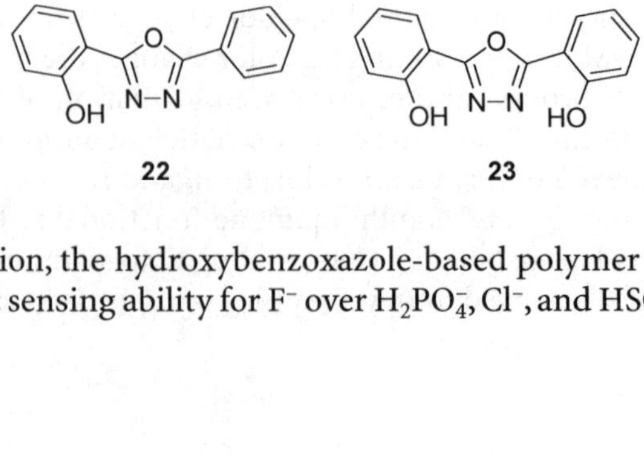

In DMF solution, the hydroxybenzoxazole-based polymer **24** showed selective colorimetric sensing ability for $F^-$ over $H_2PO_4^-$, $Cl^-$, and $HSO_3^-$ [27]. Upon ad-

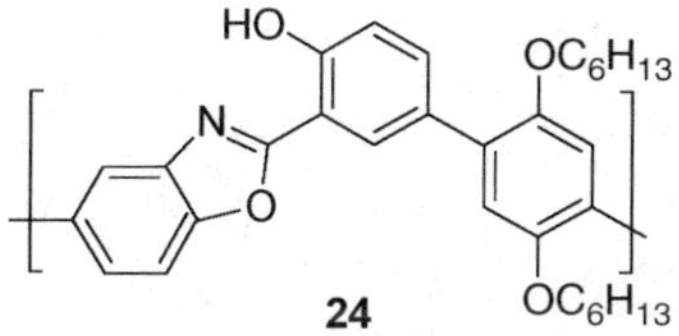

**24**

dition of $F^-$ to the solution of **24**, the color changed from colorless ($\lambda_{max}$=333 nm) to yellow ($\lambda_{max}$=420 nm) due to the binding of $F^-$ to this polymer. Therefore, compound **24** can act as a naked-eye detector for $F^-$ anion.

## 2.7
## Chromogenic Hosts Containing Dipyrroylquinoxaline Units

Sessler and coworkers have recently investigated 2,3-dipyrrol-2′-ylquinoxaline derivatives **25–27** [28] as potential anion receptors and sensors. The electronic influence of the functional groups present in the receptor played a crucial role in its recognition and sensing ability. It was found that receptor **26** had a higher affinity for $F^-$ ($K_a$=1.2×10$^5$ M$^{-1}$) than **25** ($K_a$=2×10$^4$ M$^{-1}$) in $CH_2Cl_2$. The solution of **26** underwent a dramatic color change from yellow to purple in the presence of $F^-$. Furthermore, receptor **27** showed the lack of a complete NH hydrogen bond donor. It then displayed a slight change in the optical spectrum.

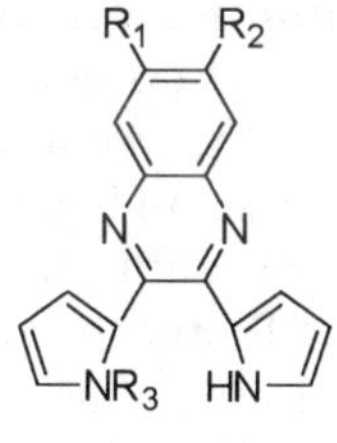

**25.** $R_1 = R_2 = R_3 = H$
**26.** $R_1 = R_3 = H$, $R_2 = NO_2$
**27.** $R_1 = R_2 = H$, $R_3 = SEM$

The fluorinated dipyrrolylquinoxaline **28** [29] underwent a sharp yellow to orange color change in the presence of both $F^-$ ($K_a$=61,600 M$^{-1}$) and $H_2PO_4^-$ ($K_a$=17,300 M$^{-1}$) anions in $CH_2Cl_2$. No changes in color were observed upon addition of $Cl^-$ ($K_a$=180 M$^{-1}$).

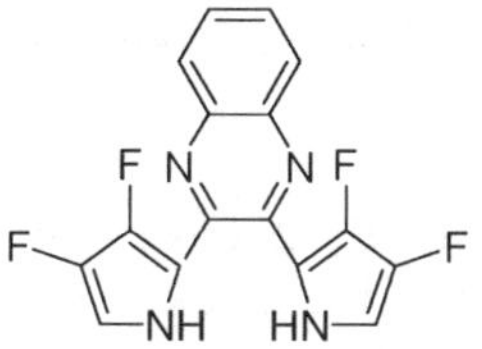

**28**

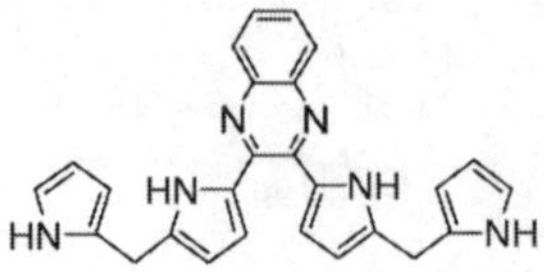

29

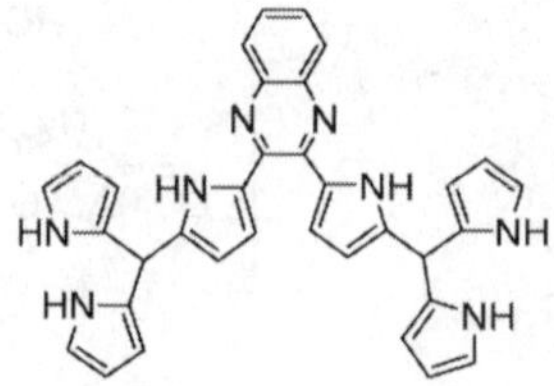

30

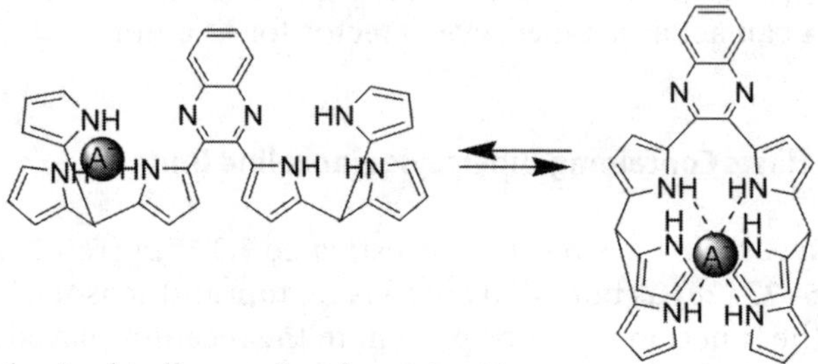

**Fig. 5** Proposed anion binding modes of receptor **30**

Sessler et al. [30] synthesized two novel quinoxaline derivatives bearing dipyrromethane **29** and tripyrromethane **30**. Both compounds were found to be much better anion receptors not only for $F^-$ ($K_a$=32,000 $M^{-1}$ for **29** and >1,000,000 $M^{-1}$ for **30**), but also for $H_2PO_4^-$ ($K_a$=4,300 $M^{-1}$ for **29** and 300,000 $M^{-1}$ for **30**) in $CH_2Cl_2$. The substantial increase in affinities seen in the case of $H_2PO_4^-$ was ascribed to the greater number of pyrrole NH donors required to bind a larger anion. In addition, $^1H$ NMR results showed that compound **29** was able to bind anions using two modes of structure as shown in Fig. 5.

Extended dipyrrolyl quinoxalines **31–34** were synthesized by Anzenbacher et al. [31] and displayed strong selectivity for $F^-$ and pyrophosphate anions ($HP_2O_7^{3-}$). Addition of these anions to the solutions of **31–34** in $CH_2Cl_2$ in all cases resulted in the decrease of absorptions at 400–450 nm concurrent with the appearance of strong bands around 500–550 nm.

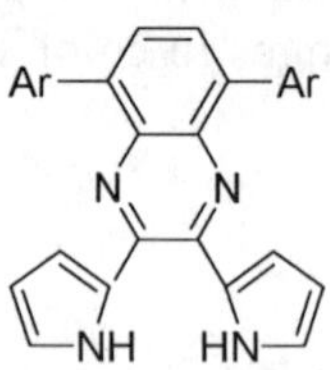

**31.** Ar = -Ph
**32.** Ar = -PhOMe
**33.** Ar = EDOT
**34.** Ar = -PhN(Me)$_2$

Recently, Sessler and coworkers [32] have synthesized a quinoxaline-bridged porphyrinoid **35** that can be used as a naked-eye sensor for $F^-$. Upon addition of $F^-$ to a 10% DMSO:$CH_2Cl_2$ solution of compound **35**, the color of the solution changed from pale orange ($\lambda_{max}$=427 nm) to red ($\lambda_{max}$=480 nm) with

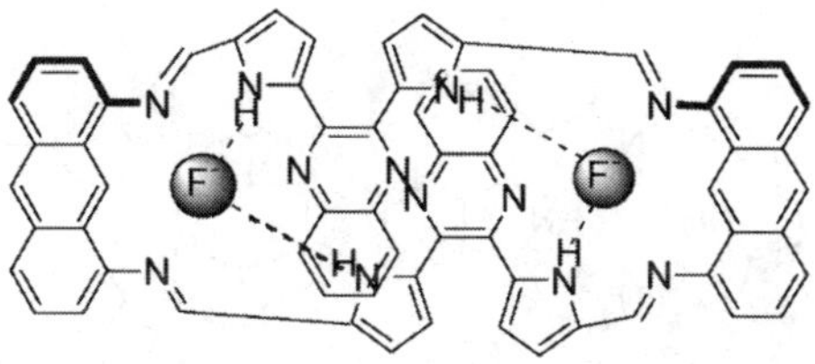


**35.2F⁻**

log $K_{a1}$=11 and log $K_{a2}$=2.2. The binding of the first fluoride anion makes the binding of the second somewhat more facile. The change in color is not observed upon addition of $Cl^-$, $Br^-$, $NO_3^-$, or $HSO_4^-$.

One example of the cavitand-based anion sensor, an open-ended host molecule with a vase-shaped cavity, was synthesized by Rebek and coworkers [33]. In this study, cavitand-based dipyrrolyl quinoxaline compound **36**, featuring four anion-detecting walls, was synthesized and showed a selective binding for $F^-$ and $AcO^-$ in toluene:acetone (4:1 v/v). A dramatic color change from yellow to dark red was observed. A band at $\lambda_{max}$=420 nm disappeared and two new bands at $\lambda$=350 and 490 nm emerged in the UV–vis spectrum.

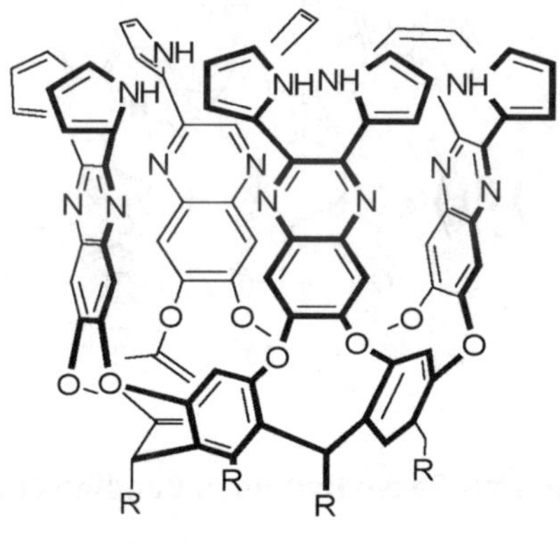


**36**, R = $(CH_2)_{10}CH_3$

## 2.8 Chromogenic Hosts Containing Porphyrin Units

Starnes et al. [34] synthesized a disulfonamide porphyrin receptor **37**. Anion binding was detected by the perturbations of the porphyrin Soret and Q bands. The Soret band of **37** at 422 nm showed a red shift upon addition of anions in $CH_2Cl_2$ and isosbestic points were observed. Compound **37** was able to bind $F^-$ via hydrogen bonding from sulfonamide NH protons in a 2:1 anion-to-ligand ratio with $K_{a1}$=255,000 $M^{-1}$ and $K_{a2}$=1,700 $M^{-1}$.

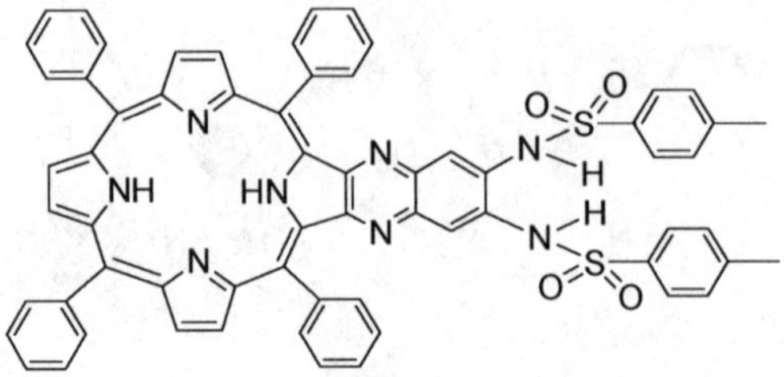

**37**

In MeCN solutions, the Soret band of receptor **38** [35] at 425 nm significantly bathochromically shifted upon addition of $H_2PO_4^-$ and hypsochromically shifted upon addition of $Cl^-$, and split into two bands at 430 and 440 nm upon addition of $HSO_4^-$. Anions formed 1:1 complexes with **38** via favorable amide hydrogen-bonding interactions and electrostatic interactions with four cobaltocenium moieties.

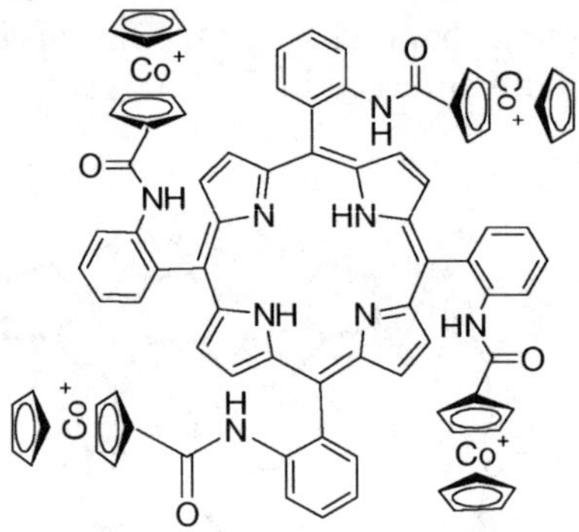

**38**

## 2.9 Nanoparticles Containing NH-Based Hydrogen Bonding Chromogenic Hosts

Nowadays, fabrication of chromophore-functionalized gold nanoparticles has emerged as one of the most exiting research areas, and their possible applications as anion sensors were also presented by Kamat et al. [36] The amide-functionalized gold nanoparticle **39** [37] was used for optical sensing of $H_2PO_4^-$, $HSO_4^-$, $AcO^-$, $NO_3^-$, $Cl^-$, $Br^-$, and $I^-$. In $CH_2Cl_2$, nanoparticle **39** showed a red-wine color with a characteristic plasmon band centered at 520 nm. When 0–0.5 equivalents of the anions were added to the solution of **39**, the plasmon band decreased in intensity with a slight red shift in wavelength. This marked decrease in the intensity of the band was ascribed to an anion-induced aggregation of **39** through hydrogen bond formation between anions and the interparticle amide groups. Further addition of anions to **39** solutions caused

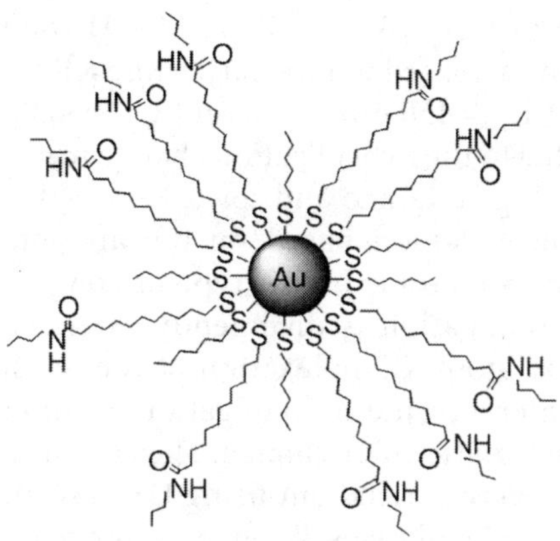

**39**

an increase in intensity of this plasmon band, which reflected the desegregation of the nanoparticle built by **39** and the corresponding anions. This nanoparticle was capable of optically sensing changes in anion concentration at a level of $10^{-6}$ M.

## 3 Lewis Acid Chromogenic Hosts

This section involves a type of receptor that requires the incorporation of a defined number and type of Lewis acids, such as B, Zn(II), and Hg(II), into a molecular skeleton with their electron-deficient sites exposed for interaction with the lone electron pair of anions.

James and coworkers [38] have synthesized an azo dye molecule with boronic acid, receptor **40**, which can act as a colorimetric sensor for detection

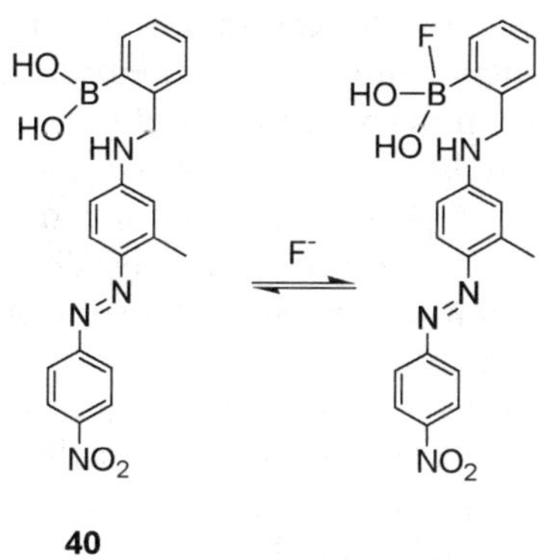

**Fig. 6** Proposed binding mode of fluoride with **40**

of $F^-$. Upon addition of KCl, KBr, and KI in MeOH solution of **40** the orange absorbance at 450 nm increased in intensity, while KF caused a dramatic color change from orange ($\lambda_{max}$=450 nm) to claret ($\lambda_{max}$=563 nm) with $K_a$=130 $M^{-1}$. The color change is due to the hybridization change from $sp^2$ to $sp^3$ when boron binds $F^-$, as shown in Fig. 6. Tamao and coworkers [39] reported a new type of multistage sensing of $F^-$ anion based on boron-containing π-conjugating systems **41–44.** The boron atoms in this type of compound possess a unique LUMO in which π-conjugation is divergently extended through the vacant p-orbital of the boron atom. Complexation of the boron π-electron systems with anions would interrupt the π-conjugation extended through the boron atom, resulting in a dramatic color change. The complexation studies of compounds **41–43** and $F^-$ were carried out using UV–vis spectrophotometry and showed $K_a$>2.6×$10^5$ $M^{-1}$ in all cases. When $F^-$ anion was added to a THF solution of each boron-containing compound, a dramatic color change occurred from colorless to orange for **41** and from yellow to colorless for compounds **42** and **43.**

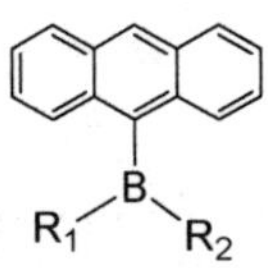


**41.** $R_1$ = $R_2$ = 9-anthryl
**42.** $R_1$ = 9-anthryl, $R_2$ = mesityl
**43.** $R_1$ = $R_2$ = mesityl

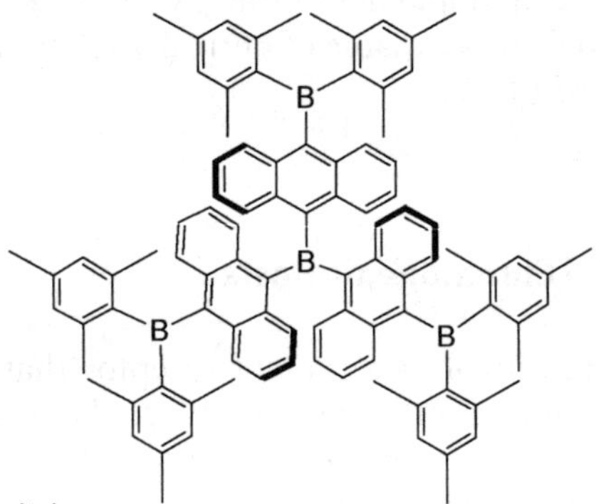


**44**

The more extended π-electron system **44** has four boron atoms in two different environments. When the concentration of $F^-$ increased, the absorption at 524 nm decreased, a new band at 474 nm emerged together with a broad shoulder band around 570 nm, and the color of the solution changed from red to orange. The UV–vis analysis implied that compound **44** was able to bind three $F^-$ ions. The three stepwise binding constants in THF solutions are 6.9×$10^4$, 9.0×$10^2$, and 2.1×$10^2$ $M^{-1}$, respectively. The results suggest that the complexation of $F^-$ in **44** may occur not with the internal boron atom but with the three peripheral boron atoms instead.

The triarylborane-porphyrin conjugate receptor **45** [40] can bind a $F^-$ ion concurrent with the change from $sp^2$ to $sp^3$ hybridization of the boron atom. The change in the hybridization interrupts both the electronic communication and the dipolar interaction between the triarylborane and porphyrin unit as well as the internal charge transfer in the presence of $F^-$. Bathochromic shifts

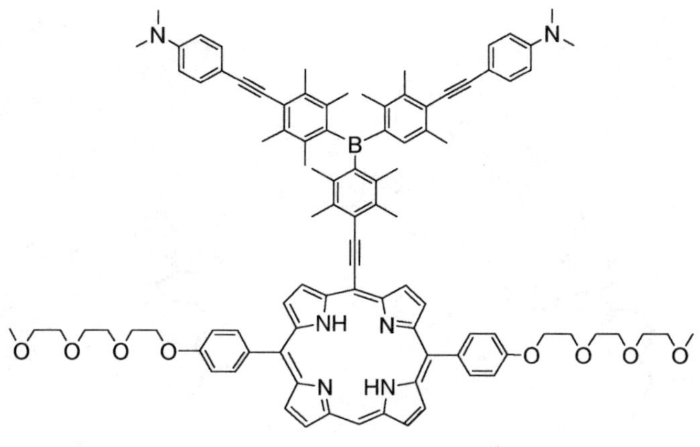

**45**

of the Soret and Q bands were observed upon addition of $F^-$. No changes in color were observed upon exposure of **45** to $Cl^-$, $Br^-$, $I^-$, $AcO^-$, or $OH^-$.

Martínez-Máñez and coworkers [41] showed the complex of Hg(II) with aza-oxa-thia macrocycle **46** can be used for the specific chromogenic sensing of $NO_3^-$ anion. Addition of $NO_3^-$ to the MeCN solution of **46** resulted in a shift of the 540 nm band to 490 nm, with a concomitant change in color from red to yellow. UV–vis titrations of **46** with $NO_3^-$ in $CH_2Cl_2$ gave a stability constant of log $K$=6.2. NMR studies of a **46**/$NO_3^-$ system showed a slight shift of signals assigned to the protons *ortho* to the azo group. This shift could be related to the presence of π-stacking interactions between the nitrate ion and the *p*-nitrophenylazobenzene group. The selective sensing of nitrate by **46** has found application in $NO_3^-$ determination in water.

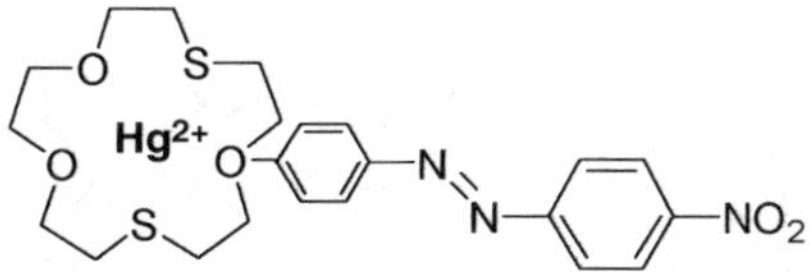

**46**

Ditopic neutral receptors composed of both a Lewis acid binding site (zinc porphyrin unit) and a Lewis base binding unit (crown ether moiety) **47** and **48** were reported by Kim and Hong [42]. Such receptors bound only NaCN in a ditopic fashion. UV–visible titrations between the receptors **47** and **48** with NaCN in $CH_2Cl_2$ resulted in a red shift of the Soret band of the zinc porphyrin units, which were the origin of the color change from red to green in all cases. The association constants of **47** and **48** complexes with NaCN in DMSO:water (9:1 v/v) are $5.7\times10^4$ and $5.1\times10^4$ $M^{-1}$, respectively.

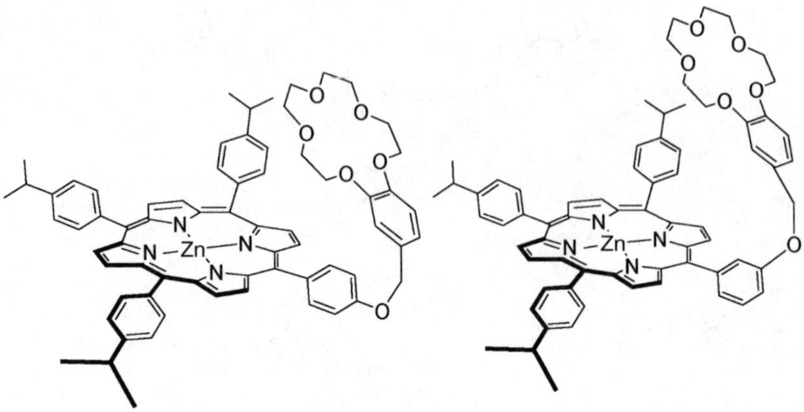

**47** **48**

New azophenolurea-containing porphyrin compounds **49–52** for the colorimetric sensing of anions were reported by Hong and coworkers [43]. These compounds contained both Lewis acid centers and urea anion binding sites. Receptors **49** and **50** show high selectivity for $AcO^-$ and $H_2PO_4^-$ in DMSO. The Soret band at 433 nm shifted to 438 nm when $H_2PO_4^-$ was added to the solution of **50**, indicating that a 1:1 complexation occurred with $K_a=2\times10^4$ $M^{-1}$. Compounds **51** and **52** showed dramatic color changes in $CH_3CN$ upon addition of $F^-$ due to an increased interaction with a nitrophenylazo phenolic OH group. Furthermore, upon addition of $H_2PO_4^-$ to the solution of **51**, the intensity of the Soret band decreased concurrent with the appearance of a new band at about 600 nm.

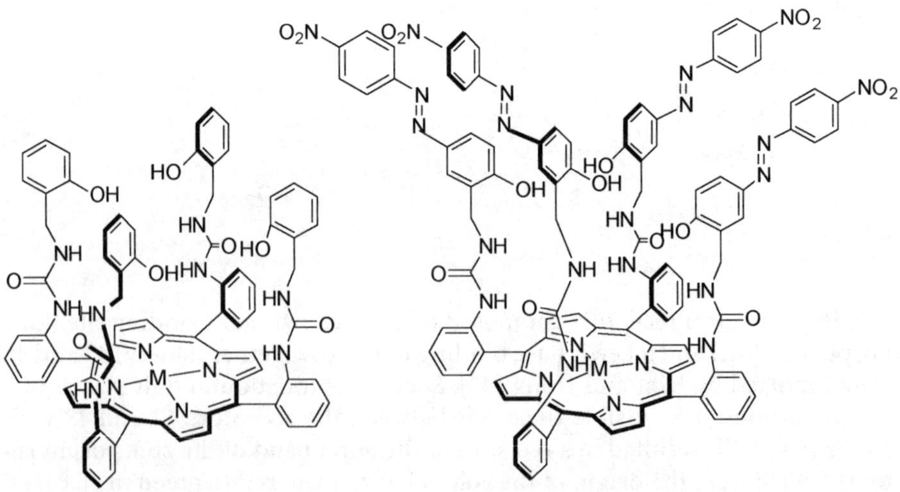

**49.** M= 2H
**50.** M = Zn

**51.** M = 2H
**52.** M = Zn

## 4
## Metal-Ion-Templated Chromogenic Hosts

The intramolecular enrollment of different binding groups can be readily achieved by a metal template. The basic concept shown in Fig. 7 requires that individual subunits contain both a metal binding region and an anion binding site. Addition of a metal ion will cause the coordination of metal binding regions and the organization of the binding site for anions [44].

Hamilton and coworkers have synthesized bipyridine-bisthiourea subunit **53** in the presence of Fe(II) or Ru(II) ions [45]. These resulted in the forma-

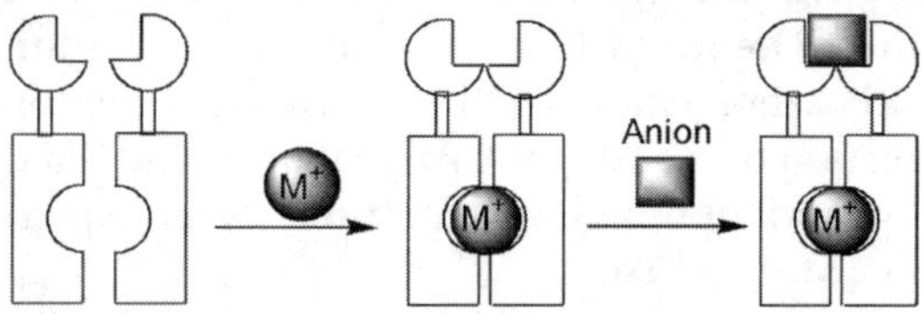


**Fig. 7** Schematic of binding site organization by metal template effects

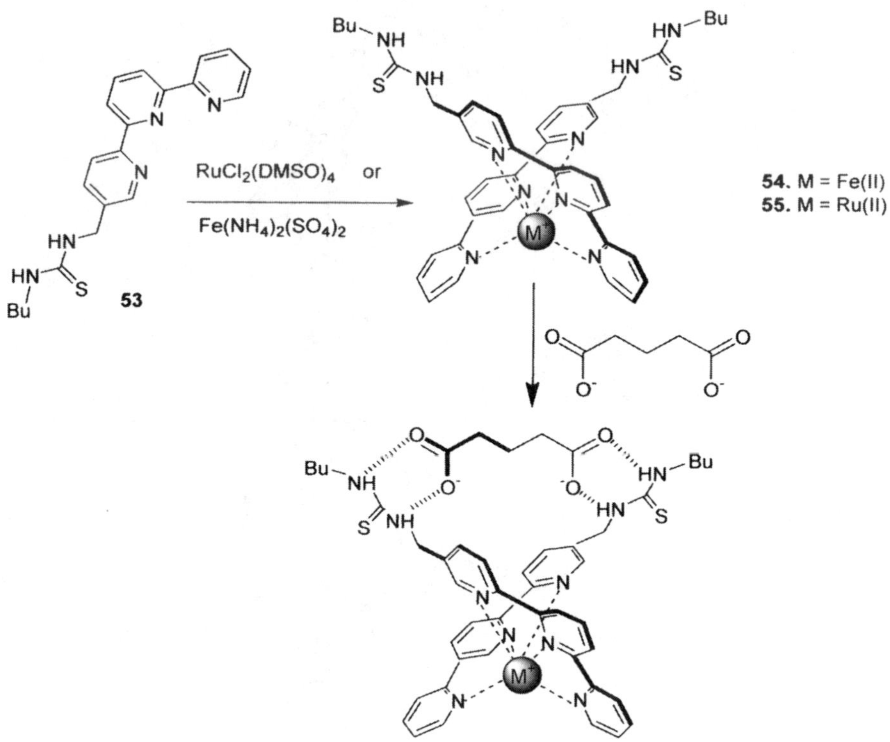


**Fig. 8** Proposed binding mode of **54** and **55** with dicarboxylates

tion of bisthiourea binding regions of complexes **54** and **55** for binding anions, as presented in Fig. 8. Preliminary results showed that complex **55** participated in binding with dicarboxylate anions such as glutarate ($K_a$=8.3×$10^3$ $M^{-1}$), adipate ($K_a$=2.9×$10^3$ $M^{-1}$), and pimelate ($K_a$=6.0×$10^3$ $M^{-1}$) to different extents.

Weiss and coworkers [46] have reported a new receptor for dicarboxylic acids, related guests to dicarboxylate anions, based on the self-assembly of functionalized *meta*-catechol **56** around *cis*-$[MoO_2]^{2+}$. UV–vis studies for dicarboxylic acid binding showed a 1:1 stoichiometry for the complex formation (Fig. 9). Receptor **57** displayed selectivity for $C_4$ and $C_5$ dicarboxylic acids. The stability constants (log *K* values) of the complex **57** with $C_4$ and $C_5$ dicarboxylic acids in $CH_3CN$ are 6.1 and 4.3, respectively. This should be explained by the free rotation of the CO–phenyl bond enabling the receptor to adapt to the length of the diacid carbon chain. Furthermore, this receptor exhibited enantioselective recognition of D and L-*N*-CBz-glutamic acid via directed multiple hydrogen bonding and lipophilic interactions with topographical control achieved by the *cis*-$[MoO_2]^{2+}$ core.

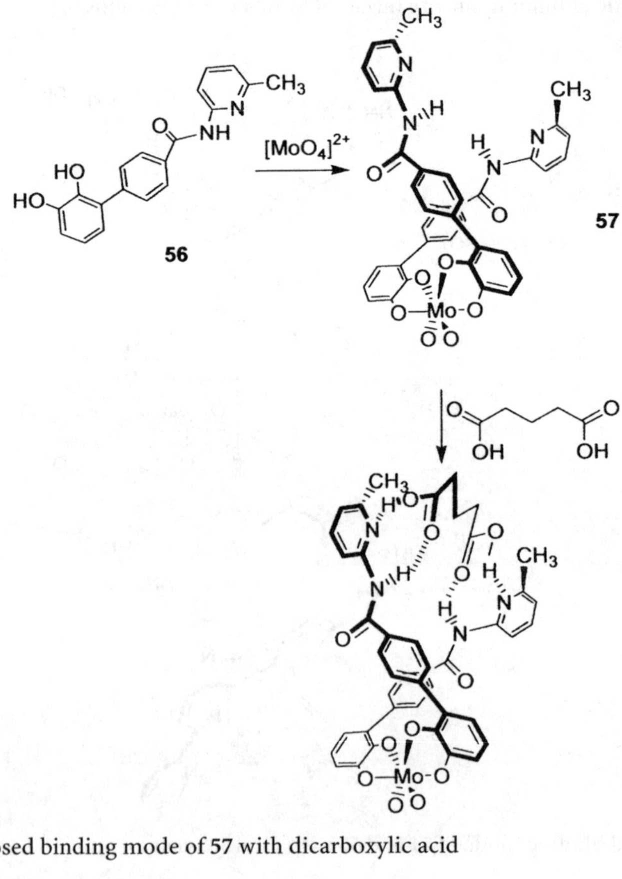


**Fig. 9** Proposed binding mode of 57 with dicarboxylic acid

Furthermore, Hamilton and coworkers [47] synthesized the dicarboxylic acid receptors **58** and **59**. These two receptors consist of a 2,9-disubstituted phenanthroline bearing one or two acylamino pyridine binding sites, which can coordinate to a Cu(I) ion. This arrangement was ideally suited to the strong hydrogen bonding complexation of dicarboxylic acids (Fig. 10), and large downfield shifts of the NH resonances were observed in an NMR study. Significant changes in the UV absorption of **58** and **59** in $CHCl_3$ also took place on addition of glutaric acid, with the color of solution changing from red-orange to orange-red for **58** and orange to red for **59**. The Job's plot showed 1:1 and 1:2 stoichiometries of complex formation between **58** or **59** and dicarboxylic acids, respectively. Binding constants for the complexes of **58** with glutaric acid and pimelic acid measured by $^1H$ NMR in $CDCl_3$ are $4.3\times10^4$ and $1.6\times10^4$ $M^{-1}$, respectively. The chromogenic effect presumably came from a change in the conjugation of the (benzoylaminopyridyl) side arm of the phenanthroline–metal system due to conformational changes on guest binding.

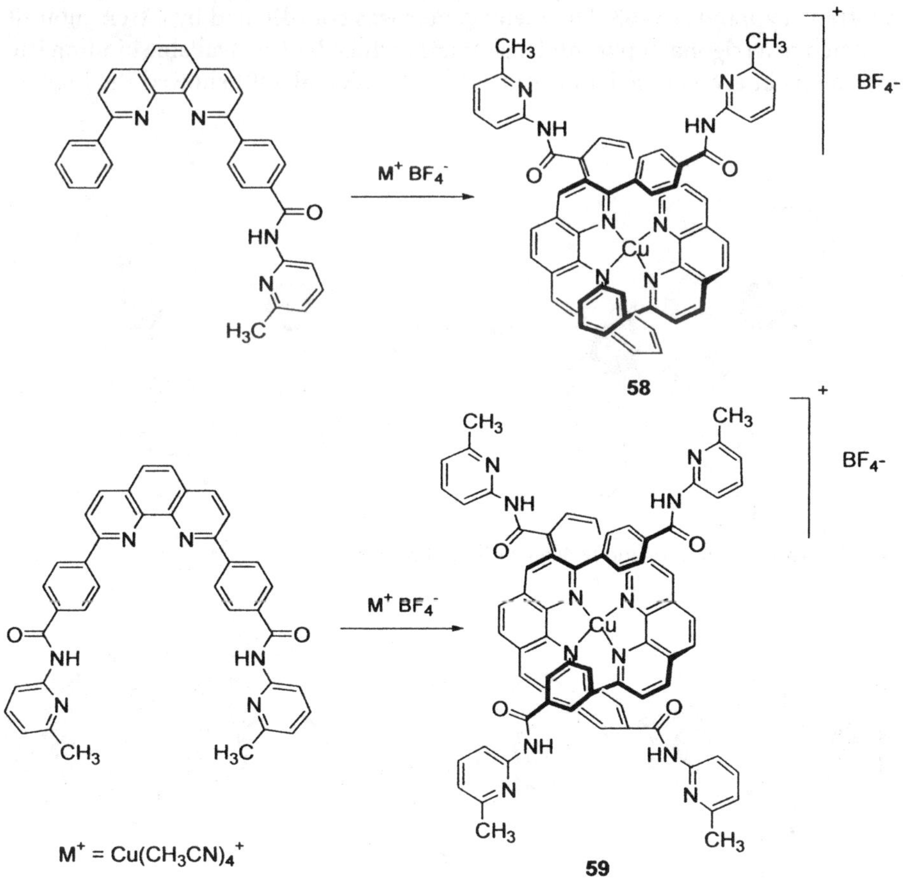


**58**

**59**

**Fig. 10** Proposed binding modes of 58 and 59 with dicarboxylates

## 5 Transition Metal Complexes as Chromogenic Hosts

This type of chromogenic sensor utilizes the coordination chemistry of transition metal complexes, which have vacant binding sites to bind specific anions or have pendant arms containing anion receptor units. Transition metal complexes already have their own specific colors due to their different electronic structures. Coordinating directly to anions or binding of anions by the pendant arms results in perturbations of their electronic structures and causes color changes.

The dicopper(II) complex **60** [48] was capable of sensing imidazolate anion over others at pH 9 with log $K$=4.7 (Fig. 11). Upon addition of imidazolate, the pale blue color of **60** ($\lambda_{max}$=640 nm) turned to a more intense blue concurrent with the absorption band being shifted to 690 nm. In addition, these results suggest that **60** can detect L-histidine which contains an imidazole moiety giving intense blue color (log $K$=5.5).

Fabbrizzi and coworkers reported the synthesis of dicopper(II) complexes of bis-tren cryptands **61–63**. The metal center was coordinated by a tren subunit oriented to a trigonal bipyramidal geometry which had an available binding site for anions at a vacant axial position. The complex **61** [49] showed the highest

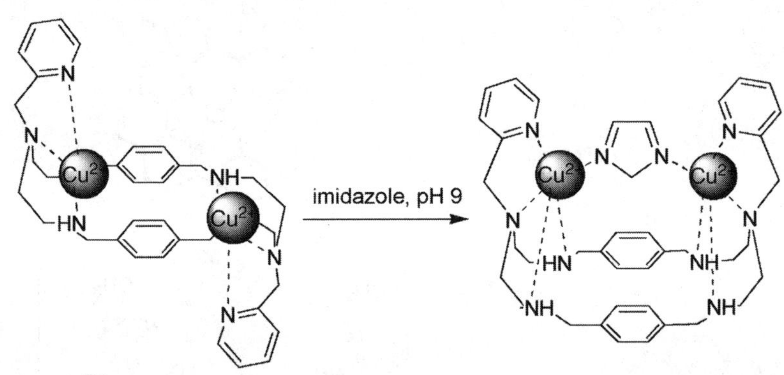


**60**

**Fig. 11** Proposed binding mode of **60** with imidazolate

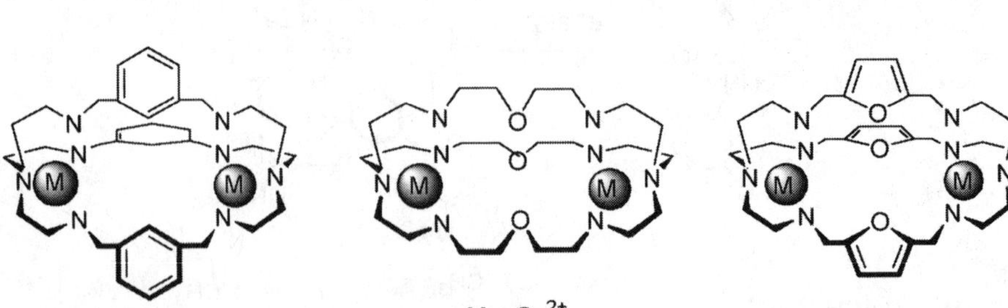

M = $Cu^{2+}$

**61** **62** **63**

selectivity for $N_3^-$ anion with log $K$=4.78 at pH 8 in a 1:1 complex stoichiometry, with a color change from blue to bright green ($\lambda_{max}$=400 nm). Receptors **62** and **63** [50] displayed a special affinity for halide ions in a 1:1 stoichiometry. Halide inclusions could be detected visually by the appearance of an intense bright yellow color above 400 nm in the spectrum ($Cl^-$, $\lambda_{max}$=410 nm; $Br^-$, 430 nm, and $I^-$, 440 nm). Compound **63** was not only able to include halide ions but also linear triatomic anions such as $N_3^-$ (log $K$=4.7) and $NCS^-$ (log $K$=4.28). The aqueous solution turned pale blue to intense olive green ($\lambda_{max}$=386 nm). This was due to the NH–$CH_2$–C–O torsion angles along the chain arm of 2,5-dimethylfuran behaving as a spring, which controlled the length of the intermetallic cavity.

Beer and coworkers [51] have combined ferrocene and mixed amide–amine moieties for a new anion sensor. The receptor **64** showed a spectacular UV–vis response to the addition of $H_2PO_4^-$ in $CH_3CN$, which resulted in hydrogen bonding to the amide units near the ferrocene and caused a large UV perturbation. The d–d band at 443 nm showed a hypsochromic shift and a hyperchromic effect. This change was visible to the naked eye with the solution becoming brighter in color.

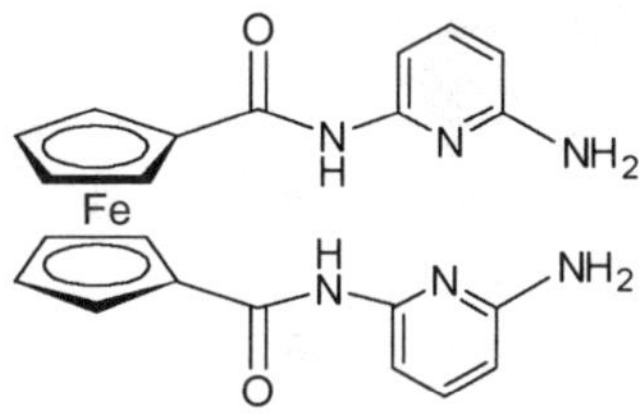


**64**

Recently, Miyaji and coworkers [52] reported the chromogenic molecular switch **65** as a new ditopic ferrocene receptor for anions in the presence of cations. Upon addition of $F^-$ to a solution of **65** in $CH_3CN$, a new absorbance appeared at 472 nm ($K_a$=9,340 $M^{-1}$). Addition of $K^+$ to the solution of **65**·$F^-$ caused a reverse behavior and the solution turned colorless. This implied that the color change was controlled by $F^-$ "switches on" and $K^+$ "switches off". This result was

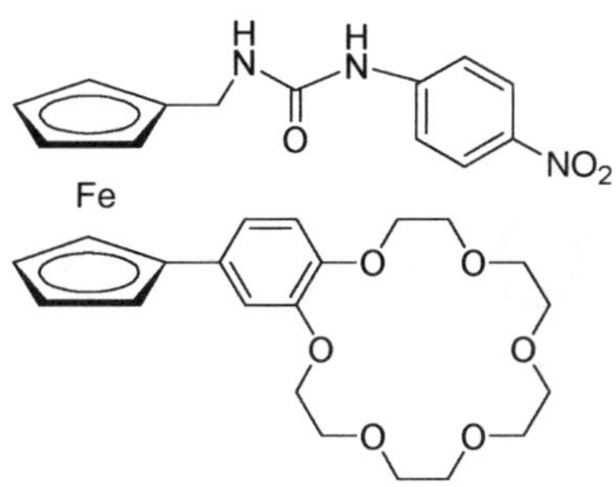


**65**

confirmed by NMR studies in $CD_3CN$. When $K^+$ was added to the solution of **65**·$F^-$, large upfield shifts of the proton of the urea-NH occurred and shifted to the position of the free receptor. The results from $^1H$ NMR and UV–vis studies, therefore, supported the assumption that it was the presence of $K^+$ in close proximity to the fluoride binding center of **65** that caused the color quenching process. This receptor may thus be applied as an optical device at the molecular level.

A luminescent Ru(II) metal complex with 2,3-di(1*H*-2-pyrrolyl)quinoxaline receptor **66** was reported by Castellano and colleagues [53] to undergo a red shift in the UV–vis spectrum upon addition of $CN^-$ in 98% $CH_2Cl_2/CH_3CN$ solution ($K_a$=428,000 $M^{-1}$). The low-lying MLCT absorption ($\lambda_{max}$=465 nm) and the high-energy π–π* ligand-based absorptions decreased monotonically throughout the addition. These absorption features signified that $CN^-$ anion bound to the metal complex through the dipyrroylquinoxaline receptor.

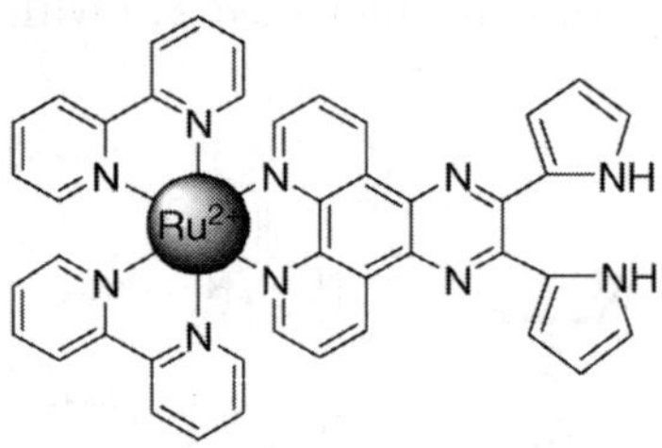


**66**

The receptor **67** [54] shows the sensing of $F^-$ in DMSO. Upon addition of $F^-$ to the solution of receptor **67** two bands at 323 and 525 nm decreased and a new band at 652 nm appeared. Furthermore, the color of the solution changed from red-pink to pale purple. These spectral changes were ascribed to the formation of a 1:1 adduct between the metal complex and $F^-$ anion, in which $F^-$ anion bound two dipyrrolylquinoxaline units ($K_a$=54,000 $M^{-1}$). Receptor **67** was also studied electrochemically by cyclic voltammetry. A clearly reversible redox signal was observed at 160 mV (vs SHE), which was assigned to the Co(III)/Co(II) reduction. The addition of $F^-$ led to a complete disappearance

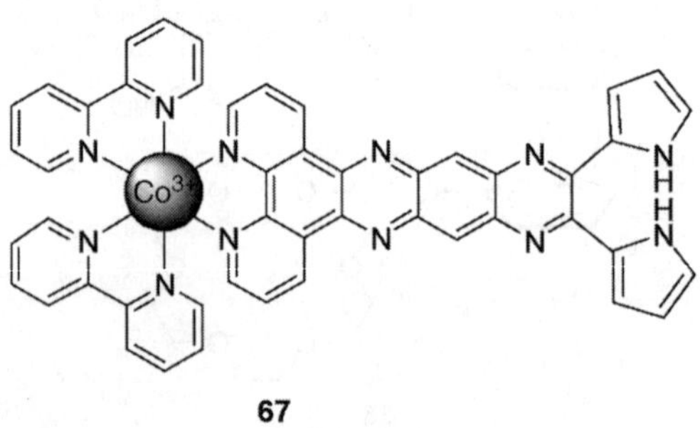


**67**

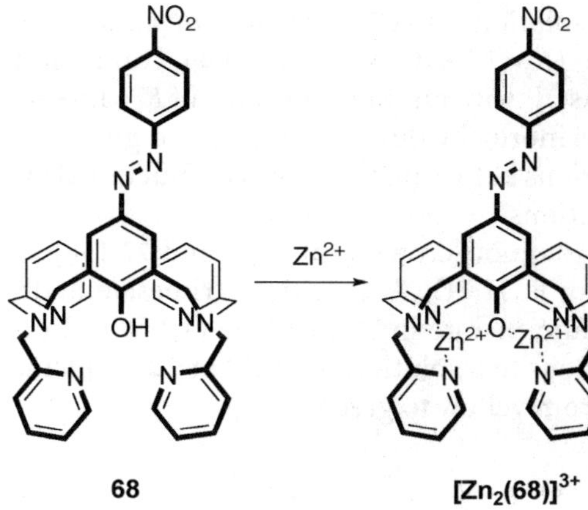

Fig. 12 Formation of $[Zn_2(\mathbf{68})]^{3+}$ complex

of the Co(III)/Co(II) reduction signal due to the formation of a redox-inactive complex.

The complex of Zn(II) with **68** [55], in which both $Zn^{2+}$ atoms coordinated to pyridine-nitrogen donors and a phenolate oxygen donor (Fig. 12), displayed a yellow color at $\lambda_{max}$=417 nm in aqueous solution at pH 7.4. Addition of $H_2PO_4^-$, $AcO^-$, $F^-$, $HCO_3^-$, $Cl^-$, $HPO_4^{2-}$, and citrate to the solution of $[Zn_2(\mathbf{68})]^{3+}$ did not induce spectral changes. However, addition of pyrophosphate causes a bathochromic shift from 417 to 463 nm and a color change from yellow to red. The result showed that a 1:1 adduct was formed between the complex and pyrophosphate anion. Pyrophosphate reacts with $[Zn_2(\mathbf{68})]^{3+}$ via oxygen to give two hexacoordinate $Zn^{2+}$ ions. The red shift occurs by the weakening of the bond between the phenolate oxygen and $Zn^{2+}$ atoms. This increases the negative charge character of the phenolate oxygen giving a bathochromic shift in the visible region.

Anslyn and coworkers [56, 57] synthesized receptors containing Cu(II) binding sites with appended ammonium **69** and guanidinium **70** groups. These receptors showed high affinities for tetrahedral anions such as $HPO_4^{2-}$ and

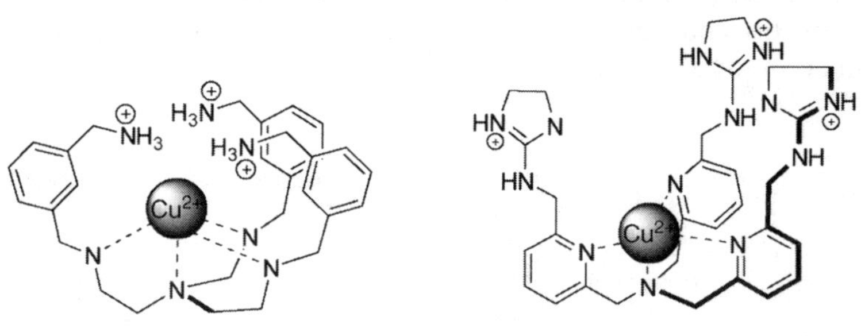

**69** **70**

$HAsO_4^{2-}$ in aqueous solution at pH 7.4. Binding constants of **69** with these guests had the same value ($K_a$=2.5×10$^4$ M$^{-1}$). However, compound **70** was found to bind $HPO_4^{2-}$ and $HAsO_4^{2-}$ with binding constants of $K_a$=1.5×10$^4$ and 1.7×10$^4$ M$^{-1}$, respectively. Furthermore, the binding of host and guest is proposed through ion-pairing interactions between the charges of functional groups and between $Cu^{2+}$ and guest reactions.

1,3,5-Triarylbenzoamido-crown ethers **71** and **72** [58] showed abilities as naked-eye sensors for $F^-$, $H_2PO_4^-$, $AcO^-$, citrate, trimesate, and isophthalate in the presence of potassium picrate in DMSO:$H_2O$ (1:1 v/v). Upon addition of the aforementioned anions to a solution of **71**·$K^+$ or **72**·$K^+$ picrate, the color of the solution changed from yellow to green.

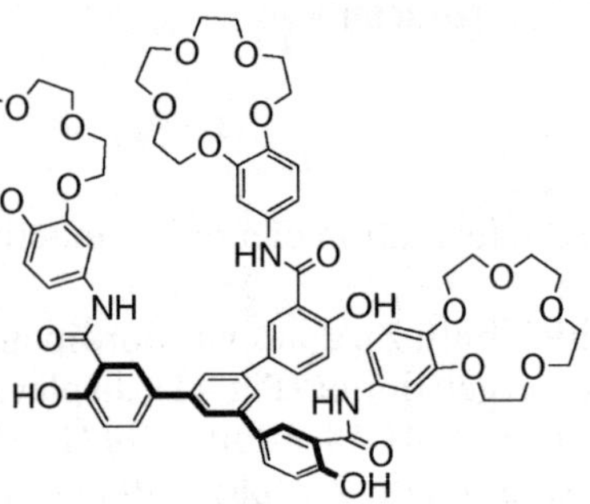

**71**

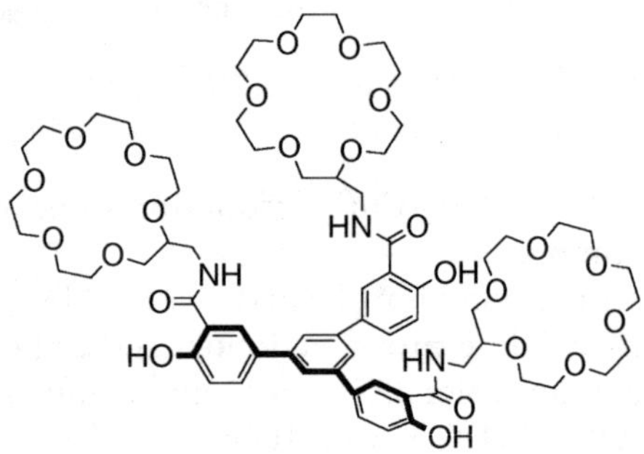

**72**

## 6 Displacement of Chromogenic Guests

An alternative method to a covalently attached chromophore is a competition between the indicator and the analyte in the receptor unit. An indicator is displaced from the binding site upon addition of an analyte, causing a signal modulation (Fig. 13). In order to create a sensitive sensing strategy, an indicator must give a large absorption change upon addition to the receptor, since the subsequent addition of an analyte must lead to a large change in the opposite direction. The advancement of this method has been described in detail by Anslyn and his colleagues [59]. Therefore, the work included in Anslyn's excellent accounts are not restated.

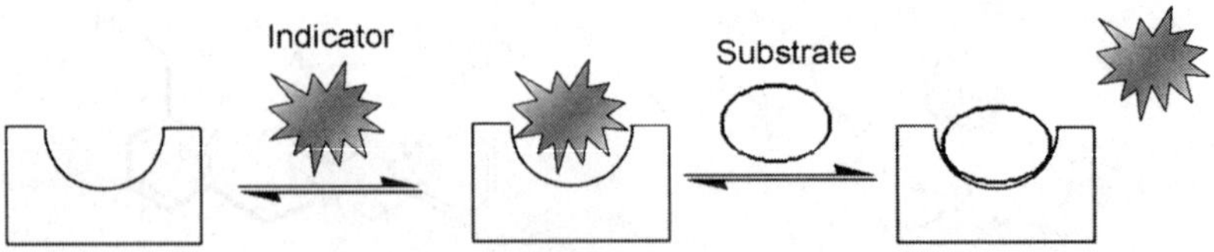


**Fig. 13** Schematic operation of displacement method for a chromogenic sensor

Gale and coworkers [60] have discovered a very simple displacement method for a calix[4]pyrrole-based anion receptor. Addition of calix[4]pyrrole **73** to a solution of tetrabutylammonium 4-nitrophenolate in $CH_2Cl_2$ caused a decrease in intensity of a UV band at 432 nm, and calix[4]pyrrole-4-nitrophenolate complex **74** was formed. Upon addition of $F^-$, 4-nitrophenolate was displaced from the complex, and the color of the solution changed from colorless to yellow due to the presence of the free nitrophenolate anion (Fig. 14).

Shao and coworkers [61] have demonstrated that a noncovalent charge-transfer complex of calix[4]pyrrole-chloranil can be used as a chromogenic sensor for $F^-$ and $H_2PO_4^-$. Upon addition of calix[4]pyrrole **75** to the solution of chloranil **76** in chloroform, the color changed from pale yellow to blue. This indicated the formation of a charge-transfer complex between calix[4]pyrrole and chloranil in chloroform. The Job's plot analysis showed a 1:1 complex with

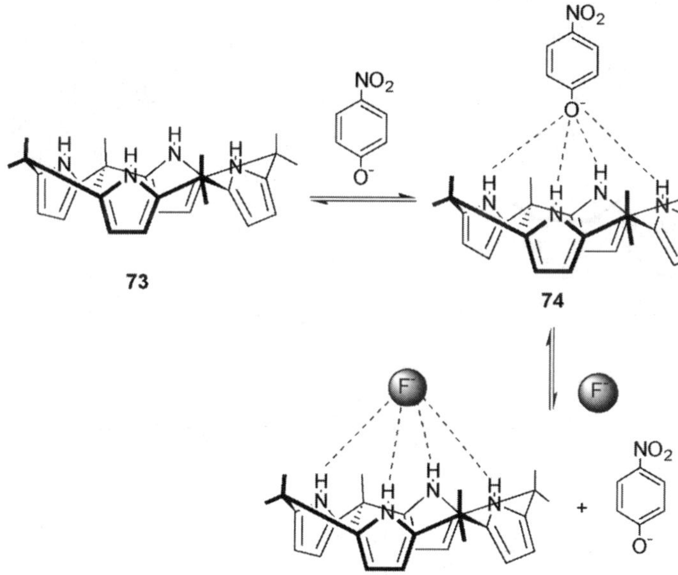


**Fig. 14** The displacement mode for detection of $F^-$ with 73

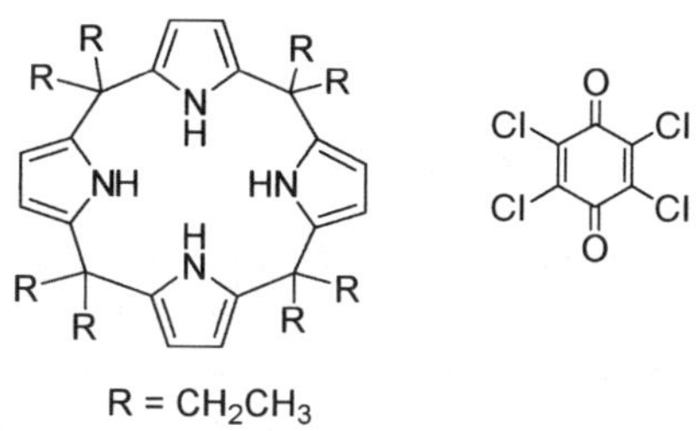


**75** **76**

$K_a$=5.8×10$^{-2}$ M$^{-1}$. In the presence of $F^-$ and $H_2PO_4^-$, the charge-transfer band, $\lambda_{max}$=622 nm, of the complex **75·76** disappeared, while a new absorption band appeared around 400–500 nm and the color of the solution turned from blue to orange-yellow.

Heparin **77**, an anticoagulation reagent, is used during cardiopulmonary bypass surgery. During surgery, heparin is commonly monitored by determining the activated clotting time (ACT). Zhong and Anslyn [62] have recently reported a new designed receptor and sensor for heparin possessing a novel amino acid with a boronic acid side chain, **78**. This new receptor showed strong binding to heparin with good selectivity. Pyrocatechol violet **79** was used as an indicator for the colorimetric sensing of heparin. The binding of **79** to the receptor **78** in MeOH/water caused a color change from yellow ($\lambda_{max}$=430 nm) to grayish purple ($\lambda_{max}$=526 nm) with $K_a$=7.1×10$^3$ M$^{-1}$. When heparin replaced **79** and was bound in the cavity of the receptor **78**, the color of the solution turned to yellow ($K_a$=3.8×10$^4$ M$^{-1}$).

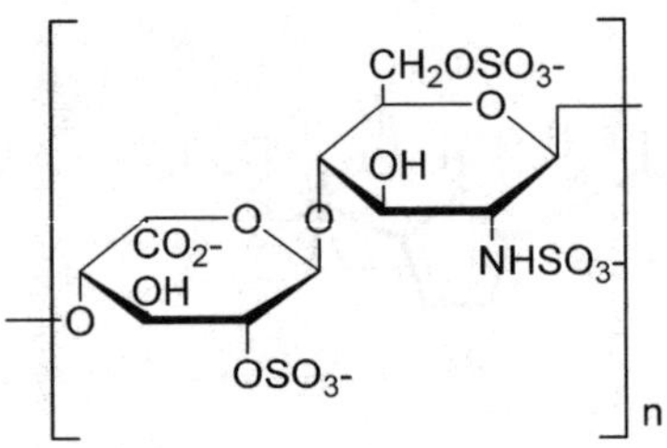


77

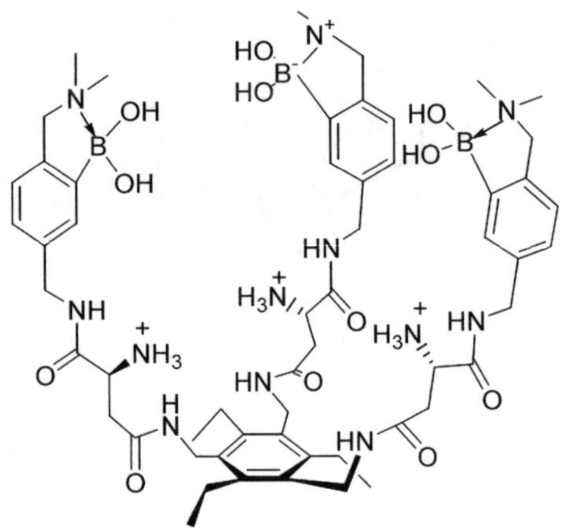


78

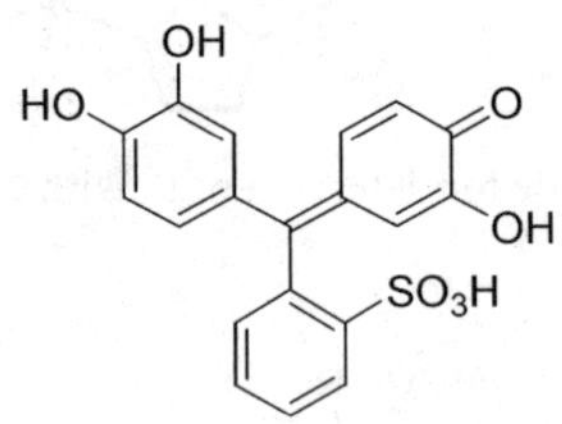


79

The detection of anions such as $HPO_4^{2-}$ in water is a challenging task due to the competing solvation effect between water and anions. Han and Kim [63] have recently reported a colorimetric sensor that can detect $HPO_4^{2-}$ in aqueous solution at neutral pH. The dinuclear Zn(II) complex of 2,6-bis{[bis(2-pyridylmethyl)amino]methyl}-4-methylphenol (H-bpmp) **80** was synthesized.

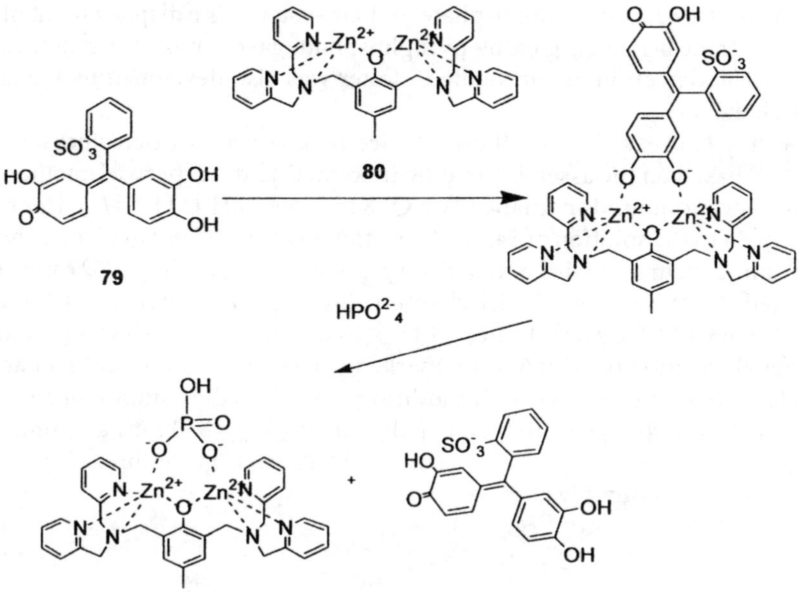


**Fig. 15** Schematic representation of the displacement of **79** from **80** by $HPO_4^{2-}$

The Zn(II) H-bpmp is colorless and has good water solubility. In addition, it can bind $HPO_4^{2-}$ very strongly. The authors used pyrocatechol violet **79**, a catechol-type pH-sensitive dye, as a chromogenic indicator for the sensor. When the solution of complex **80** was titrated into the solution of the indicator, the color changed from yellow ($\lambda_{max}$=444 nm) to blue ($\lambda_{max}$=624 nm) signifying the binding of **79** to the zinc complex. The addition of $HPO_4^{2-}$ to the aqueous solution of the ensemble resulted in a change of the color from blue to yellow, indicating free pyrocatechol violet that was replaced by the anion (Fig. 15). Moreover, other anions failed to cause the color change.

A similar system for detecting pyrophosphate optically in water was created by Fabbrizzi et al. [64]. The researchers used the Cu(II) complex of the bisdiene macrocycle 3,6,9,16,19,22-hexaazatricyclo[22.2.2.2(11.14)]-triacontal(26),-11(12),13,24,27,29-hexaene **81** as a receptor for binding anions, dyes such as

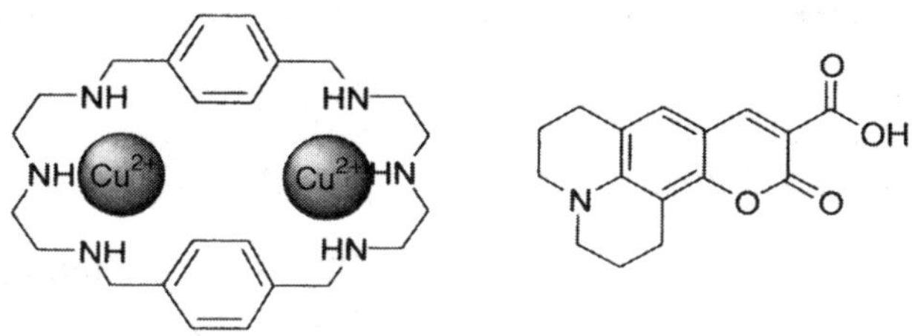

**81** **82**

coumarin 343, **82**, and pyrophosphate in their system. The displacement of the dye from the Cu(II) complex by pyrophosphate gave a naked-eye detectable change. The detection of carbonate in water was also developed by the same researchers.

Recently, Guo et al. [65] synthesized selective colorimetric detectors for $PO_4^{3-}$ and $CO_3^{2-}$ based on the assembly of dihydroxymethyl-di-(2-pyrroyl)methane **83** and tetracyanoquinodimethane (TCNQ) **84** in MeCN:$H_2O$ (1:1 v/v). Upon addition of **83** to the solution of **84**, an absorption peak at 397 nm evidently shifted to 317 nm concurrent with a new absorption band appearing at 627 nm. This indicated the formation of a 1:1 charge-transfer complex between **83** and **84** with $K_a$=463.4 $M^{-1}$. Upon addition of $PO_4^{3-}$ to the solution of **83·84**, the charge-transfer absorption band at 627 nm markedly decreased and the color changed from blue to orange. Similarly, the addition of $CO_3^{2-}$ to the solution of **83·84** induced the color change from blue to pale purple ($\lambda_{max}$=550 nm) and finally to orange-yellow after 20 h. These results were caused by the break-up of the charge-transfer assembly **83·84**.

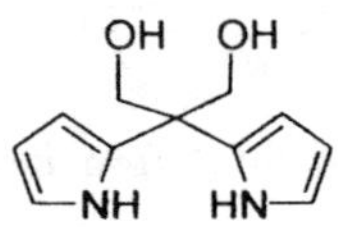


**83**

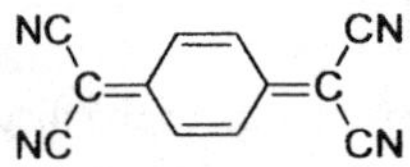


**84**

## 7 Chromoreactands: Concurrent Recognition and Reaction

Besides the covalent attachment of the chromophore to the receptor units or displacement of a chromogenic guest, one can employ the advantages of the selective reactivity between a particular anion and chromogenic hosts, which will display dramatic color changes. There have been at least nine examples of these chromoreactands published recently.

The mixture of **85** and methylene blue **86** [66] at pH 3 showed the characteristic blue color ($\lambda_{max}$=665 nm) of the dye. Upon addition of $F^-$, the oxidation potential of the redox-active ferrocene unit decreased. Under these conditions, the ferrocene group was able to reduce the dye as detected by the disappearance of the band at 665 nm. Other anions had no effect on the ferrocene and dye units. Therefore, this system can be used as a colorimetric detector for fluoride.

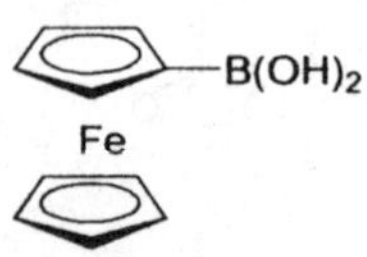


**85**

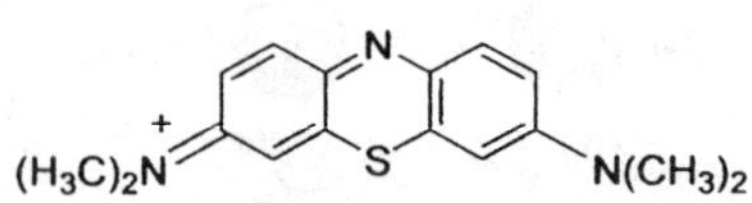


**86**

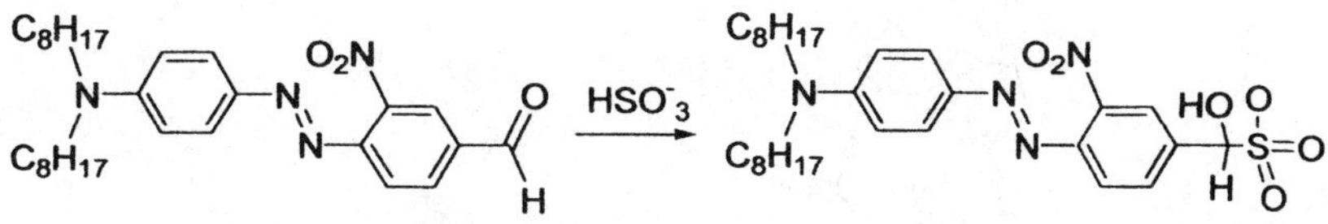

**87**

**Fig. 16** The chemical structure of receptor **87** and the corresponding bisulfite adduct

Mohr has used 4-*N*,*N*-dioctylamino-4′-formyl-2′-nitroazobenzene **87** as a chromoreactand for selective sensing of hydrogen sulfite ($HSO_3^-$) [67]. The aldehyde moiety in **87** is known to interact with hydrogen sulfite by forming a bisulfite adduct, as shown in Fig. 16. The conversion of the aldehyde into the bisulfite adduct can bring about a change in the electron acceptor strength. Consequently, a color change of the formyl dye upon interaction with hydrogen sulfite can be noticed. The investigator has demonstrated that in a sensor layer composed of the chromoreactand and dioctadecylmethylamine in PVC plasticized by 2-nitrophenyloctyl ether, first with plain buffer of pH 4.9 and then with buffer containing 30 mM hydrogen sulfite, a shift in absorbance maximum from 524 to 484 nm was observed.

Ros-Lis et al. [68] have taken advantage of the selective reactivity toward a particular reaction that certain anions display. The nucleophilic characteristics of $CN^-$ allow a reaction with a squaraine dye derivative **88** that results in a distinct color change. The chromogenic sensing ability of **88** was studied in acetonitrile in the presence of various anions: $F^-$, $Cl^-$, $Br^-$, $I^-$, $NO_3^-$, $H_2PO_4^-$, $HSO_4^-$, $AcO^-$, $BzO^-$, $CN^-$, and $SCN^-$. Only $CN^-$ was able to decolorize **88**. $^1H$ NMR data for **88** in $CD_3CN$ upon $CN^-$ addition were consistent with $CN^-$ attack on a carbon of the four-atom squaraine ring next to the phenyl group. This caused the rupture of the electronic decolorization with the consequent disappearance of the 641-nm charge-transfer band.

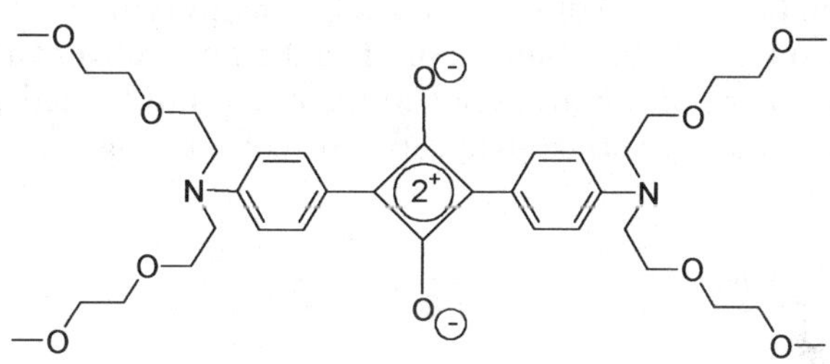

**88**

The transformation of a reactant to another form by changing the pH of the solution can be used as a chromogenic sensor. The 1,3,5-triaryl-1,5-pentanedione derivative **89** [69] can be readily transformed into the corresponding pyrylium ion, as shown in Fig. 17. Addition of nitric acid to a solution of **89** in

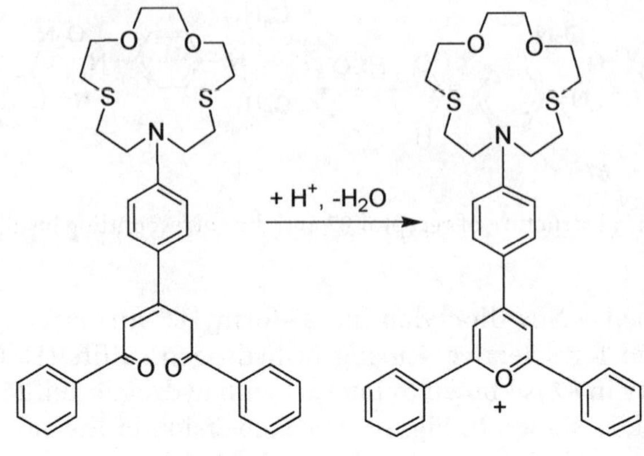

89

**Fig. 17** Transformation of receptor **89** into the pyrylium cation

1,4-dioxane/water caused a dramatic change in color from yellow to magenta. The changes in absorption spectrum of **89** at neutral pH upon adding anions such as chloride, bromide, sulfate, phosphate, ATP, ADP, and GMP have also been studied. However, the most remarkable effect was observed in the presence of ATP, whereby a bright magenta color was fully developed. Therefore, **89** is a selective chromogenic reagent for sensing ATP.

Sancenón et al. [70] have synthesized a number of colorimetric probes to discriminate isomeric dicarboxylate anions. A dioxane/water (70:30 v/v) solution of compound **90** at ca. pH 6 using 0.01 M HEPES buffer is yellow. Upon addition of oxalate and malonate anions, solutions of compound **90** turned to red-magenta. Other dicarboxylate anions such as succinate, glutarate, adipate, pimelate, and suberate did not change the color of the solutions, which remained yellow. The remarkable color change was a result of the anion-induced selective cyclization of **90** to give the colored 2,4,6-triphenylpyrylium cation similar to the one shown in Fig. 17. The solution of **91** in the same solvent and buffer was yellow in the presence of fumarate (a *trans* isomer), but turned to magenta in the presence of maleate (a *cis* isomer). Specific color change occurred because

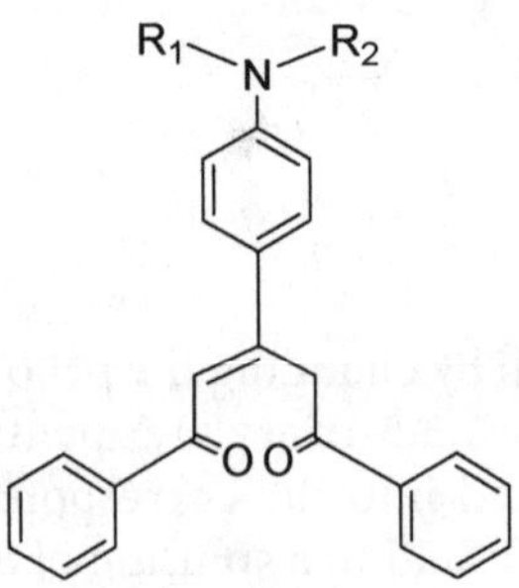

**90**. $R_1 = CH_2CH_2OH$, $R_2 = (CH_2CH_2O)_4CH_2CH_2OH$
**91**. $R_1 = R_2 = CH_3$

the tweezer-like diacid (*cis* isomer and oxalate) was able to act as a chelating ligand toward the hydroxyl group whereas other diacids did not. This shape-induced recognition enhanced the ease of cyclization to the magenta pyrylium cation.

A sensor containing a pyrylium–aniline backbone, **92** [71], reacted toward $SH^-$ to give the pyrylium cycle (Fig. 18). Further addition of an acid induced cyclization and resulted in the corresponding anilinethiopyrylium derivative **93**. This transformation induced a charge-transfer color change from magenta ($\lambda_{max}$=540 nm) to blue ($\lambda_{max}$=580 nm) in water–MeCN (1:1 v/v) that allowed visual recognition of $SH^-$.

Bachas and coworkers [72] have reported a highly selective optical sensor for chloride in blood, based on the multidentate Lewis acid ionophore [9]mercuracarborand-3, **94**. This compound incorporates three electron-deficient mercury centers within a macrocyclic cavity and, therefore, is selective for chloride by such a preorganized cavity. In addition, the compound is chemically stable and hypophilic and, thus, suitable for using as a sensor for $Cl^-$ in blood. This sensor was constructed by embedding the mercuracarborand, a suitable acidic lipophilic dye **95**, and lipophilic cationic sites in a plasticized polymeric membrane film. The sensing mechanism was based on anion binding by **94**, followed by a concomitant protonation of the acid dye to ensure electroneutrality. Such protonation of the proton chromoionophore led to a large change in the absorbance of the optical film.

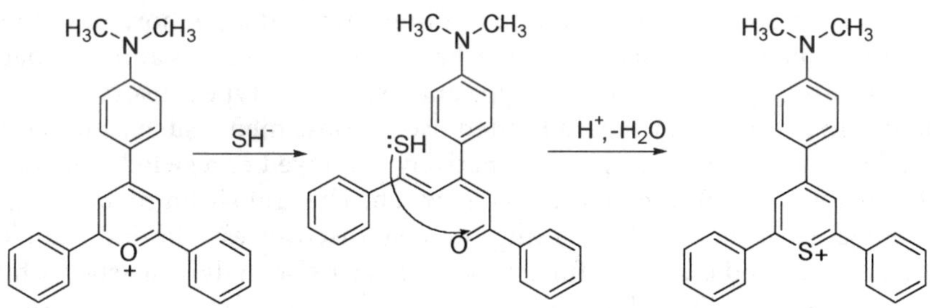


**92** **93**

**Fig. 18** Transformation of receptor **92** into the thiopyrylium form in the presence of $SH^-$

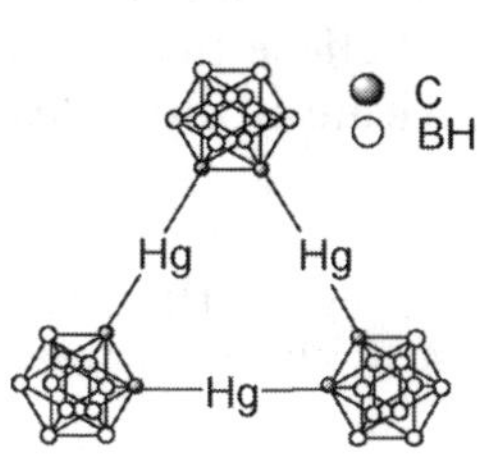


**94**

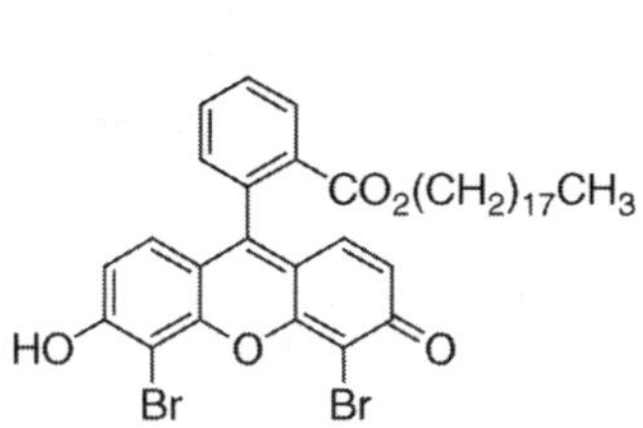


**95**

Leclerc et al. [73] have demonstrated that a cationic polymer which consists of polythiophene derivatives can be a colorimetic sensor for iodide. In aqueous solutions, polymer **96** is yellow ($\lambda_{max}$=406 nm) due to a random coil conformation of the polythiophene derivative that leads to a decrease in effective conjugation through the polythiophene chain. Addition of $F^-$, $Cl^-$, $Br^-$, $SO_4^{2-}$, $CO_3^{2-}$, $HCO_3^-$, $H_2PO_4^-$, $HPO_4^{2-}$, $EDTA^{4-}$, $(C_6H_5)B^-$, and $AcO^-$ induced negligible changes in the UV–visible spectrum of **96**. However, addition of $I^-$ gave a red shift of the band centered at 406 to 543 nm (yellow to red-violet). This color variation was ascribed to iodide promoting the aggregation and planarization of polymer **96**. Furthermore, fluorimetric detection of iodide anion was also possible because the fluorescence spectrum of **96** in the yellow random coil form ($\lambda_{ex}$=420 nm, $\lambda_{em}$=550 nm) was quenched in the planar aggregated form.

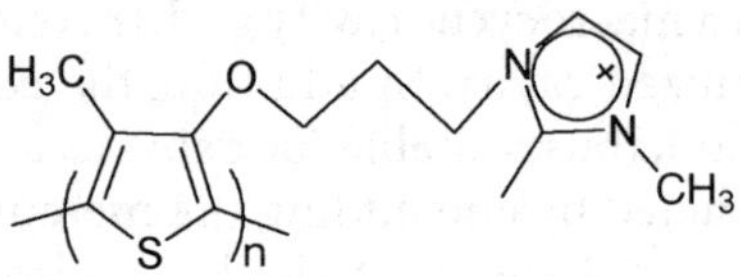


**96**

## 8
## Conclusion

We have summarized six types of chromogenic anion sensors. Many anion receptors containing different anion binding sites and a plethora of chromophores have been described. Each type has its own advantages and drawbacks. The NH-based hydrogen bonding-type and Lewis acid-type sensors are simple and easy to use, but they have a limited use in some media such as water. The metal ion template type can be alternatively employed as a switchable "on" or "off" sensor utilizing a metal ion as a switch. The transition metal complex chromogenic hosts are still waiting for new discoveries. The indicator displacement method is a powerful approach in terms of analytical aspects, but it is sometimes complicated to use. Therefore, this type of chromogenic sensor, in particular, needs more development. There should be a lot of space to fill in this area. The chromoreactand has only recently emerged and definitely needs more investigation. Overall, chromogenic anion sensing is still a young field waiting for new disclosures. It is yet a long way to bring these synthetic sensors or systems to practical use. We optimistically believe that there is a way to combine the aforementioned approaches to fabricate an ultimate chromogenic sensor for a particular anion.

**Acknowledgments** This work was financially supported by the Thailand Research Fund (RSA4680013 and PHD/00123/2545). CS is a Ph.D. student supported by the Royal Golden Jubilee Program.

## References


1. Lehn J-M (1988) Angew Chem Int Ed Engl 27:90
2. Schmidtchen FP, Berger M (1997) Chem Rev 97:1609
3. Harley JH, James TD, Ward CJ (2000) J Chem Soc Perkin Trans I 3155
4. Gale PA (2000) Coord Chem Rev 199:181
5. Gale PA (2001) Coord Chem Rev 213:79
6. Gale PA (2003) Coord Chem Rev 240:191
7. Beer PD, Gale PA (2001) Angew Chem Int Ed 40:486
8. Lavigne JJ, Anslyn EV (2001) Angew Chem Int Ed 40:3118
9. Fabbrizzi L, Poggi A (1995) Chem Soc Rev 24:197
10. Suksai C, Tuntulani T (2003) Chem Soc Rev 32:192
11. Martínez-Máñez R, Sancenón F (2003) Chem Rev 103:4419
12. Miyaji H, Sessler JL (2001) Angew Chem Int Ed 40:154
13. Kato R, Nishizawa S, Hayashita T, Teramae N (2001) Tetrahedron Lett 42:5053
14. Piatek P, Jurczak J (2002) Chem Commun 2450
15. Camiolo S, Gale PA, Hursthouse MB, Light ME (2003) Org Biomol Chem 1:741
16. Miyaji H, Sato W, Sessler JL, Lynch VM (2000) Tetrahedron Lett 41:1369
17. Lee DH, Lee KH, Hong J-I (2001) Org Lett 3:5
18. Lee DH, Lee HY, Lee KH, Hong I-J (2001) Chem Commun 1188
19. Lee DH, Lee HY, Hong J-J (2002) Tetrahedron Lett 43:7273
20. Gunnlaugsson T, Kruger PE, Lee TC, Parkesh R, Pfeffer FM, Hussey GM (2003) Tetrahedron Lett 44:6575
21. Gunnlaugsson T, Kruger PE, Jenson P, Pfeffer FM, Hussey GM (2003) Tetrahedron Lett 44:8909
22. Cho EJ, Moon JW, Ko SW, Lee JY, Kim SK, Yoon J, Nam KC (2003) J Am Chem Soc 125:12376
23. Kondo S, Nagamine M, Yano Y (2003) Tetrahedron Lett 44:8801
24. Jiménez D, Martínez-Máñez R, Sanceñon F, Soto J (2002) Tetrahedron Lett 43:2823
25. Miyaji H, Sato W, Sessler JL (2000) Angew Chem Int Ed 39:1777
26. Tong H, Zhou G, Wang L, Jing X, Wang F, Zhang J (2003) Tetrahedron Lett 44:131
27. Lee JK, Na J, Kim TH, Kim YS, Park WH, Lee TS (2004) Mater Sci Eng C 24:261
28. Black CB, Andrioletti B, Try AC, Ruiperez C, Sessler JL (1999) J Am Chem Soc 121:10438
29. Anzenbacher P Jr, Try AC, Miyaji H, Jursíková K, Lynch VM, Marquez M, Sessler JL (2000) J Am Chem Soc 122:10268
30. Sessler JL, Maeda H, Mizuno T, Lynch VM, Fututa H (2002) Chem Commun 862
31. Aldakov D, Anzenbacher P Jr (2003) Chem Commun 1394
32. Sessler JL, Maeda H, Mizuno T, Lynch VM, Furuta H (2002) J Am Chem Soc 124:13474
33. Lücking U, Rudkevich DM, Rebek J Jr (2000) Tetrahedron Lett 41:9547
34. Starnes SD, Arungundram S, Saunders CH (2002) Tetrahedron Lett 43:7785
35. Beer PD, Drew MGB, Hesek D, Jagessar R (1995) J Chem Soc Chem Commun 1187
36. Thomas KG, Kamat PV (2003) Acc Chem Res 36:888
37. Watanabe S, Sonobe M, Arai M, Tazume Y, Matsuo T, Nakamura T, Yoshida K (2002) Chem Commun 2866
38. Ward CJ, Patel P, James TD (2001) Chem Lett 406
39. Yamaguchi S, Akiyama S, Tamao K (2001) J Am Chem Soc 123:11372
40. Kubo Y, Yamamoto M, Ikeda M, Takeuchi M, Shinkai S, Yamaguchi S, Tamao K (2003) Angew Chem Int Ed 42:2036
41. Sancenón F, Martínez-Máñez R, Soto J (2002) Angew Chem Int Ed 41:1416
42. Kim Y-H, Hong J-I (2002) Chem Commun 512

43. Lee C, Lee DH, Hong J-I (2001) Tetrahedron Lett 42:8665
44. Linton B, Hamilton AD (1997) Chem Rev 97:1669
45. Goodman MS, Jubian V, Hamilton AD (1995) Tetrahedron Lett 36:2551
46. Goodman MS, Hamilton AD, Weiss J (1995) J Am Chem Soc 117:8447
47. Prévot-Halter I, Smith TJ, Weiss J (1997) J Org Chem 62:2186
48. Fabbrizzi L, Pallavicini P, Parodi L, Perotti A, Taglietti A (1995) J Chem Soc Chem Commun 2439
49. Fabbrizzi L, Pallavicini P, Parodi L, Taglietti A (1995) Inorg Chim Acta 238:5
50. Amendola V, Bastianello E, Fabbrizzi L, Mangano C, Pallavicini P, Perotti A, Lanfredi A, Ugozzoli F (2000) Angew Chem Int Ed 39:2917
51. Beer PD, Graydon AR, Johnson AOM, Smith DK (1997) Inorg Chem 36:2112
52. Miyaji H, Collinson SR, Prokeŝ I, Tucker JHR (2003) Chem Commun 64
53. Anzenbacher P Jr, Tyson DS, Jursíková K, Castellano FN (2002) J Am Chem Soc 124:6232
54. Mizuno T, Wei W-H, Eller LR, Sessler JL (2002) J Am Chem Soc 124:1134
55. Lee DH, Im JH, Son SU, Chung YK, Hong J-I (2003) J Am Chem Soc 125:7752
56. Tobey SL, Jones BD, Anslyn EV (2003) J Am Chem Soc 125:4026
57. Tobey SL, Anslyn EV (2003) J Am Chem Soc 125:14807
58. Morey J, Orell M, Barceló MA, Deyà PM, Costa A, Ballester P (2004) Tetrahedron Lett 45:1261
59. Wiskur SL, Ait-Haddou H, Lavigne JJ, Anslyn EV (2001) Acc Chem Res 34:963
60. Gale PA, Twyman LJ, Handlin CI, Sessler JL (1999) Chem Commun 1851
61. Shao S, Guo Y, He L, Jiang S, Yu X (2003) Tetrahedron Lett 44:2175
62. Zhong Z, Anslyn EV (2002) J Am Chem Soc 124:9014
63. Han MS, Kim DH (2002) Angew Chem Int Ed 41:3809
64. Fabbrizzi L, Marcote N, Stomeo F, Taglietti A (2002) Angew Chem Int Ed 41:3811
65. Guo Y, Shao S-J, Xu J, Shi Y-P, Jiang S-X (2004) Inorg Chem Commun 7:333
66. Yamamoto H, Ori A, Ueda K, Dusemund C, Shinkai S (1996) Chem Commun 407
67. Mohr G (2002) Chem Commun 2646
68. Ros-Lis JV, Martínez-Máñez R, Soto J (2002) Chem Commun 2248
69. Sancenón F, Descalzo AB, Martínez-Máñez R, Miranda MA, Soto J (2001) Angew Chem Int Ed 40:2640
70. Sancenón F, Martínez-Máñez R, Miranda MA, Seguí M-J, Soto J (2003) Angew Chem Int Ed 42:647
71. Jiménez D, Martínez-Máñez R, Sancenón F, Ros-Lis JV, Benito A, Soto J (2003) J Am Chem Soc 125:9000
72. Badr IHA, Johnson RD, Diaz M, Hawthorne MF, Bachas LG (2000) Anal Chem 72:4249
73. Ho HA, Leclerc M (2003) J Am Chem Soc 125:4412

Top Curr Chem (2005) 255: 199–229
DOI 10.1007/b101167


# Abiotic Guanidinium Receptors for Anion Molecular Recognition and Sensing

Ronald J. T. Houk · Suzanne L. Tobey · Eric V. Anslyn (✉)

Department of Chemistry and Biochemistry, University of Texas at Austin,
1 University Station A5300, Austin, TX 78712-0165, USA
*anslyn@ccwf.cc.utexas.edu*



**Abstract** The guanidinium functionality has been utilized by molecular recognition chemists as a strong and selective binder for anionic guests. Guanidiniums, because of their high p$K_a$, offer both hydrogen-bonding and charge-pairing interactions over a wide range of pH. Strong binding interactions in competitive media have been reported for receptors containing this functionality. Due to the advantageous properties of the guanidinium group, it has been used in receptor designs for almost every class of small anionic target molecules. This review article will record the history of this group in supramolecular systems by cataloging the target analyte classes for which synthetic guanidinium-based anion receptors have been developed.

**Keywords** Guanidinium · Molecular recognition · Anion receptor · Sensing · Supramolecular

**Abbreviations**

| | |
|---|---|
| AMP | Adenosine monophosphate |
| ATP | Adenosine triphosphate |
| Bpy | Bipyridine |
| cAMP | Cyclic AMP |
| CD | Circular dichroism |
| d(ApA) | 2′-Deoxyadenylyl(3′→5′)-2′-deoxyadenosine |
| EPR | Electron paramagnetic resonance spectroscopy |
| FT-IR | Fourier-transform infrared spectroscopy |
| GABA | γ-Aminobutyric acid |
| GTP | Guanine triphosphate |
| HIV-1 | Human immunodeficiency virus type 1 |
| HPNPP | 2-Hydroxypropyl-*p*-nitrophenyl phosphate |
| ITC | Isothermal titration calorimetry |
| MLCT | Metal-to-ligand charge transfer |
| mRNA | Messenger RNA |
| NOESY | Nuclear Overhauser effect spectroscopy |
| PAGE | Polyacrylamide gel electrophoresis |
| PEG | Polyethylene glycol |
| PET | Photoinduced electron transfer |
| ROE | Rotating Overhauser effect |
| RRE | Rev response element |
| SN | Staphylococcal nuclease |
| TPB | Tetraphenylboron |
| TREN | Tris(2-ethylamino)amine |
| UMP | Uracil monophosphate |
| XPS | X-ray photoelectron spectroscopy |

## 1 Introduction

Until relatively recently, the field of supramolecular chemistry, more specifically molecular recognition, has been mainly concerned with the complexation of cationic guests. In the past 15 years or so, however, there has been a boom in the literature of synthetic receptors designed to bind interesting anions or zwitterions. The guanidinium group, which is a carbon bound to two amines and an iminium, has been shown to be one of the most versatile functionalities for binding anions via strong noncovalent interactions with several biologically and commercially interesting targets, such as carboxylates, phosphates, carbonates, nitrates, flavins, and sulfates. The idea of using guanidiniums for these receptors arises directly from nature. Enzymes often use the guanidinium-containing side chain of arginine in their active sites to bind anionic substrates [1–4].

Several factors contribute to the guanidinium's strong ability to bind anionic guests. The wide pH range through which the group remains protonated allows for binding to occur through both hydrogen bonding and charge pairing. The $pK_a$ value of a typical guanidinium lies between 12 and 13, which means that at physiological pH, 7–8, the guanidinium remains charged [5, 6]. The second important factor is the geometry of the group, which allows for better alignment of the hydrogen bonds with the diffuse anions listed above than a simple ammonium. Also, the more delocalized nature of the positive charge in the guanidinium group allows for better interactions with softer anions such as sulfates and phosphates than does the hard, localized charge of an ammonium. With the numerous advantages presented by the guanidinium group for binding a wide variety of anionic guests, there has been a steady rise in the use of the group in anion receptor designs. Recently, two reviews on this topic have been published which focus on the diversity of the receptor design motifs into which guanidiniums have been incorporated [7, 8]. This review will attempt to demonstrate the universality of the guanidinium group as an anion binder by cataloging the myriad anionic targets of guanidinium-containing receptors and the data and usefulness gleaned from such studies.

## 2 Phosphates

By far the most extensively sought after class of analytes using guanidiniums is phosphate derivatives such as inorganic phosphates, synthetic phosphodiesters, nucleotides, and nucleic acids. The prevalence of the phosphodiester linkage in biological systems makes the phosphate a very attractive target anion for molecular recognition chemistry. As mentioned above, because the positive charge of a guanidinium is rather delocalized and soft, it has a much higher potential for binding big, soft anions, of which phosphates are a prime example, than does a charge-localized ammonium group. Interestingly, as will be seen later, very few instances have been reported of heightened affinity for sulfate binding with guanidiniums. Because most of the phosphate targets for which receptors have been designed have some biological significance, the binding studies have been conducted at physiological pH. In this pH range, a typical phosphate guest is dianionic. As such, many of the guanidinium-based receptors designed to bind phosphate guests consist of two guanidinium moieties connected by a linker, which allows them to orient around the tetrahedral geometry of the anion. For many of the receptors designed to bind phosphodiester guests, binding studies are only a preliminary step to the real objective of the design, catalysis of hydrolytic cleavage. Our discussion of this analyte class will begin with simple inorganic phosphate and small organic phosphates and move up to binding and/or hydrolysis of synthetic phosphodiesters and nucleic acids.

### 2.1 Inorganic Phosphates ($H_nPO_4^{x-}$ and $P_2O_7^{4-}$)

Some of the earliest work using guanidiniums for trapping anionic guests dealt with binding inorganic phosphate. Lehn and coworkers synthesized the three guanidinium-containing macrocycles **1**, **2**, and **3**, and characterized their binding to tribasic phosphate anion by analyzing pH titration curves [9]. They found that stability constants, $K_s$, for the macrocycles were considerably greater than those for their acyclic control compounds.

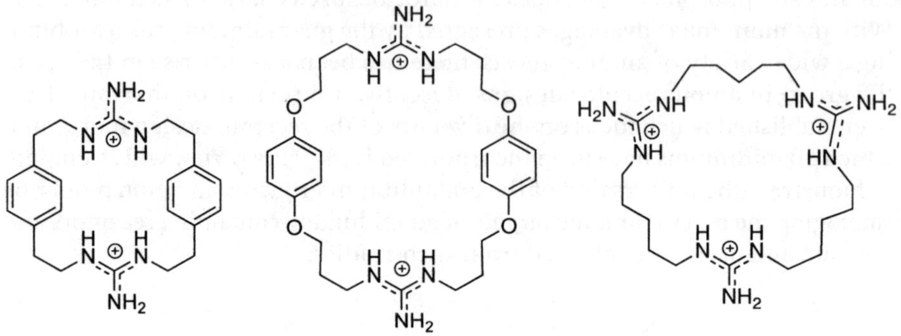

**1** **2** **3**

Later work by Schmidtchen's group using a bicyclic guanidinium core, which will be seen in many of the later receptor scaffolds, and four appendant arms capable of hydrogen bond donation, were used to bind a variety of different guests [10]. Using $^1$H NMR to watch the shift in the N–H signal, they were able to determine that **4** bound phosphate with a 1:1 stoichiometry and a very strong dissociation constant in both acetonitrile and chloroform. Hard evidence of the 1:1 stoichiometry of this host with its analytes was given by a crystal structure of **4** with acetate. This study was particularly important, in that it demonstrated the versatility of the guanidinium, by showing high binding affinities for not only phosphates, but also carboxylates and sulfates.

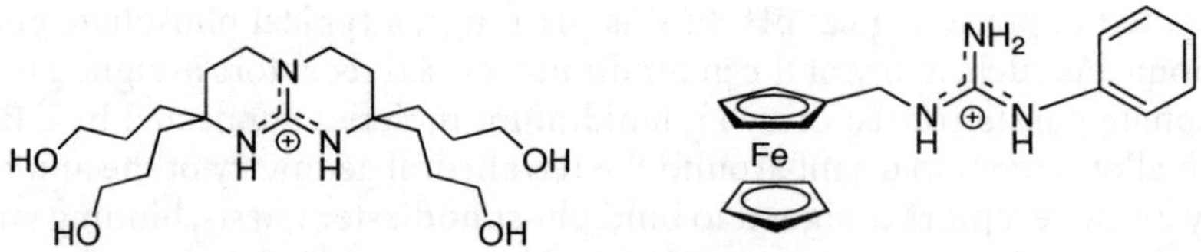

**4** **5**

Another interesting study on a receptor which bound phosphate was conducted almost a decade later by Beer and coworkers. Using the ferrocene-appended guanidinium **5**, they studied the binding of monobasic phosphate and tetrabasic pyrophosphate by $^1$H NMR titration [11]. They were unable to fit the monobasic phosphate binding to a recognizable stoichiometric curve; however, pyrophosphate was found to bind in a 2:1 (host:guest) stoichiometry with an affinity of $4.6\times10^3$ $M^{-2}$ in 50/50 $H_2O$/MeOH. Interestingly, receptor **5** was extremely selective to the phosphates over sulfates. The novelty of this study was the experiments conducted on the effect that binding the phosphates had on the redox potential of the receptor. Cyclic voltammetry studies showed that only pyrophosphate induced any significant enhancement of the oxidation potential of the iron, thus showing the selectivity of **5** for this bidentate guest.

Another receptor with strong, selective 2:1 complexation to inorganic phosphate was prepared by Teramae [12]. The pyrene fluorophore of **6** allowed for quantitation of binding through fluorimetric assays. This analysis was completed by observing the change in the ratio of excimer to monomer emission upon 2:1 complex formation. By fluorimetry and $^1$H NMR in methanol, a 2:1 binding constant of $1.2\times10^8$ $M^{-2}$ was observed with a 1:1 $K_a$ value of $1\times10^4$ $M^{-1}$ for receptor **6** with pyrophosphate. However, no emission changes were observed with monobasic phosphate anion. This study and studies with host **5** delineate two different ways to distinguish between phosphate and pyrophosphate.

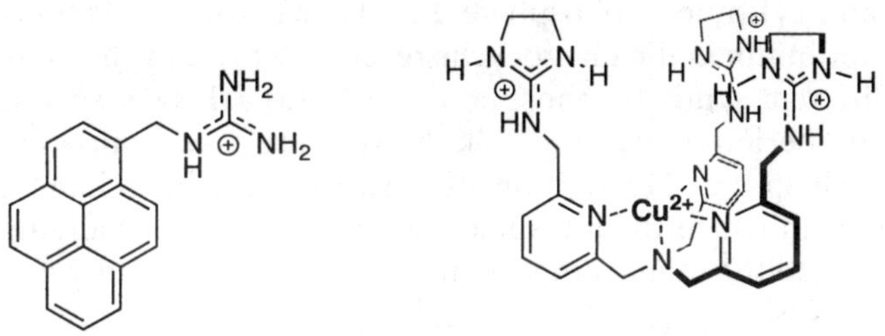


**6** **7**

Finally, metallo-receptor **7** and a tris(ammonium) TREN-based analog were reported by Anslyn and coworkers as selective binders for phosphate and arsenate in water [13]. UV/vis titrations and isothermal titration calorimetry (ITC) experiments were used to determine the binding energetics of the receptors. Binding affinities in 98:2 water/methanol for both receptors to phosphate and arsenate were similar, ~$10^4$ $M^{-1}$ for both analytes. The guanidinium receptor **7** showed slightly higher selectivity, however the binding was stronger for the ammonium host. ITC studies showed that the bulk of the free energy associated with binding the anions to these hosts resulted from complexation with the Cu(II) center, while the cationic arms fine-tuned the selectivity. Binding of phosphate to **7** was driven by both favorable enthalpy and entropy changes, whereas the ammonium host association was completely entropy driven. This disparity was attributed to solvation effects, where the ammoniums bind a tighter solvent sphere than the guanidiniums.

## 2.2 Organic Phosphates

This section will cover a wide variety of compounds from small organic phosphates and phosphodiesters, to nucleotides, to fully formed DNA and RNA chains. In addition to binding these guests, many of the receptors have been found to act as RNase mimics, in that they catalyze hydrolytic cleavage of the phosphodiester functionality of their guests. Most of the catalysis has been reported for nucleotides or synthetic replicas of nucleic acids; however, several receptors have been tested with actual RNA or DNA strands.

### 2.2.1 Organic Phosphates and Phosphodiesters

In an early study, Schmidtchen employed the bicyclic guanidinium scaffold to study the binding of tetrahedral oxoanions by a ditopic receptor. The result was the urethane-linked bis(guanidinium) compound **8**, which was found to selectively bind dianionic tetrahedral guests in water [14]. While simple, planar bis(guanidinium) guests showed no binding in water, host **8** was able to encompass the shape of such biologically relevant phosphates as *p*-nitrophenyl phosphate and cytidine 5′-phosphate. In a later study, de Mendoza used the bicyclic guanidinium in the macrocyclic receptor **9** to study the binding modes of several different synthetic phosphodiesters [15]. $^{1}$H NMR spectra of **9** with phosphate counterions of varying bulk showed single resonances at room temperature which split at 213 K, indicating two rapidly interchanging binding conformations in which the phosphate counterion was exchanging from one side of the ring to the other rather than sitting inside the cavity as predicted.

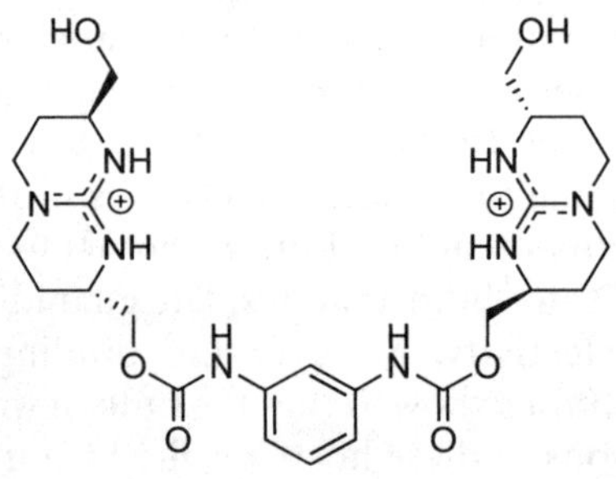

**8**

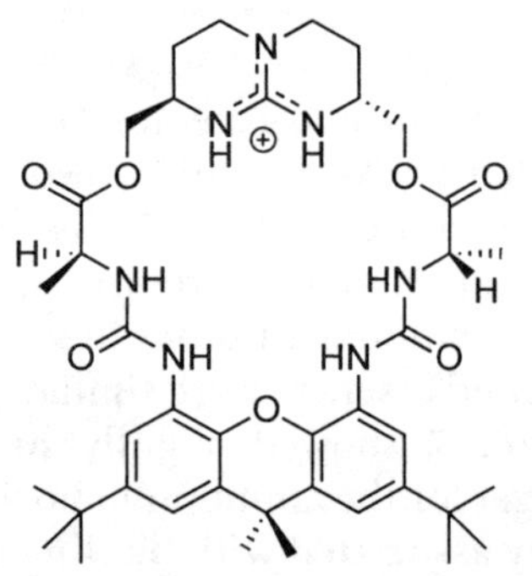

**9**

The indicator-displacement assay exploited in the Anslyn group uses an indicator with a weak binding affinity for the designed host which undergoes a spectrophotometric change upon binding to that host. The target guest is then titrated into a solution containing the host–indicator complex, and the absorption of the indicator is monitored as it is displaced from the binding cavity of the host. This indicator-displacement assay was utilized to study the binding of the tris(phosphate) secondary messenger inositol triphosphate with compound **10**, which contains six guanidiniums directed toward the center of the binding

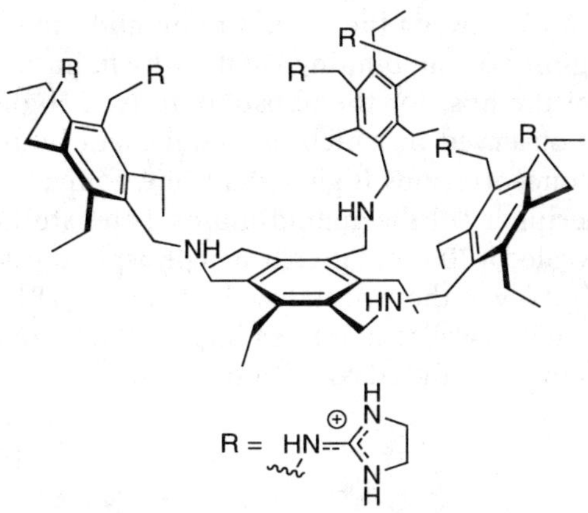

10

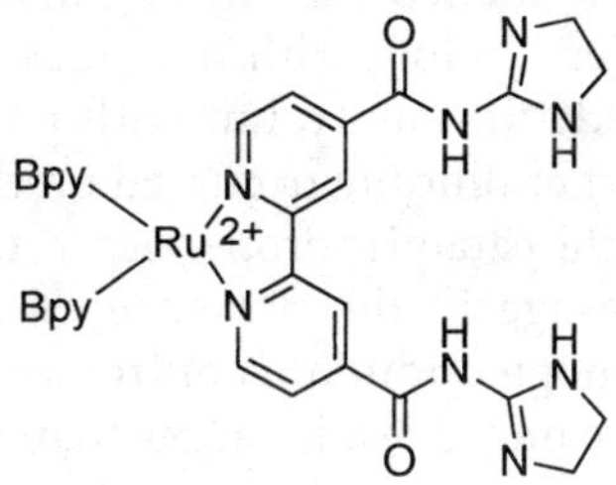

11

cavity [16]. The 2,4,6-triethylbenzene scaffold employed, and several other receptors shown below, bias the substituents in the 1, 3, and 5 positions via steric gearing to preferentially remain on the same face of the aromatic platform. This facial selectivity preorganizes the binding substituents – in this case the three appendant triethylbenzenes – into a rigid binding cavity. In studies on the binding affinities of **10** and a hexakis(ammonium) analog with various anionic guests, the absorbance of 5-carboxyfluorescein was measured as it was displaced from the binding cavity of the host. Binding constants for both **10** and the ammonium analog were found to be similar; however, the ammonium compound showed less selectivity, and the binding fell off more strongly as ionic strength increased.

Another interesting spectroscopic assay using a guanidine-based receptor for organic phosphate guests was developed in Watanabe's laboratory. The binding of neutral and anionic phosphoric acid diesters to ruthenium complex **11** was studied by monitoring the MLCT absorbance and fluorescence changes upon binding [17]. Large fluorescence emission changes were seen upon complexation. Interestingly, anionic guests caused an increase in the luminescence intensity, whereas neutral phosphoric acid binding resulted in a decrease in intensity. Binding of anionic guests probably introduced an increased rigidity to the metal complex. However, because the hydrogen-bonding modes are different for the binding of neutral guests, intracomplex proton transfer probably accounts for the reduction in luminescence.

Many of the guanidinium receptors used for binding organic phosphate guests were designed not only as receptors, but also as enzyme mimics to study the catalysis of hydrolytic cleavage of phosphodiesters. Many of these receptors contain two guanidiniums separated by a spacer, which allows them to orient in a way which mimics the active site of staphylococcal nuclease (SN). Hamilton and coworkers reported the isophthaloyl bis(guanidinium) receptor **12** which bound phosphodiester guests with high affinity. $^1$H NMR analysis determined that an intramolecular hydrogen bond between the guanidinium and the isophthaloyl oxygen increased the rigidity of the binding cavity, which, in turn, strengthened the binding ability of the host for the phosphodiesters. Higher order binding with one guest was observed in which one equivalent bound inside the cavity, but with guest concentrations higher than one equivalent, further binding occurred on the periphery of the guanidiniums. Later studies on the catalytic properties of **12** revealed a 700-fold increase in phosphodiester cleavage in the presence of catalyst over the uncatalyzed scenario [18]. A monoguanidinium control was studied as well, and interestingly, catalytic rates were only 2.5-fold larger than rates in uncatalyzed conditions.

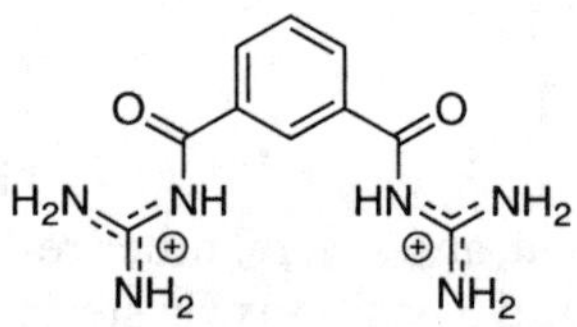

**12**

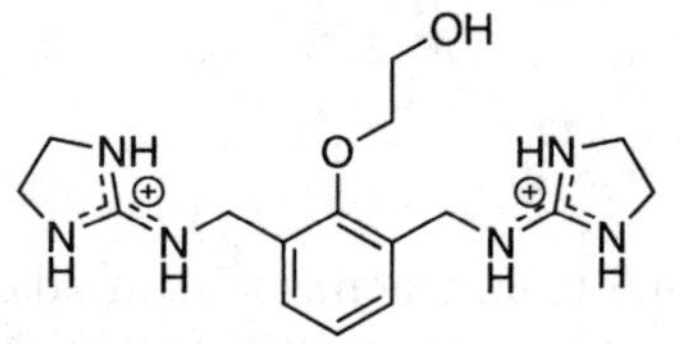

**13**

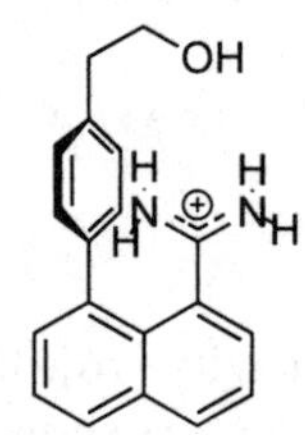

**14**

Göbel and coworker used the phenyl alcohols **13** and **14** to study binding and catalysis of hydrolysis of synthetic phosphodiester guests [19]. In the absence of base, binding constants for the two compounds were obtained, with **13** ($K_a$=2.9×$10^3$ $M^{-1}$) having a much higher affinity for the chosen cyclic phosphate guest than **14** ($K_a$=190 $M^{-1}$). Kinetics measurements using $^{31}$P NMR showed that **13** catalyzed the phosphorylation under basic conditions of the appendant alcohol 800 times faster than **14** and 380,000 times faster than phenylethanol, a result which directly shows the ability of the guanidinium to enhance the binding and thus the reactivity of phosphate guests.

Hamilton and coworkers expounded on the idea of placing additional pendant functionalities on the receptor for hydrolysis of organic phosphates by studying a series of compounds similar to **15** [20]. They found that appending amines to the guanidinium groups to act as general bases to deprotonate the free alcohol of 2-hydroxypropyl-*p*-nitrophenyl phosphate (HPNPP) could effect large rate enhancements for intramolecular phosphate transesterification. It

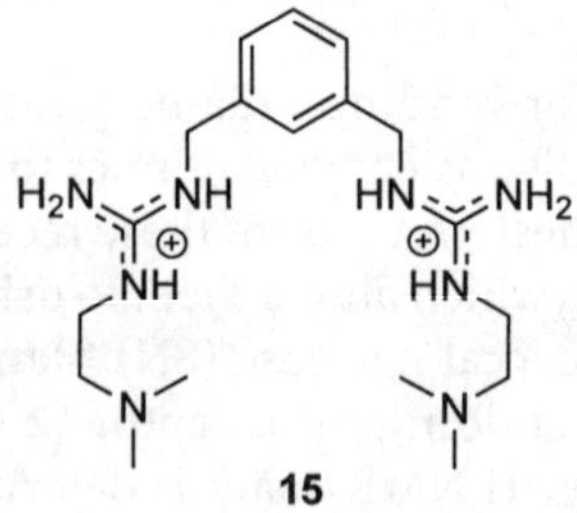

**15**

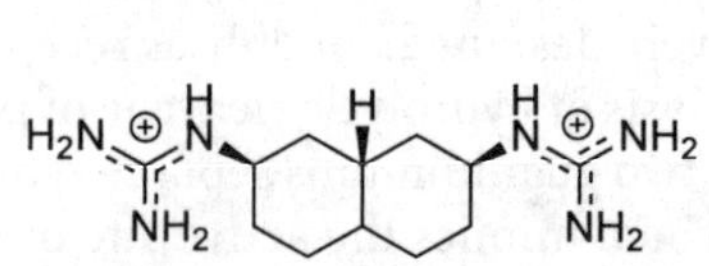

**16**

was found spectroscopically that the trialkylamine **15** was optimal for binding and catalysis of HPNPP. Receptor **16** was shown to have good binding to phosphodiester anion guests [21]. The *trans*-decalin spacer, however, proved too rigid to effectively catalyze hydrolysis of the diester.

Another SN mimic was designed concurrently in the Anslyn group using an octahydroacridine spacer between two aminoimidazoline groups [22]. Binding studies in competitive media (2:1 DMSO–$D_2O$) of **17** with dibenzyl phosphate gave a binding affinity of $3.6\times10^2$ $M^{-1}$, whereas no association was observed with guanidinium chloride. Variation of the linker to a less rigid pentane moiety in a later study showed that the constraining octahydroacridine linker imparted 1.8 kcal $mol^{-1}$ of stabilization to the binding of the host to uridine 5′-monophosphate [23]. This work was expanded with studies on the individual stereoisomers of **17** and an analog containing a hexahydrodicyclopenta[*b*,*e*]pyridine. It was found that guest binding was affected by the guest counterion on the spatial orientation of the binding groups. With small counterions like chloride, the *d*,*l* forms of the host bound better due to a specific interaction within the binding cavity, as shown by the crystal structure. The *meso* isomer, as predicted, was preferred for larger counterions such as tetraphenylboron [24]. In a final study, polyacrylamide gel electrophoresis (PAGE) was used to study the catalysis of mRNA cleavage in the presence of **17**. Hydrolysis was 20 times faster in the presence of the catalyst than without [25].

The calix[6]arene-based receptor **18** was reported [26] as a synthetic acetylcholinesterase by Mandolini and de Mendoza. Designed as a receptor for dioc-

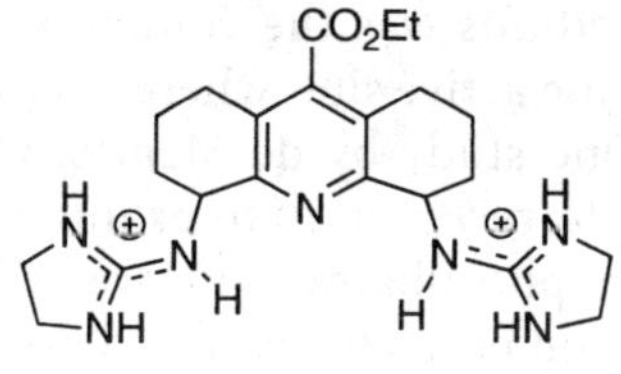


**17**

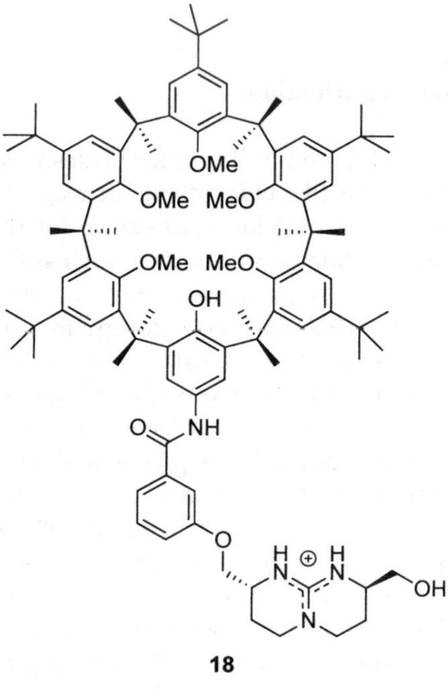


**18**

tanoyl-L-α-phosphatidylcholine, a transition state analog for the hydrolysis of esters and carbonates, the host binds a carbonate guest bearing a trimethylammonium through the calixarene and the phosphate of the target phosphatidylcholine through the appendant guanidinium. Methanolysis of *p*-nitrophenyl carbonate was successfully achieved in the presence of **18** in 1% methanol–chloroform in the absence of the phosphate. In the end, the catalyst was able to act as a successful mimic, though with neither great selectivity nor high activity.

The bis(guanidinium)-appended steroid **19** was designed as an enzyme mimic for RNA cleavage. The study utilized 12 very different guanidinium-containing scaffolds to study the hydrolysis of HPNPP [27]. Of these, compound **19** was found to have a tenfold greater rate than any of the others for cleavage of HPNPP and was three times faster at hydrolyzing uridylyl(3′,5′)-uridine.

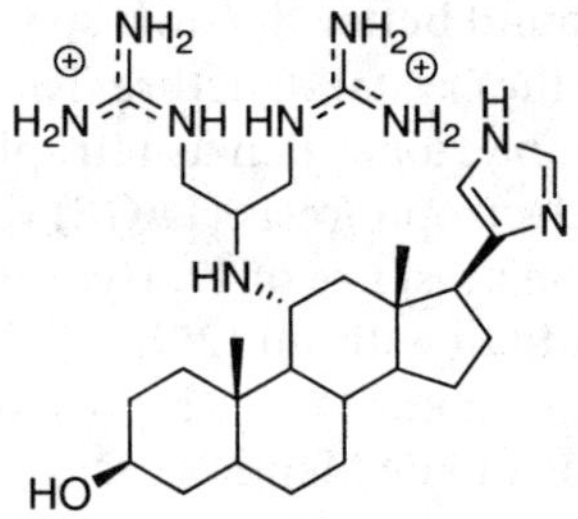


**19**

## 2.2.2 Biological Nucleotides

Early work with binding nucleotides was generally conducted in noncompetitive solvents like chloroform, or as extractions from aqueous solutions into organic solvents. The justification for these methods was the common view that intracellular environments such as the enzyme active sites where reactions occur can be nonaqueous and even aprotic. One study by de Mendoza and coworkers used a series of compounds similar to receptor **20** to examine the selective extraction of various adenosine monophosphates from water into chloroform [28]. They found a range of extraction efficiencies from 0 to 50% for several analogs of **20**, with efficiencies much greater at higher pH where the phosphates are doubly deprotonated. This pH effect was most pronounced for 3′-AMP. Ironically, their most selective binder, **20**, was not lipophilic enough to use in the extraction experiments. The naphthoyl group of **20** allows for π-stacking interactions with the nucleic base, while the uridine moiety allows for complementary hydrogen bonding with the adenosine of the phosphates. $^1$H NMR titrations in DMSO revealed almost no interaction between **20** and 3′,5′-cAMP, whereas there was strong binding with 3′-AMP as their models suggested.

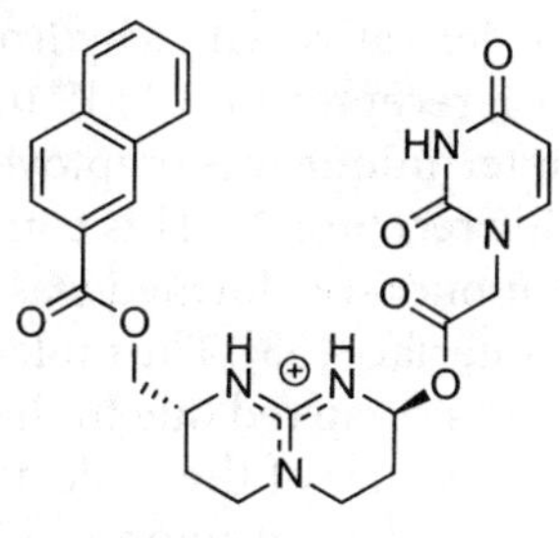

**20**

Joint work by de Mendoza, Bruix, and Rebek's groups resulted in the lipophilic, guanidinium-containing host **21**, which was expected to bind well to dideoxy dinucleotides. The carbazole arms, derived from Kemp's triacid, are expected to complement the purine nucleic bases, especially adenine, through concomitant Hoogsteen and Watson–Crick base pairing. The bicyclic guanidinium core was added to bind the phosphate linker of the dinucleotide. In extraction experiments once again, it was found that **21** bound 2′-deoxyadenylyl(3′→5′)-2′-deoxyadenosine (d(ApA)) very strongly and extracted a full equivalent of d(ApA) [29]. NOE data showed strong intermolecular coupling between the phosphate guest and the carbazole arms. Later studies led to a similar host with only one arm designed to bind cAMP under similar conditions. 2′,3′-cAMP and 3′,5′-cAMP were both extracted in full equivalency into the organic layer with slight selectivity for 2′,3′-cAMP [30]. Separately, these carbazole-based receptors were utilized as membrane transport agents for adenine mono- and dinucleoside monophosphates across model liquid membranes. Also, it was found that **21** was effective in extracting longer oligonucleotides, as long as they contained adenosine, and the efficiency rose with increasing ApA occurrence in the oligonucleotide [31]. In a final study with these recep-

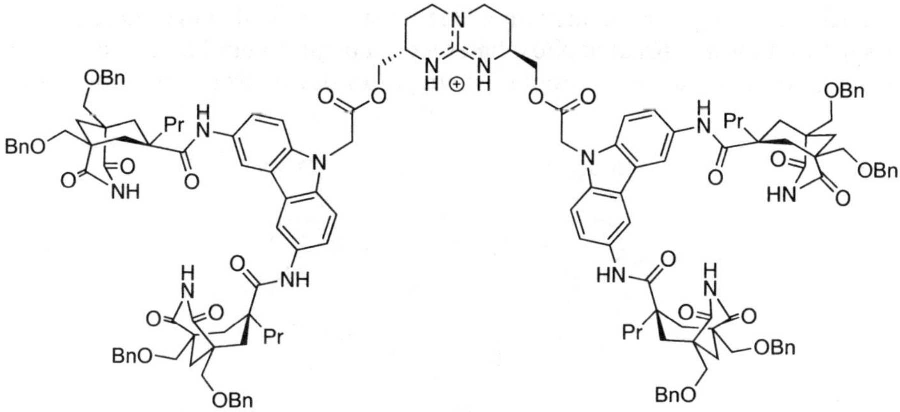

**21**

tors, another monocarbazole derivative with a hydrophilic group in place of the second arm was found to be a receptor for cAMP in water [32].

A twist on the extraction technique was employed in Kunitake's laboratory using the extremely lipophilic receptor **22**. This compound forms a monolayer on a water surface, and the monolayer formed was studied for the binding of nucleotides at the air–water interface [33]. The molecular packing of the monolayer was monitored as changes occurred due to the binding of ATP and AMP. The resulting $\pi$–$A$ curve – a measure of the molecular area expansion or contraction of the monolayer – condensed upon binding of ATP, whereas AMP caused the curve to expand. Thus, the host was found to be selective and distinguishable for the two phosphate guests based on monolayer signal change. Binding constants to this receptor in the solid phase were determined by XPS analysis of Langmuir–Blodgett films drawn from substrate-saturated monolayers. Binding of ATP occurred in a 1:3 ATP:amphiphile ratio with a binding constant of $1.7\times10^7\ M^{-1}$, whereas AMP bound in a 1:1 fashion with an affinity of $3.2\times10^6\ M^{-1}$. FT-IR studies on the films led to the conclusion that binding occurred through phosphate–guanidinium interactions and that the counterion played very little role in the association. Later studies with **22** found enhanced binding with UMP over AMP due to secondary binding interactions between the uracil carbonyl groups and the guanidiniums of the amphiphile [34].

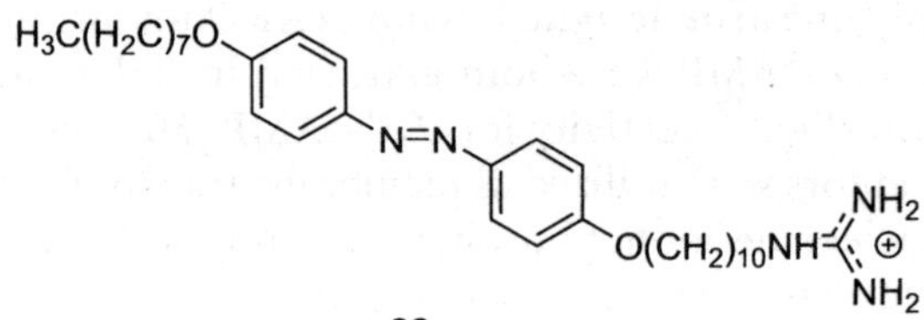


**22**

Other hosts for binding nucleotides in aqueous, or at least protic, media have been designed. Receptor **23** has been used to bind several different mono- and dinucleotides in water and methanol [35]. It was thought that in switching from methanol to water, organic phosphates would begin to bind better than inorganic phosphate due to hydrophobic effects and/or hydrogen bonding; how-

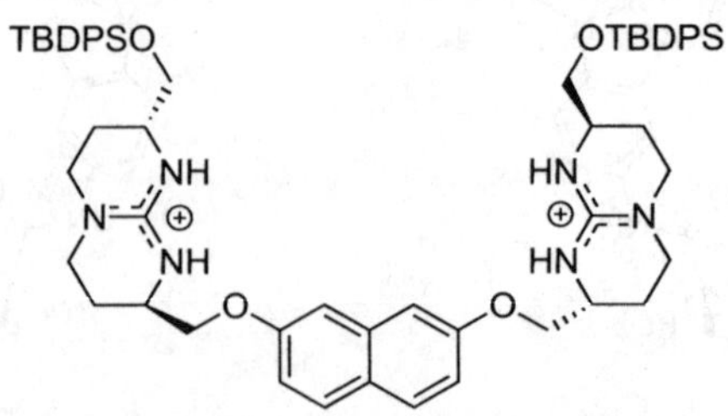


**23**

ever, this was not the case. Dianionic inorganic phosphate was the best binder, with $ATP^{4-}$ running a close second. This study also brings into sharp focus the extreme effects that solvents have on binding, even when switching between supposedly similar solvents. The binding affinity of 5′-AMP with **23** in water was nearly 40 times less than in methanol.

One final study, using a combinatorial library of immobilized receptors based on the 2,4,6-triethylbenzene scaffold **24**, was conducted in the Anslyn group [36]. The core guanidinylated triethylbenzene was bound to a PEG resin and the combinatorial library was grown by placing peptides on the free arms. Finally, fluorophores were appended to the peptide arms, and the library was fluorometrically screened for ATP binding with UV light through inspection. Potential binders were sequenced, resynthesized, then screened more rigorously. The analog of **24** containing two Ser-Tyr-Ser tripeptide arms gave strong selectivity for ATP over AMP and GTP. Later studies found that an unscreened combinatorial array based on the **24** scaffold was suitable for differentially recognizing structurally similar analytes (i.e., ATP, GTP, and AMP) [37]. It was found that the highest response factors were for analogs containing serine and aromatic amino acids.

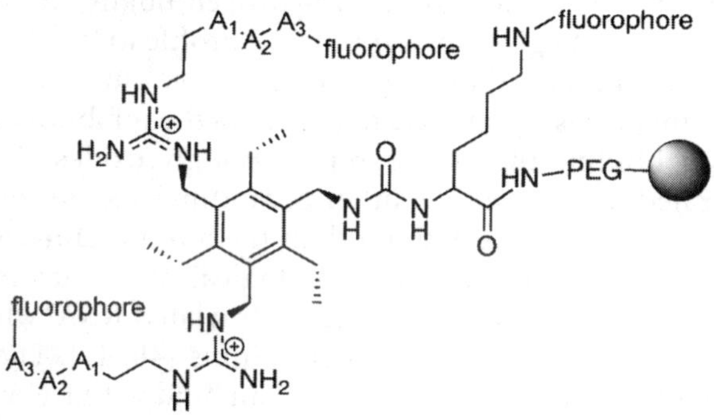


**24**

### 2.2.3
### Oligonucleotides, RNA, and DNA

We have already seen two receptors which have been used to bind or catalyze the cleavage of oligonucleotides and RNA. A 20-fold increase in the rate of mRNA cleavage was found in the presence of Anslyn's host **17** over background hydrolysis, and receptor **21** was found to bind well to oligonucleotides containing a high percentage of ApA units [24, 29]. Several other guanidinium-based receptors have been reported to have binding and/or catalytic properties with long phosphate polymers. The tetraguanidinium structure **25** and a related *meta*-linked compound were designed to bind to the DNA minor groove [38]. Association constants measured by both CD titration and ultrafiltration assays with calf thymus DNA were found to be $\sim 10^6$ and $\sim 10^7$ $M^{-1}$ for **25** and the *meta*-

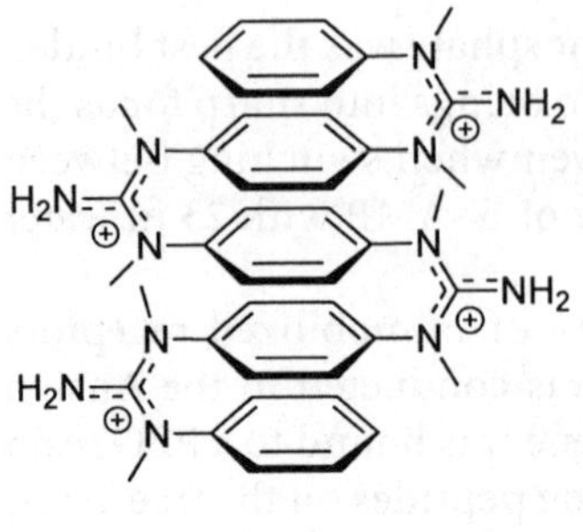

25

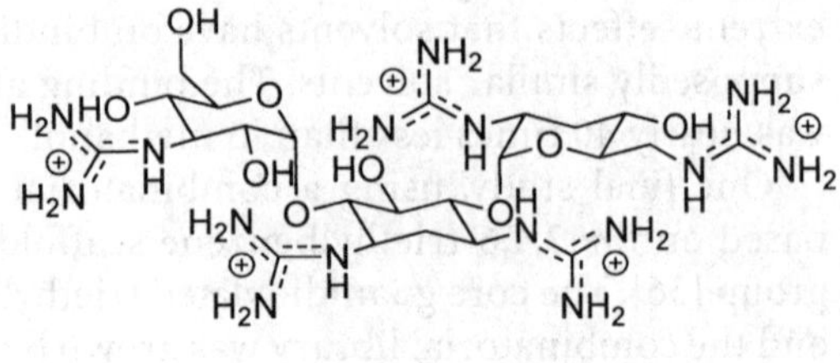

26

linked analog, respectively. Sequence specificity for these compounds was determined by NOESY analysis. Both **25** and the analog were found to preferentially bind 3′-GAA-5′ regions.

Several studies have used the strategy of appending guanidiniums to biological molecules to effect higher selectivity. Receptor **26**, a tobramycin derivative, and several other guanidinoglycosides were reported by Tor as possible alternatives to natural aminoglycosides for increased binding affinity to RNA [39]. Guanidiniums, with their directed hydrogen bonds and diffuse charge density, were expected to give a better binding profile to the RNA phosphodiester linkages than ammoniums. Using fluorescence anisotropy, it was found that the guanidinoglycoside had five to ten times the inhibitory power toward HIV-1 RRE-Rev binding than the parent aminoglycosides. The researchers also demonstrated that **26** and the other guanidinylated glycosides could be tuned toward greater RNA selectivity than the ammonium-containing parents. Two other studies on modifying biological molecules to create novel molecular recognition scaffolds used guanidinylated RNA, where the phosphodiester links were replaced by guanidiniums [40, 41]. The result of the substitution is a polycationic strand which can bind with DNA and RNA, not only through the ion-pairing interactions, but also through nucleic base complementarity.

# 3
# Carboxylates

After phosphates, the most highly prized guests for guanidinium-based receptors are carboxylates. Due to the similarity in size and charge distribution, carboxylates and guanidiniums complement each other very well. A wide variety of carboxylates, from small organic monocarboxylates, to amino acids, to polypeptides, to full proteins have been targeted by receptors with guanidinium cores. With the advent of proteomics in chemistry and biochemistry, the demand for better, stronger, more selective carboxylate binders is expected to skyrocket, and the phosphate may be forced to take a back seat.

## 3.1 Simple Carboxylates

This section will cover receptors and studies which deal with binding small organic mono-, di-, tri-, etc. carboxylates. One of the earliest receptors used to study carboxylate binding has already been discussed as an inorganic phosphate binder. Receptor **4**, reported by Schmidtchen and coworkers, was studied with a variety of anionic guests [10]. Strong binding for *p*-nitrobenzoate was characterized by $^1$H NMR titration, and had a dissociation constant of less than $10^{-4}$ M. As mentioned above, the 1:1 binding stoichiometry was confirmed by X-ray crystal structure analysis of acetate bound to **4**. A year later, Lehn and de Mendoza reported the bis(naphthoyl) host **27** as a strong carboxylate binder and the first anion receptor showing chiral discrimination [42]. The naphthoyl side arms were incorporated to provide π-stacking interactions with aromatic guests. Quantitative extraction of *p*-nitrobenzoate from water to chloroform was achieved using **27**, and $^1$H NMR titrations quantified the binding with this and other aromatic carboxylates. The strongest binding was found with *p*-nitrobenzoate, having a stability constant of 1,609 $M^{-1}$ in chloroform. In extraction experiments using optically pure **27**-*S*,*S* or -*R*,*R* and racemic carboxylate salts such as mandelate, naproxenate, and *N*-acetyltryptophan, the researchers observed slight diastereomeric excesses in the extracted salts.

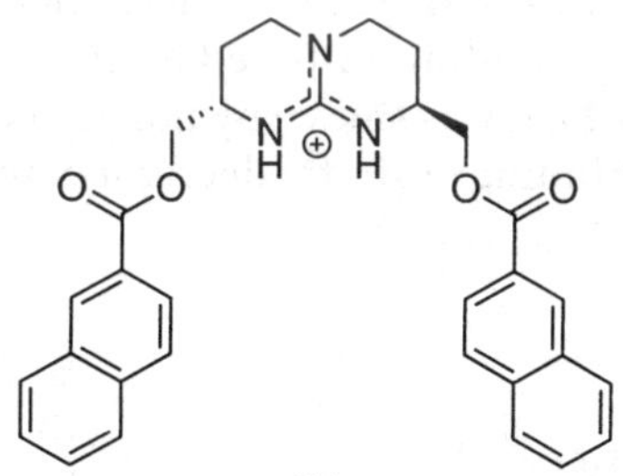


**27**

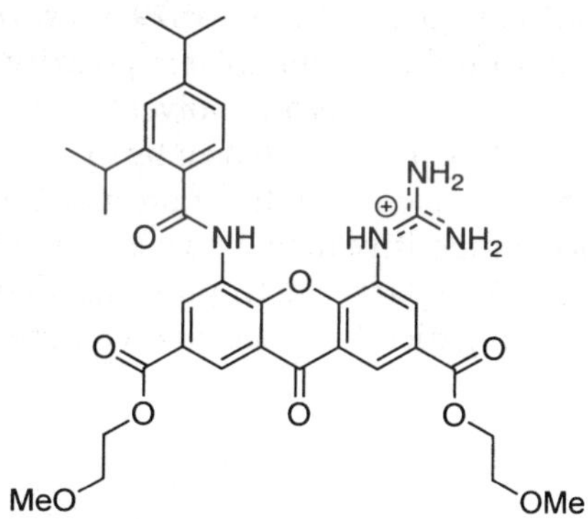


**28**

Morán and coworkers reported the unique xanthone-based receptor **28** for the binding of monocarboxylates [43]. Apart from the guanidinium, the xanthone core includes extra binding interactions through hydrogen bond donation from the amide NH and the diisopropylbenzoate substituent for positive dipole–π interactions. $^1$H NMR analysis of the binding of **28** to acetate showed that acetate would displace methanesulfonate, but none of the characteristic shifts which indicate host–guest binding with guanidiniums could be observed. It was determined that proton transfer from the guanidinium to the carboxylate guest was occurring rather than binding. To test this theory, a series of carboxylates with steadily decreasing p$K_a$ values were titrated into the free-base guanidine form of the host. The system was found to be extremely sensitive to carboxylate p$K_a$. For example, decanoic acid (p$K_a$ 4.9) bound with $K_a$=6.0×10$^4$ M$^{-1}$ while monochloroacetic acid (p$K_a$ 2.86) had an association constant of 6.5×10$^7$ M$^{-1}$: a 1,000-fold increase in binding with only a 100-fold increase in acidity.

More complex carboxylates, especially di- and tricarboxylates such as tartrate and citrate, have often been targeted due to their prevalence in biological systems as a product of various metabolic pathways. Hamilton and coworkers examined the binding energetics of mono- and dicarboxylates with various phenyl-based receptors including **29** [44]. Binding for both mono- and dicarboxylates increased with increasing acidity of the recognition element from urea to thiourea to guanidinium. Whereas the urea and thiourea derivatives bound glutarate with affinities of 6×10$^2$ and 1×10$^4$ M$^{-1}$ respectively, binding with the bis(guanidinium) **29** was too strong for determination by NMR analysis in pure DMSO. Using 25% $D_2O$/DMSO, a binding constant of 4.8×10$^2$ M$^{-1}$ was found. In later studies, isothermal titration calorimetry (ITC) experiments with the urea, thiourea, and guanidinium compounds further quantified the binding thermodynamics with dicarboxylates [45]. Based on the earlier report, ITC studies with solvent mixtures of increasing polarity were conducted to determine the thermodynamic effects of solvent on binding. With the urea and thiourea receptors, associations were always exothermic or thermoneutral, while the entropy becomes more favorable with increasing solvent polarity due to solvent release. Receptor **29**, however, switched from exothermic to endothermic binding as the solvent polarity increased, showing extremely favorable entropy in polar protic solvents. This result suggests that binding in noncompetitive media is driven by hydrogen bonding and electrostatics, whereas solvent release accounts for the association to guanidinium-based receptors in more competitive media. In a later study with **29**, it was shown by nonaqueous affinity capillary electrophoresis that the binding affinity decreased with decreasing dicarboxylate chain length [46].

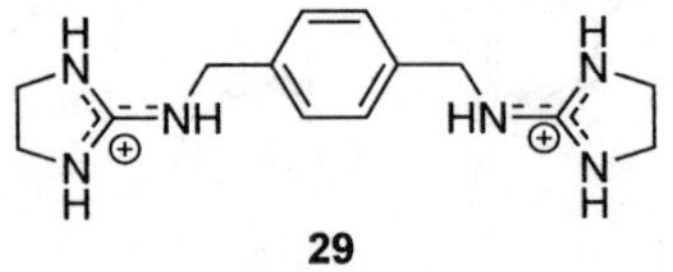

**29**

Air–water interface binding studies with dicarboxylates were conducted with octadecyl (**30**) and dioctadecylguanidinium receptors [47]. A monolayer of **30** gave greater molecular area expansion than the dioctadecylguanidinium amphiphiles. The binding of **30** to a range of dicarboxylates was monitored once again by the change in the π–*A* isotherm. Generally, the monolayer area increased with increasing chain length from oxalate through to adipate. Interestingly, the curve changes were once again very sensitive to guest structure,

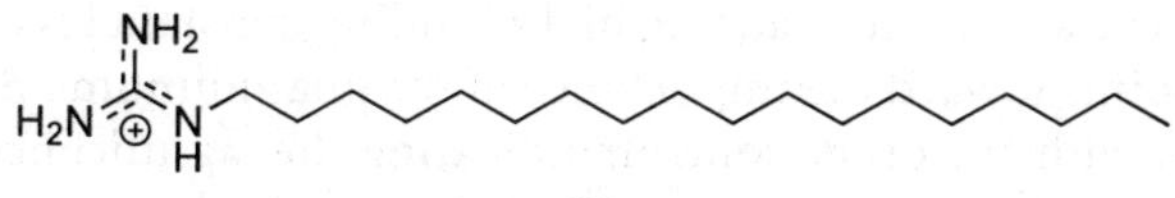

**30**

with oxalate and phthalate condensing the monolayer while *cis*-1,2-cyclohexanedicarboxylate and 1,1-cyclohexanediacetate gave expansion.

Though not multicarboxylates, uronic acid derivatives are important biological small molecules, and have received some attention for molecular recognition. A bis(deoxycholic acid) guanidinium receptor, **31**, was synthesized and tested for binding with glucuronate and galacturonate [48]. Though **31** did indeed bind the uronic acids with a slight selectivity for glucuronate, modifications of the host design showed that the steroid arms contributed little to the association of the guests. Other modifications proved that binding was best when the linker between the guanidinium core and the steroid arms was shortest.

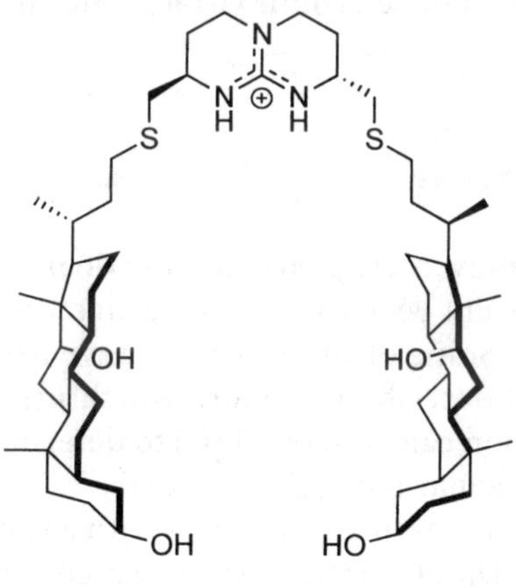

**31**

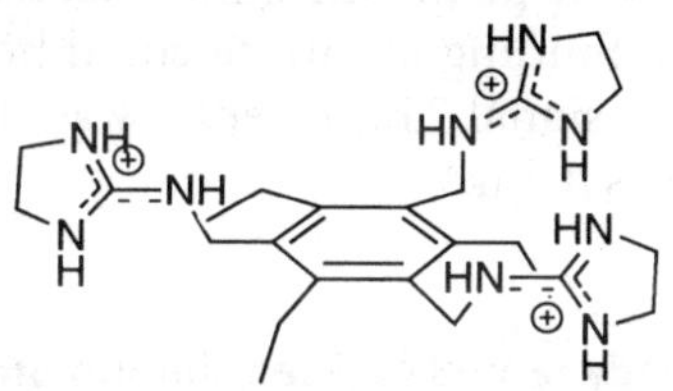

**32**

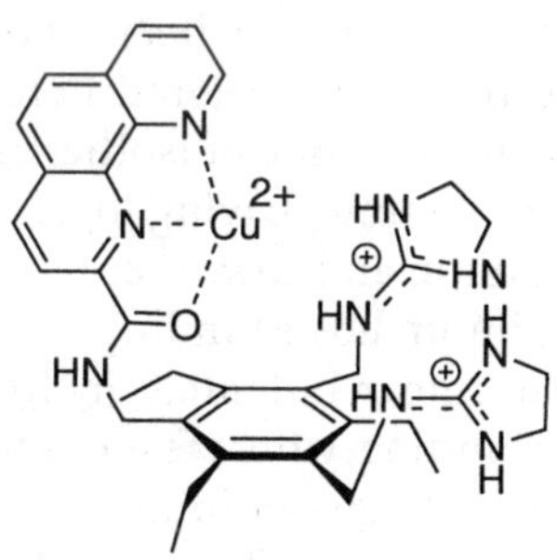

**33**

Citrate has been the target of two compounds developed in the Anslyn group. Receptors **32** and **33** are both based once again on the triethylbenzene platform. The orientation of the aminodihydroimidazolium groups in **32** was expected to be well suited for the binding of the tricarboxylate guest. In pure water, the binding constant of **32** with citrate was found to be $7\times10^3\ M^{-1}$ [49]. Two control experiments were conducted using an analog with ammonium groups instead of guanidiniums and an analog lacking the ethyl groups of the core scaffold. The ammonium analog had a binding constant less than half that of **32**, which points toward the importance of the guanidiniums. Similar results were obtained with the other control, indicating the significance of the rigid binding cavity created by the steric gearing. X-ray crystal structures confirmed the complexation orientation of the association. In a later study, **32** was used to successfully quantify citrate in sports drinks through an indicator-displacement assay [50]. Finally, ITC studies covering a wide range of concentrations and ionic strengths were conducted with **32** to determine the driving forces for the extremely strong citrate binding [51]. At high ionic strength, 1:1 binding was found to be driven by large positive entropy and negative enthalpy changes. At low ionic strength, higher order complexes were seen that were driven by a large positive entropy, whereas the enthalpy became unfavorable. Interestingly, it was determined that the higher order complexes resulted from the dilution of the 1:1 host–guest complexes which then formed the 2:1 complexes.

Receptor **33** was also designed to target citrate. The phenanthroline–copper complex was added not only as a binding element but also as a signaling motif [52]. The phenanthroline fluorescence was quenched upon complexing with copper. Binding of citrate could be detected by the reemergence of the fluorescence signal. This receptor was also used to determine citrate concentrations in sports drinks.

### 3.1.1 Thermodynamics of Guanidinium Binding

As observed in several of the examples above, designed receptors do not always give the results expected of them. In recent years, several investigations as to why this occurs have been undertaken. Some of these reports have been discussed briefly above, and here we will describe two more which utilize the powerful technique of isothermal titration calorimetry (ITC) to determine the thermodynamic driving forces for molecular recognition events.

Hamilton and coworker conducted a series of experiments with a variety of guanidinium hosts binding to acetate in DMSO [53]. Receptors which were able to form linear bidentate hydrogen bonds with acetate showed moderate to strong binding, whereas more highly derivatized hosts showed no or very weak binding. TPB guanidinium and aminoimidazolium both formed 2:1 guest/host complexes, which ITC was able to characterize while NMR titration could not. It was also found that the guanidinium's counterion played a role in the binding energetics, with small coordinating counterions such as chloride damping the association while large counterions, e.g., iodide or tetraphenylboron, coordinated less and gave rise to more accurate binding thermodynamics. For most of the guanidiniums in DMSO, binding was characterized by favorable enthalpy and entropy changes. However, it was found for one host that could only form bidentate coordination through geminal hydrogen bond donation that association was endothermic, with the driving force arising from favorable entropy. This scenario was seen for all guanidiniums in more competitive media (methanol), indicating that solvent release rather than the formation of strong hydrogen bonds was responsible for binding.

In a later study conducted in Schmidtchen's laboratory, ITC measurements on a series of bicyclic guanidiniums with varying peripheral arms akin to **4** were conducted using benzoate as the guest in acetonitrile [54]. The ultimate result of this study was a more complete, thermodynamic understanding of why the basic lock-and-key concept for receptor design fails in many cases. Receptors with bulky, organic peripheries which were less tightly solvated showed more favorable enthalpy changes, while entropy due to solvent release was diminished. Better solvated hosts saw smaller enthalpies of binding, but greater solvent dissipation produced more favorable entropy. As in the previously discussed study, similar counterion dependencies were encountered here. The binding increased by more than an order of magnitude from chloride to tetrafluoroboron. Again, it was seen that the more coordinating the counterion, the lower the enthalpy change and the greater the effect of the positive entropy change. These results demonstrate well the emerging theory of enthalpy–entropy compensation in competitive media molecular recognition. Hosts designed by the lock-and-key method tend to take only the enthalpic interactions of binding into account, but groups that are good enthalpically for the guest to bind also increase competition from the solvent or counterion. Thus, the design can fail or give less desirable results if enthalpy–entropy compensation is not taken into account.

## 3.2 Amino Acids and Peptides

Amino acids, peptides, and proteins are very attractive targets for molecular recognition studies for the obvious reason of their biological abundance and functions. In the past two years, the vast majority of publications on guanidinium-based receptors have dealt with recognizing peptidic species. As mentioned above, the desire for better peptide receptors derives mainly from the growing field of proteomics. New proteins are being discovered every day, and receptor design is required to keep up. The following sections will discuss several studies dealing with receptors designed to bind various types of amino acids and peptides, such as *N*-acylated or zwitterionic amino acids, oligopeptides, and proteins.

### 3.2.1 Single Amino Acids

In 1992, de Mendoza and coworkers reported the synthesis of the bifunctionalized chiral bicyclic guanidinium **34** for the purpose of enantioselectively binding zwitterionic aromatic amino acids (i.e., tryptophan and phenylalanine) [55]. By including binding elements for the carboxylate and ammonium of the amino acids which were noncomplementary, they hoped to avoid the potential problem of self-coordination which would prevent substrate binding. Hence, the azacrown ether was used to bind the ammonium while the guanidinium was expected to complex the carboxylate. The naphthoyl arm added a third binding interaction with π-stacking to the aromatic side chains. Receptor **34** was found to selectively bind phenylalanine and tryptophan over valine in competition extraction experiments. More complex extraction experiments with 13 different amino acids showed an enhanced binding preference for phenylalanine. Enantioselectivity of the host was also studied by competition assays. By $^1$H NMR, >95% ee values were found for the *R,R*-**34**/D-AA and the *S,S*-**34**/L-AA complexes. The high enantioselectivity was attributed to the three-point binding profile of the system. In a later study, **34** and several other similar structures were used to study the rate of amino acid transport across a model liquid membrane [56]. It was found that a derivative of **34** containing amide rather than ester linkages gave the fastest and most efficient transport, though it was less enantioselective.

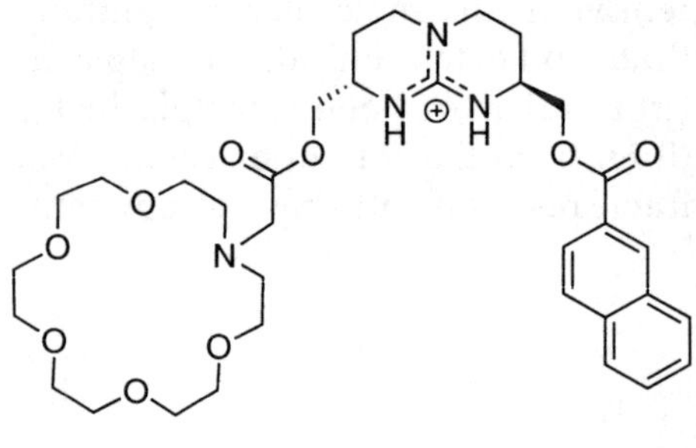

**34**

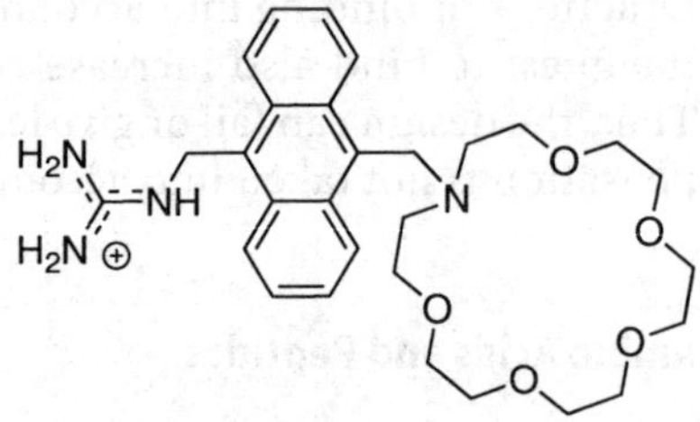

**35**

In another study attempting to bind zwitterionic guests, receptor **35** was designed as a host for the neurotransmitter γ-aminobutyric acid (GABA) [57]. Once again, the noncomplementary azacrown ether and guanidinium moieties were utilized along with the anthracene spacer as the signaling motif. The hope was that binding of the ammonium portion of the guest would disrupt the photoinduced electron transfer (PET) quenching of the fluorescence of the anthracene by the azacrown amine. An increase in fluorescence intensity was indeed detected upon introduction of GABA. Control experiments with an analog of **35** lacking the guanidinium group showed no fluorescence change in the presence of either GABA or its biological precursor, glutamic acid.

Another example of a receptor design using PET enhancement as the signaling motif was recently reported. Host **36** was studied by monitoring the fluorescence upon the binding of various essential amino acids, GABA, and alkyl ammoniums [58]. Strong fluorescence enhancement was seen with GABA and lysine, while weaker binding was observed with glycine and 1-aminobutane. Other amino acids tested showed no strong response, and the fluorescence spectrum closely resembled that of the free receptor.

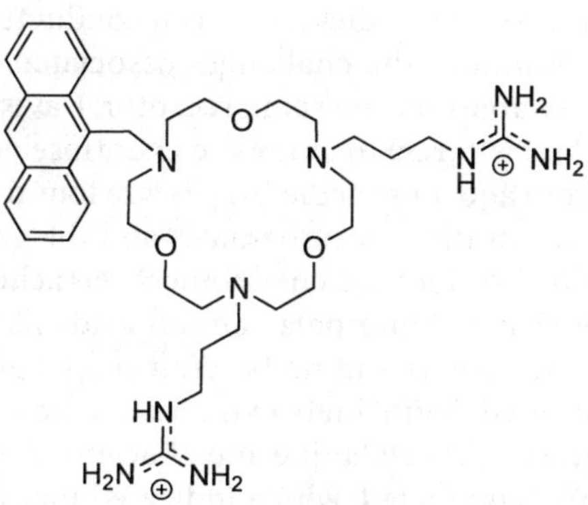

**36**

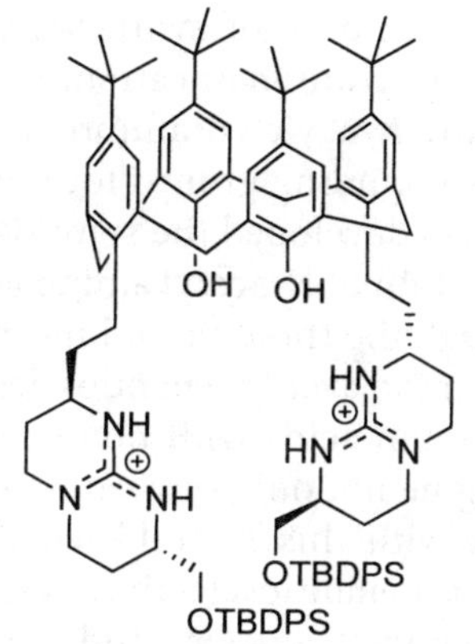

**37**

Calixarene-based host **37** was utilized in the chiral recognition of various zwitterionic amino acids [59]. In competitive extraction experiments, a 9:1 enantioselectivity was observed for L-phenylalanine over the D-isomer with the bis(*S*,*S*-guanidinium) **37**.

The 2,2′:6,2″ terpyridine-based ligand **38** was developed by Anslyn and collaborators to bind doubly anionic amino acids [60]. The goal was to achieve strong binding through a three-point profile with the two guanidiniums and the complexed zinc. The host–guest interactions were studied using an indicator-displacement assay as described above, using pyrocatechol violet as the indicator. While binding of the indicator to **38** showed higher order binding than simple 1:1, the displacement by the analyte was characterized by a sharp isosbestic point in the UV/vis titration, indicating that interaction only occurred between the singly bound indicator complexes. Fortunately, a very large absorbance shift (ca. 200 nm) was observed for the indicator upon displacement from the metal center by the analyte. The best binding in 50/50 water/methanol was observed with aspartate ($K_a=1.5\times10^5\ M^{-1}$) while other amino acids, including glutamate, were less strongly bound by an order of magnitude or more. While most of the binding affinity for amino acids derives from the amine–zinc complexation, the selectivity for aspartate was attributed to better interaction with the appendant guanidinium arms.

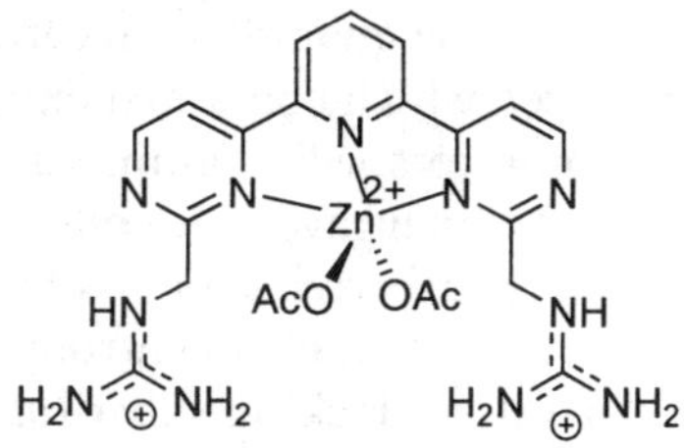

**38**

The binding of zwitterions inevitably raises more problems than simply binding an anion or cation. Therefore, several studies have been conducted on binding *N*-acylated amino acids to minimize the challenges associated with placing binding groups for opposite charges on the same receptor. Davis and Lawless developed the steroid-based receptor **39** to study the enantioselective extraction of *N*-acetyl amino acids from aqueous media [61]. It was found that, through the three-point binding profile of the two carbamates and the guanidinium, an average enantioselectivity of 80% for the L enantiomers was achieved for amino acids with nonpolar side chains. More polar amino acids like asparagine not only showed no enantioselectivity, but no binding at all. Further work with this scaffold led to derivatives with high extraction efficiencies and 10:1 enantioselectivities for *N*-acetyl-phenylalanine and *N*-acetyl-alanine. Membrane transport studies were also conducted, which indicated that these receptors showed consistent chiral recognition properties over an extended time period and with improved enantiomer ratios [62].

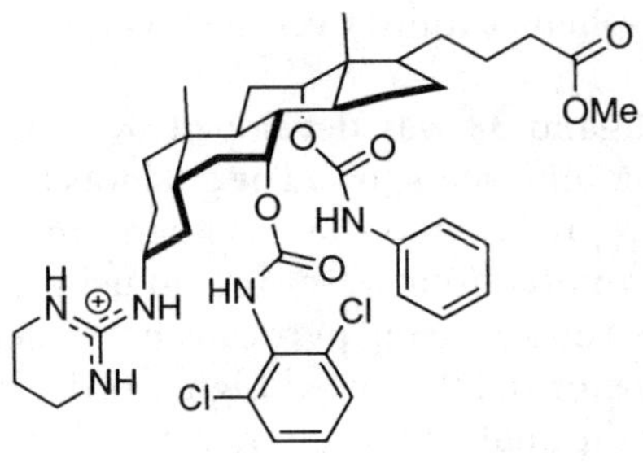

**39**

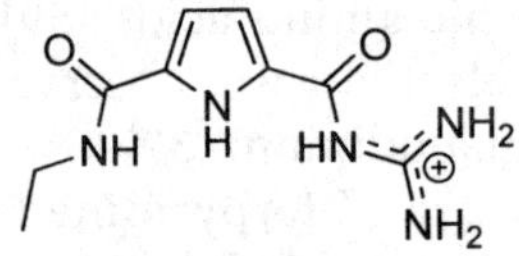

**40**

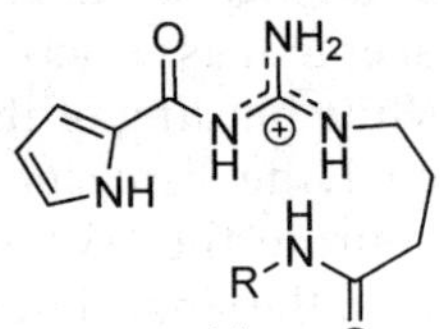

**41**

A series of receptors for *N*-acylated amino acids have been reported in several publications by Schmuck. Attempts to attach more recognition elements close to the guanidinium resulted in the pyrrole-appended acyl-guanidinium receptor **40** [63]. Binding studies in 40% water–DMSO revealed binding affinities ranging from 360 to 1,700 $M^{-1}$. Selectivity for *N*-Ac-phenylalanine was attributed to π-stacking between the aromatic side chain and the pyrrole of the receptor. In general, binding with **40** was found to be orders of magnitude higher than with the parent *N*-acetyl guanidinium cation. In a later study, several analogs of **40** were produced which contained extra recognition elements appended to the pyrrole to bind to specific amino acid side chains. A chiral derivative was also synthesized which gave moderate enantioselectivities [64]. An analog of **40** containing a free carboxylate opposite the guanidinium on the pyrrole was found to self-assemble through an entropy-driven oligomerization [65]. Finally, *N*-alkylated derivatives lacking substitution at position 5 of the pyrrole (**41**) were produced and found to bind hydrophobic *N*-acetyl amino acids with very high affinity, even in water. The appendant recognition units imparted extra selectivity to the binding based on the side chain of the amino acid bound [66].

### 3.2.2
### Peptides and Proteins

Proteins, due to their wide variety of surface features, present special challenges for molecular recognition researchers. Different surface structures (i.e., α-helices, β-sheets, β-turns) require different binding motifs to accommodate them. Most work using abiotic guanidinium receptors has focused on binding helical peptides and proteins. Hamilton and coworkers designed the bicyclic bis(guanidinium) structure **42** to bind helical peptides with high concentrations of carboxylic amino acids [67]. Several 16-mer peptides were synthesized with varying distances between aspartate residues to measure the specificity of the receptor. The best binding occurred when the aspartates were four residues apart ($K_a$=2.2×10$^3$ M$^{-1}$ in 10% water/methanol), and the binding dropped off as the distance increased. Also, circular dichroism titrations showed increased helicity upon binding of **42**. Controls with cyclopentyl guanidinium chloride gave nearly immeasurably low binding by NMR and showed no change in the peptide helicity. In later studies, several analogous hosts showed decreased association with the peptides [68].

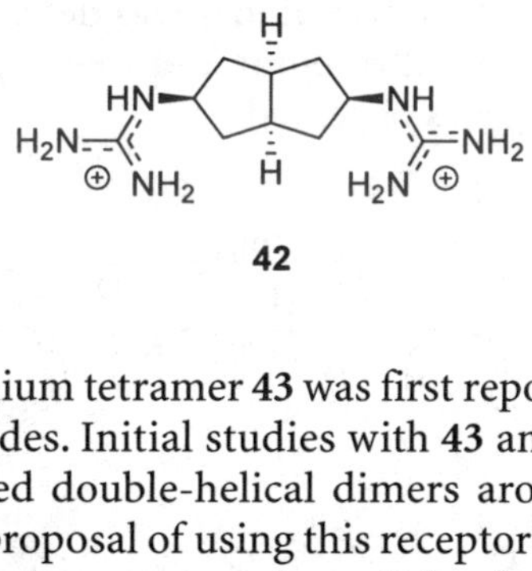

**42**

The bicyclic guanidinium tetramer **43** was first reported as a possible binder for helical oligonucleotides. Initial studies with **43** and sulfate anions showed that the tetramer formed double-helical dimers around sulfate counterions [69]. Despite the initial proposal of using this receptor for membrane transport of oligonucleotides, all subsequent work on **43** has been conducted on helical peptides. For example, the binding of **43** with several synthetic peptides caused an increase in the helicity and helical stability of the peptides in 10% water/methanol [70]. The peptide containing four Asp derivatives showed the

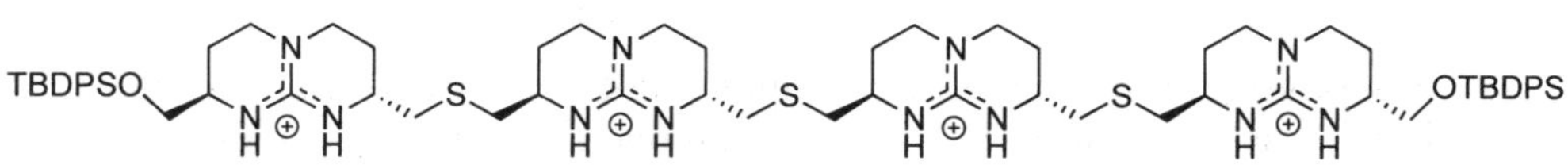

**43**

strongest binding ($K_a$=1.6×10$^5$ M$^{-1}$) and a 40% increase in helicity over the free peptide chain. A later study attempted to look at this system in pure water. While NOE and ROE spectra gave evidence of increased helicity of the peptide in the presence of **43**, CD titrations showed no detectable change [71]. Binding constants in water were not reported. A related study on the thermodynamics of binding based on the side chain length of the anionic residues (Asp vs. Glu) showed that as the average length of the side chain increased, the binding became strongly entropy driven. Furthermore, with the introduction of Glu in place of Asp residues, less induced helicity was observed upon binding [72]. This report is yet another example of the role of enthalpy–entropy compensation in molecular recognition events, as the apparent binding constants did not show much change with increasing side chain length. Finally, **43** was studied as a receptor for the tetramerization domain of protein P53 [73]. Whereas in all the previous studies, binding was best with every fourth residue being anionic, the primary binding site of **43** with P53 was with a slightly extended domain. Sterics and the presence of the longer glutamate residues were thought to be the cause of this elongated binding site.

The tweezer-like compound **44** was reported as another receptor for peptides [74]. This study incubated a 1,000-member library of resin-bound tripeptide guests with the dansylated receptor. After washing, the beads were examined for fluorescence, and approximately 3% of the library was found to bind the host. A 95% selectivity for Val-C-terminating peptides was observed as well as a 40% selectivity for *t*-BuO-Glu in the central position of the tripeptide. A binding constant of 4×10$^5$ M$^{-1}$ for their best binder was determined by ITC. A reverse experiment with a resin-bound receptor and free tripeptides was attempted, but no significant binding was observed.

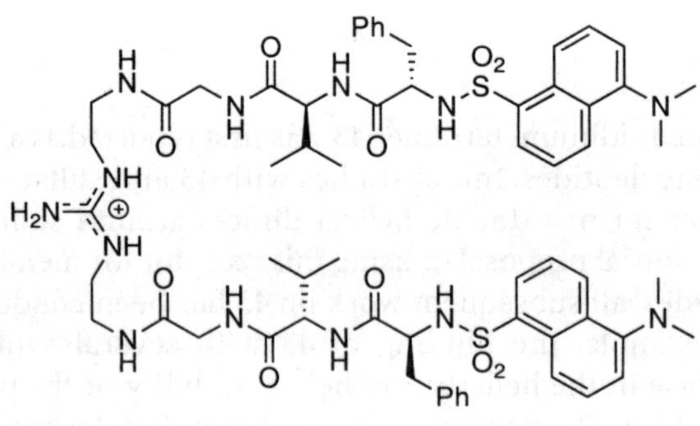


**44**

In another study by Kunitake, the amphiphile **45** was used to bind dipeptides at the air–water interface [75]. Using mixed monolayers of **45** and a dioctadecyl glycylglycinamide amphiphile, a Langmuir isotherm for the binding of Gly-Leu was produced. The result was a cooperative binding event between the two

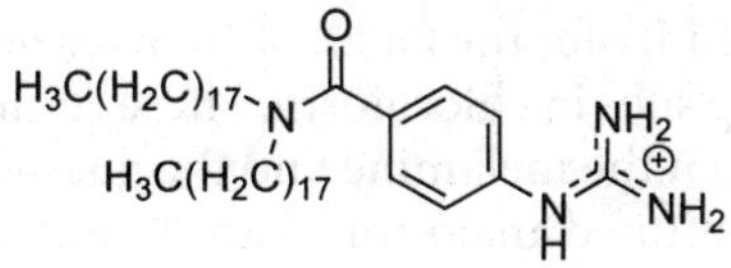

**45**

components of the monolayer and the dipeptide with a binding constant of $6.4\times10^3\ M^{-1}$. Reversal of the dipeptide to Leu-Gly resulted in a 1/3 decrease in binding affinity, indicating the specificity of the mixed monolayer recognition profile. Similar studies with single-component monolayers showed highly decreased association affinities.

## 4
## Miscellaneous Targets

This section will cover some of the less popular analytes for which guanidinium-based receptors have been reported. These analytes include flavins, nitronates, carbonates, sulfates, squarates, and lactones. Some of the receptors discussed above have been used to study these analytes as well; however, their primary purpose was for recognition of the targets listed above. For instance, even though the first report of receptor **43** dealt with monitoring the induced helicity upon binding of sulfate counterions, the purpose of the design was to bind helical biomacromolecules.

Flavin-based coenzymes are particularly attractive targets for molecular recognition chemists because of the way they interact with the enzymes they work with. The hydrogen-bonding interactions between flavins and proteins hold the flavin in place and regulate the reactivity of the coenzyme. Thus, it might be possible to design enzyme mimics if that hydrogen-bonding profile could be emulated. Yano and coworkers designed the melamine-based receptor **46** to bind to 6-azaflavin through a five-point hydrogen-bonding profile after their initial attempts with three-point binding systems failed to have strong interactions [76]. Addition of the guanidinium to the melamine core saw

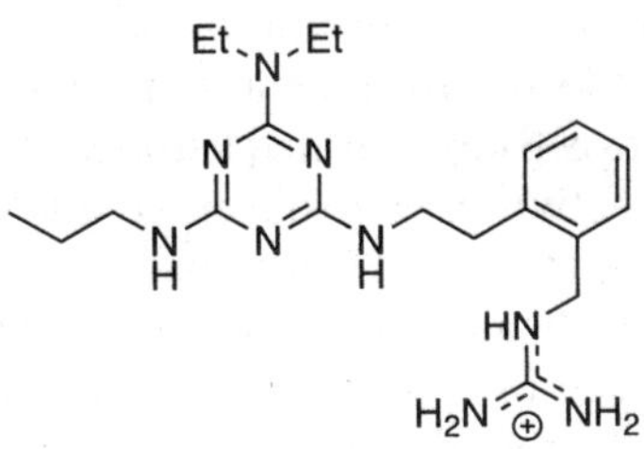

**46**

an increased binding affinity for the flavin of $10^3$ over receptors containing only three hydrogen-bonding sites in chloroform. These results indicate a large binding cooperativity between the melamine and the guanidinium arm. In another study, an *N*-phenyl-substituted analog of **46** was found to bind the semiquinone radical anions of 6-azaflavin with an association constant of $7.7\times10^5\ M^{-1}$, which was 100 times greater than that for the binding of neutral 6-azaflavin [77]. The radical anions were greatly stabilized and were detectable by EPR spectroscopy even after 48 hours. Finally, later studies with this receptor scaffold incorporated a variety of different functionalities at the guanidinium center to probe the effect of hydrogen bonding by the guanidinium group to different acceptors on the flavin [78]. Kinetic studies of the oxidation of benzyl-1,4-dihydronicotinamide and thiophenol by 6-azaflavin in the presence of the receptor analogs were carried out. The greatest rate enhancement was induced by receptors which hydrogen bonded to the N1 position.

As seen above with the phosphate receptors, researchers are often not satisfied with only binding their targets, but they also want to mimic enzymes and produce some catalytic activity from their receptors. De Mendoza and collaborators used the bicyclic guanidinium **47** to effect the addition of pyrrolidine to α,β-unsaturated lactones [79]. Using a 10% catalyst loading in chloroform, an eightfold reduction in the half-life of the starting materials was observed by $^1$H NMR. No similar catalysis was observed for α,β-unsaturated esters due to the *s-cis* conformation of the ester. In a later study, analogs of **47** with aromatic substituents on the bicyclic guanidinium were synthesized. These compounds were shown to increase catalysis, but no chiral recognition was observed [80].

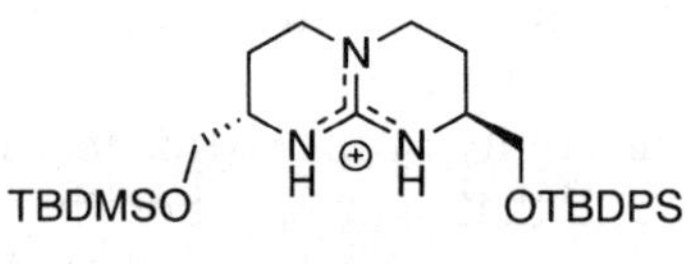


**47**

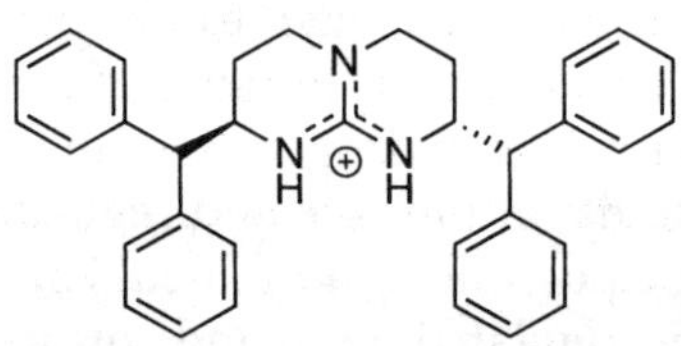


**48**

Another host design which was eventually probed for catalytic properties is receptor **48**. This structure was found to bind well to nitronate anions in solution and gave enantioselective recognition of carboxylate guests like naproxenate [81]. The binding of **48** to nitronate anions was exploited to catalyze the addition of nitroalkanes to α,β-unsaturated ketones. While the catalysis worked, disappointing enantioselectivities were observed due to the sterically hindered binding cavity. In an attempt to quantify the guanidinium–nitronate association, an unsubstituted analog of **48** along with another receptor previously reported by Hamilton (**29**) were used in a later study [82]. The binding of nitronate to guanidinium receptors was found to be greater than that to thiourea analogs, though not as strong as that of carboxylates. Kinetic studies with the unsubstituted bicyclic guanidinium receptor showed a decrease in nitronate activity allowing for an increase in C-alkylation over O-alkylation.

Continuing in their pursuit of the physical understanding of molecular recognition events, Schmidtchen and coworkers presented two studies using squarates as generalized oxoanions for binding studies with various receptors. In the first study, expanding upon the ditopic receptor design seen previously (**23**), a series of receptors akin to **49** were synthesized [83]. The purpose of the study was to examine the effect of rigidity of the spacer of a ditopic host to ultimately determine the cost effectiveness of building receptors with stiff preorganization. Unfortunately, there was no observable correlation found, since receptors with similarly rigid spacers varied widely – by almost two orders of magnitude – in their binding affinities for the squarate guests. The conclusion was that the free energy due to solvation well outweighed any differences in rotational or vibrational energies. The next study used the electroneutral receptor **50**, which incorporates a nonhydrophilic *closo*-borane dianion to compensate for the two guanidinium cations [84]. Binding studies with several dianionic guests gave in many cases convoluted high-order associations. However, 1:1 binding constants were obtained for several of the targeted guests, and it was found that cyclic oxoanions such as squarate and croconate exhibited the strongest binding. Further studies with **50** and analogs used ITC to measure the thermodynamic parameters of binding different dianions. It was found that, based on the anion studied and the solvent used, binding moved from enthalpy driven to having large favorable entropy as the driving force [85, 86].

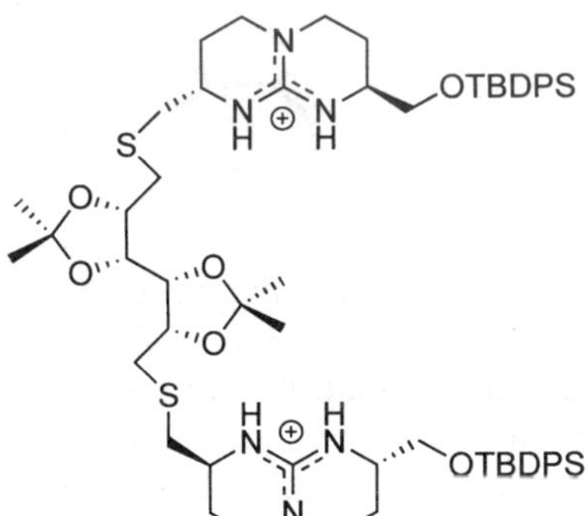


**49**

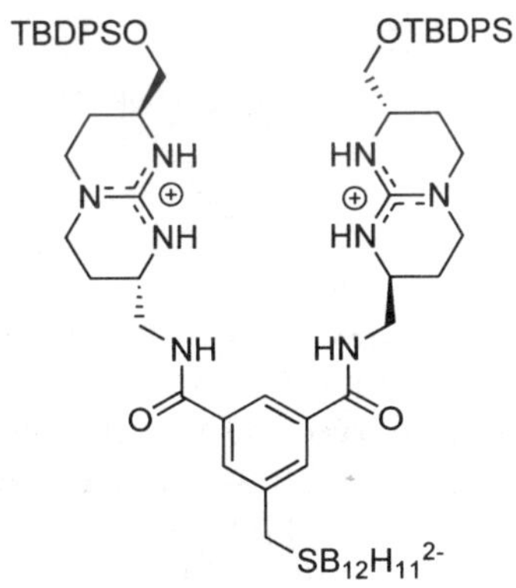


**50**

Carboxylates containing vicinal diol functionalities such as tartrate, malate, and gallate are attractive commercial targets for molecular recognition. These carboxylates are present in all aged alcoholic beverages. Tartrate and malate are essential compounds for wine production since their concentrations over time can be traced to the proper aging of the wine. Gallate and similar cyclic carboxylates are found in the oak used in the aging barrels in Scotch whisky production and are extracted into the Scotch during the aging process. Two receptors reported by Anslyn and coworkers were specifically designed to bind to vicinal diols which contain carboxylates. The triethylbenzene-based receptor **51** consists of two aminodihydroimidazolium groups for carboxylate binding and a phenylboronic acid which can form reversible borate esters with a vicinal diol [87]. Tartrate, with two carboxylate functionalities as well as a vicinal diol, was chosen as the target for this receptor. This host was studied using an indicator-displacement assay with alizarin complexone because this indicator contains the same complementary functionalities as tartrate. Qualitative naked-eye experiments showed distinct color changes upon introduction of tartrate to a **51**–indicator solution. Quantitative analyses revealed that binding of tartrate ($K_a$=5.5×10$^5$ M$^{-1}$) in 25% water/methanol was selective over all other analytes except malate ($K_a$=4.8×10$^5$ M$^{-1}$). The host was successfully used to quantify the concentration of tartrate–malate in grape-based beverages such as wine. In a later study, the bis(boronic acid) analog **52** was studied for binding with gallate and similar structures [88]. Indicator-displacement assays with pyrocatechol violet found binding constants between 100 (4-hydroxycinnamic acid) and 10,000 (gallic acid) for these targets. Interestingly, when the assay was applied to determination of the ages of Scotch whiskies, only the complex binding profile of all the analytes gave an accurate age-to-concentration ratio.

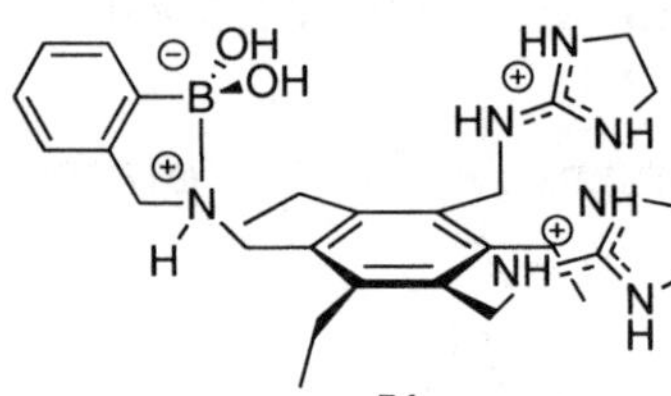

**51**

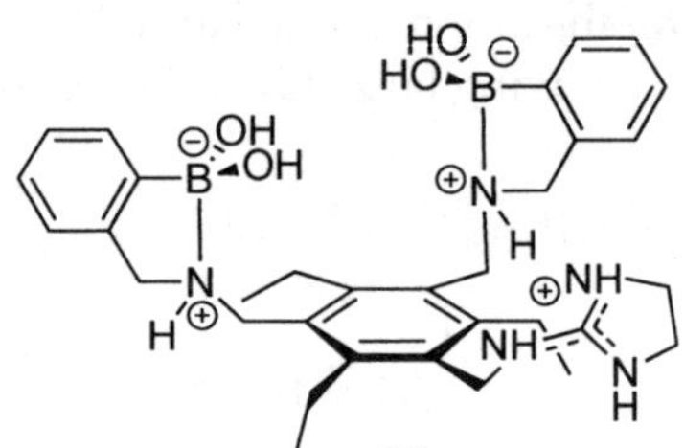

**52**

One final class of analytes which has been studied with guanidinium-containing receptors is carbonates. The neutral bis(iminoylthiourea) host **53** was found to undergo a 53-fold fluorescence enhancement upon binding of $CO_3^{2-}$ [89]. Binding fell off with bicarbonate (39-fold enhancement), and phosphate and sulfate showed poor binding. The corresponding 1,2,4-thiadiazole derivative showed only a 12-fold fluorescence enhancement upon binding of carbonate anion. This study highlights the use of truly electroneutral hosts for anion recognition.

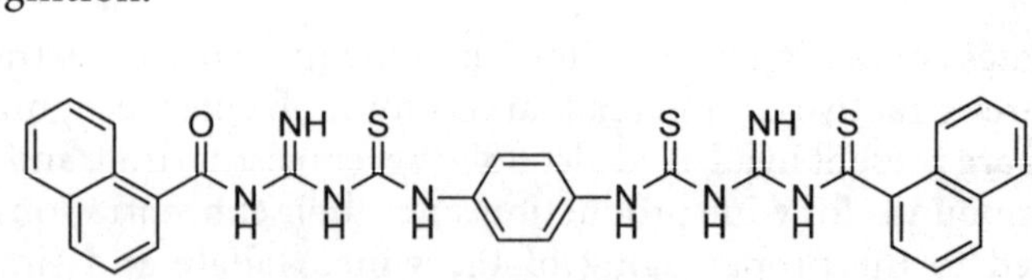

**53**

## 5 Conclusions and Outlook

Guanidinium groups have been used to study a wide variety of anionic substrates and have been incorporated into many different molecular scaffolds. Even with the addition of other recognition elements into many of these scaffolds, the guanidinium group still continues to offer a large driving force for the binding of most of these analytes. Major advancements have been made in understanding the physical properties behind recognition events with guanidinium-based receptors. This insight has been used to design receptors with stronger binding, higher catalytic activity, and structural stabilization properties. However, many of the goals toward which molecular recognition chemists have been striving still go unmet. Creating synthetic enzymes with similar selectivity and activity as natural enzymes, designing receptors with strong host–guest interactions in water, and creating strongly enantioselective receptors for relevant racemic substrates are all still lacking. Further research in these areas and in the deeper understanding of the driving forces of host–guest interactions are essential to the progress of the field and remain open for the future.

## References


1. Christianson DW, Lipscomb WN (1989) Acc Chem Res 22:62
2. Cotton FA, Hazen EE Jr, Legg MJ (1979) Proc Natl Acad Sci USA 76:2551
3. Hannon CL, Anslyn EV (1993) In: Dugas H, Schmidtchen FP (eds) Bioorganic chemistry frontiers. Springer, Berlin Heidelberg New York, p 194
4. Kim EE, Wyckoff HW (1991) J Mol Biol 218:449
5. Angyal SJ, Warburton WK (1951) J Chem Soc 2492
6. Dean JA (1985) Lange's handbook of chemistry. McGraw-Hill, New York
7. Best MD, Tobey SL, Anslyn EV (2003) Coord Chem Rev 240:3
8. Tobey SL, Anslyn EV (2004) In: Atwood JL, Steed JW (eds) Encyclopedia of supramolecular chemistry. Marcel Dekker, New York, p 1
9. Dietrich B, Flyes TM, Lehn J-M, Pease LG, Fyles DL (1978) J Chem Soc Chem Commun 934
10. Müller G, Riede J, Schmidtchen FP (1988) Angew Chem Int Ed Engl 27:1516
11. Beer PD, Drew MGB, Smith DK (1997) J Organomet Chem 543:259
12. Nishizawa S, Kato Y, Teramae N (1999) J Am Chem Soc 121:9463
13. Tobey SL, Anslyn EV (2003) J Am Chem Soc 125:14807
14. Schmidtchen FP (1989) Tetrahedron Lett 30:4493
15. Alcázar V, Segura M, Prados P, de Mendoza J (1998) Tetrahedron Lett 39:1033
16. Niikura K, Metzger A, Anslyn EV (1998) J Am Chem Soc 120:8533
17. Watanabe S, Onogawa O, Komatsu Y, Yoshida K (1998) J Am Chem Soc 120:229
18. Jubian V, Dixon RP, Hamilton AD (1992) J Am Chem Soc 114:1120
19. Muche M-S, Göbel MW (1996) Angew Chem Int Ed Engl 35:2126
20. Jubian V, Veronese A, Dixon RP, Hamilton AD (1995) Angew Chem Int Ed Engl 34:1237
21. Gross R, Bats JW, Göbel MW (1994) Liebigs Ann Chem 205
22. Ariga K, Anslyn EV (1992) J Org Chem 57:417

23. Perreault DM, Chen X, Anslyn EV (1995) Tetrahedron 51:353
24. Kneeland DM, Ariga K, Lynch VM, Huang C-Y, Anslyn EV (1993) J Am Chem Soc 115: 10042
25. Perreault DM, Cabell LA, Anslyn EV (1997) Bioorg Med Chem 5:1209
26. Cuevas F, Stefano SD, Magrans JO, Prados P, Mandolini L, de Mendoza J (2000) Chem Eur J 6:3228
27. Oost T, Filippazzi A, Kalesse M (1997) Liebigs Ann Chem 1005
28. Galán A, Pueyo E, Salmerón A, de Mendoza J (1991) Tetrahedron Lett 32:1827
29. Galán A, de Mendoza J, Bruix CM, Deslongchamps G, Rebek J Jr (1991) J Am Chem Soc 113:9424
30. Deslongchamps G, Galán A, de Mendoza J, Rebek J Jr (1992) Angew Chem Int Ed Engl 31:61
31. Andreu C, Galán A, Kobrio K, de Mendoza J, Park TK, Rebek J Jr, Salmerón A, Usman N (1994) J Am Chem Soc 116:5501
32. Kato Y, Conn MM, Rebek J Jr (1994) J Am Chem Soc 116:3279
33. Sasaki DY, Kurihara K, Kunitake T (1991) J Am Chem Soc 113:9685
34. Sasaki DY, Kurihara K, Kunitake T (1992) J Am Chem Soc 114:10994
35. Schliessl P, Schmidtchen FP (1994) J Org Chem 59:509
36. Schneider SE, O'Neil SN, Anslyn EV (2000) J Am Chem Soc 122:542
37. McCleskey SC, Griffin MJ, Schneider SE, McDevitt JT, Anslyn EV (2003) J Am Chem Soc 125:1114
38. Fukutomi R, Tanatani A, Kakuta H, Tomioka N, Itai A, Hashimoto Y, Shudo K, Kagechika H (1998) Tetrahedron Lett 39:6475
39. Luedtke NW, Baker TJ, Goodman M, Tor Y (2000) J Am Chem Soc 122:12035
40. Barawkar DA, Linkletter B, Bruice TC (1998) Bioorg Med Chem Lett 8:1517
41. Kojima N, Szabo IE, Bruice TC (2002) Tetrahedron 58:867
42. Echavarren A, Galán A, Lehn J-M, de Mendoza J (1989) J Am Chem Soc 111:4994
43. Martín M, Almaraz M, Hernández J, Tejeda A, Caballero MC, Morán JR (1999) Heterocycles 50:47
44. Fan E, Van Armen SA, Kincaid S, Hamilton AD (1993) J Am Chem Soc 115:369
45. Linton BR, Goodman MS, Fan E, van Armen SA, Hamilton AD (2001) J Org Chem 66:7313
46. Peddicord MB, Weber SG (2002) Electrophoresis 23:431
47. Kamino A, Koyano H, Ariga K, Kunitake T (1996) Bull Chem Soc Jpn 69:3619
48. Segura M, Alcázar V, Prados P, de Mendoza J (1997) Tetrahedron 53:13119
49. Metzger A, Lynch VM, Anslyn EV (1997) Angew Chem Int Ed Engl 36:862
50. Metzger A, Anslyn EV (1998) Angew Chem Int Ed 37:649
51. Rekharsky M, Inoue Y, Tobey SL, Metzger A, Anslyn EV (2002) J Am Chem Soc 124:14959
52. Cabell LA, Best MD, Lavigne JJ, Schneider SE, Perreault DM, Monahan M-K, Anslyn EV (2001) J Chem Soc Perkins Trans 2:315
53. Linton B, Hamilton AD (1999) Tetrahedron 55:6027
54. Haj-Zaroubi M, Mitzel NW, Schmidtchen FP (2002) Angew Chem Int Ed 41:104
55. Galán A, Andreu D, Echavarren AM, Prados P, de Mendoza J (1992) J Am Chem Soc 114:1511
56. Breccia P, Gool MV, Pérez-Fernández R, Martín-Santamaría S, Gago F, Prados P, de Mendoza J (2003) J Am Chem Soc 125:8270
57. de Silva AP, Gunaratne HQN, McVeigh C, Maguire GEM, Maxwell PRS, O'Hanlon E (1996) Chem Commun 2191
58. Sasaki S-I, Hashizume A, Citterio D, Fujii E, Suzuki K (2002) Tetrahedron Lett 43:7243
59. Fang L, Lu G-Y, He W-J, Wang Z-S, Zhu L-G (2001) Chin J Chem 19:317
60. Aït-Haddou H, Wiskur SL, Lynch VM, Anslyn EV (2001) J Am Chem Soc 123:11296

61. Lawless LJ, Davis AP (1999) Chem Commun 9
62. Baragaña B, Blackburn AG, Breccia P, Davis AP, de Mendoza J, Padrón-Carrillo JM, Prados P, Reidner J, de Vries JG (2002) Chem Eur J 8:2931
63. Schmuck C (1999) Chem Commun 843
64. Schmuck C (2000) Chem Eur J 6:709
65. Schmuck C (2001) Tetrahedron 57:3063
66. Schmuck C, Bickert V (2003) Org Lett 5:4579
67. Albert JS, Goodman MS, Hamilton AD (1995) J Am Chem Soc 117:1143
68. Albert JS, Peczuh MW, Hamilton AD (1997) Bioorg Med Chem 5:1455
69. Sánchez-Quesada J, Seel C, Prados P, de Mendoza J (1996) J Am Chem Soc 118:277
70. Peczuh MW, Hamilton AD, Sánchez-Quesada J, de Mendoza J, Haack T, Giralt E (1997) J Am Chem Soc 119:9327
71. Haack T, Peczuh MW, Salvatella X, Sánchez-Quesada J, de Mendoza J, Hamilton AD, Giralt E (1999) J Am Chem Soc 121:11813
72. Salvatella X, Peczuh MW, Gairí M, Jain RK, Sánchez-Quesada J, de Mendoza J, Hamilton AD, Giralt E (2000) Chem Commun 1399
73. Salvatella X, Martinell M, Gairí M, Mateu MG, Feliz M, Hamilton AD, de Mendoza J, Giralt E (2004) Angew Chem Int Ed 43:196
74. Davies M, Bonnat M, Guillier F, Kilburn JD, Bradley M (1998) J Org Chem 63:8696
75. Ariga K, Kamino A, Cha X, Kunitake T (1999) Langmuir 15:3875
76. Tamura N, Kajiki T, Nabeshima T, Yano Y (1994) J Chem Soc Chem Commun 2583
77. Kajiki T, Moriya H, Kondo S, Nabeshima T, Yano Y (1998) Chem Commun 2727
78. Kajiki T, Moriya H, Hoshino K, Kuroi T, Kondo S, Nabeshima T, Yano Y (1999) J Org Chem 64:9679
79. Alcázar V, Morán JR, de Mendoza J (1995) Tetrahedron Lett 36:3941
80. Portugués MM, Alcázar V, Prados P, de Mendoza J (2002) Tetrahedron 58:2951
81. Davis AP, Dempsey KJ (1995) Tetrahedron Asymmetry 6:2829
82. Linton BR, Goodman MS, Hamilton AD (2000) Chem Eur J 6:2449
83. Peschke W, Schmidtchen FP (1995) Tetrahedron Lett 36:5155
84. Berger M, Schmidtchen FP (1996) J Am Chem Soc 118:8947
85. Berger M, Schmidtchen FP (1998) Angew Chem Int Ed 37:2694
86. Berger M, Schmidtchen FP (1999) J Am Chem Soc 121:9986
87. Lavigne JJ, Anslyn EV (1999) Angew Chem Int Ed 38:3666
88. Wiskur SL, Anslyn EV (2001) J Am Chem Soc 123:10109
89. Hennrich G, Sonnenschein H, Resch-Genger U (2001) Tetrahedron Lett 42:2805

# Author Index Volumes 251–255

*Author Index Vols. 26–50 see Vol. 50*
*Author Index Vols. 51–100 see Vol. 100*
*Author Index Vols. 101–150 see Vol. 150*
*Author Index Vols. 151–200 see Vol. 200*
*Author Index Vols. 201–250 see Vol. 250*

*The volume numbers are printed in italics*

# Subject Index